ALSO BY CRAIG NELSON

Rocket Men: The Epic Story of the First Men on the Moon

Thomas Paine: Enlightenment, Revolution, and the Birth of Modern Nations

The First Heroes: The Extraordinary Story of the Doolittle Raid—America's First World War II Victory

Let's Get Lost: Adventures in the Great Wide Open

THE AGE OF RADIANCE

THE EPIC RISE AND DRAMATIC FALL OF THE ATOMIC ERA

CRAIG NELSON

SCRIBNER

New York London Toronto Sydney New Delhi

SCRIBNER
A Division of Simon & Schuster, Inc.
1230 Avenue of the Americas
New York, NY 10020

First Scribner hardcover edition March 2014

Book design by Ellen R. Sasahara
Jacket design by Steve Attardo
Back jacket photograph © Time Life Pictures/Getty Images

Manufactured in the United States of America

1 3 5 7 9 10 8 6 4 2

Library of Congress Control Number: 2013042192

ISBN 978-1-4516-6043-2
ISBN 978-1-4516-6045-6 (ebook)

For Stuart—

You are the best in the world at what you do.

The most beautiful and deepest experience a man can have is the
sense of the mysterious. It is the underlying principle of religion
as well as of all serious endeavor in art and science.
He who never had this experience seems to me,
if not dead, then at least blind.

—Albert Einstein

Nothing in life is to be feared, it is only to be understood.
Now is the time to understand more, so that we fear less.

—Marie Curie

CONTENTS

PART ONE: THE OLD WORLD

1. Radiation: What's in It for Me? *3*
2. The Astonished Owner of a New and Mysterious Power *8*
3. Rome: November 10, 1938 *55*
4. The Mysteries of Budapest *73*

PART TWO: THE NEW WORLD

5. The Birth of Radiance *109*
6. The Secret of All Secrets *140*
7. The First Cry of a Newborn World *196*
8. My God, What Have We Done? 206

PART THREE: WORLD'S END

9. How Do You Keep a Cold War Cold? *225*
10. A Totally Different Scheme, and It Will Change the Course of History *247*
11. The Origins of Modern Swimwear *270*
12. The Delicate Balance of Terror *276*

PART FOUR: POWER AND CATACLYSM

13. Too Cheap to Meter *303*
14. There Fell a Great Star from Heaven, Burning as It Were a Lamp *312*
15. Hitting a Bullet with a Bullet *327*
16. On the Shores of Fortunate Island *340*
17. Under the Thrall of a Two-Faced God *368*

Heartfelt Thanks *381*
Notes *383*
Sources *399*
Photo Credits *417*
Index *419*

PART ONE

THE OLD WORLD

1

Radiation: What's in It for Me?

YESTERDAY you dashed your breakfast eggs with dried spices that had been irradiated against bacteria, germination, and spoilage. The secret ingredient in the microwave oven reheating that morning's to-go coffee was radioactive thorium, first isolated from hearty Scandinavian minerals and named for their tempestuous lord. Your kitchen's smoke + CO_2 detector then started beeping every thirty seconds since it had to be replaced—its 0.9 micrograms of americium-241 had expired. The Brazil nuts in your cereal, meanwhile, had a thousand times more radium than any other food eaten by modern humans. Your banana's potassium was radioactive, as was the body of the person you slept next to the night before. All night long you returned the favor, two lovers irradiating each other across snores, dreams, and twitching REM eyes.

Since you live at sea level, you get an annual cosmic-ray shower of thirty millirems (*r*oentgen *e*quivalent *m*an (or *m*ammal), a measure of the cancerous effects of radiant emanations, with one rem meaning .055 percent chance of cancer), but moving to Denver with its increased elevation will double that, while an aviation career equals 1 mrem for every thousand miles soaring at thirty thousand feet. Living in a home of masonry—stone, brick, adobe—doses you with 7 mrems annually, and in a city of buildings made from those same types of earth, 10 mrems, with an extra shot every time you pass through halls of granite, such as New York's Grand Central Terminal or Washington's Capitol (which is so vibrant it would fail the Nuclear Regulatory Commission's licensing conditions for a reactor site).

Visiting the dentist, you are draped in a lead apron but, even so, get another 80 mrems, as you would with a chest X-ray, CAT scan, or nuclear stress test. There's uranium in your dental work, added to porcelain for long-lasting whiteness and fluorescence, and if you walk too closely by certain policemen after a dental or medical procedure, you'll set off their Geiger counters (otherwise engaged in the hunt for mythic dirty bombs). On the way home, you bought a balloon for your daughter, the helium that made it float produced by all-natural, all-organic radioactivity within the planet's mantle. Your delinquent nephew still smokes, inhaling 12,000 to 16,000 mrem every year from radon isotopes trapped in tobacco's delicious leaves. As you watch the glittering skyline of New York City on your way home that night, 30 percent of its power draws from a nuclear reactor that, if it suffered a Chernobyl-like failure, would mean the evacuation of 10 million people.

Your family is radioactive; your friends are radioactive; your pets are radioactive; and the earth itself throws off a gaseous froth of radon, 200 mrem for each of us, as uranium and thorium decay within the planet's restive loam. Your vacation at a health spa includes daily bathing in mineral waters, but doesn't include the information that hot springs are hot in two senses—the water is heated by rocks burning from those same uranium and thorium emissions, the force that also powers earthquakes to tremor, and volcanoes to erupt. Physicist Paul Preuss: "What spreads the sea floors and moves the continents? What melts iron in the outer core and enables the Earth's magnetic field? Heat. Geologists have used temperature measurements from more than 20,000 boreholes around the world to estimate that some 44 terawatts (44 trillion watts) of heat continually flow from Earth's interior into space. Where does it come from? Radioactive decay of uranium, thorium, and potassium in Earth's crust and mantle is a principal source."

Radiation is so organic that in 1972, geologic evidence of fourteen naturally occurring nuclear reactors, 1.5 billion years old, were found in the Oklo mines of Gabon, Central Africa. When groundwater leaked into a radiant vein of pitchblende ore unusually dense with that rare and quick-to-fission isotope uranium-235, the water slowed the energies of the free-range neutrons that make uranium radioactive, until their bouncing around like multiple strikes on a hundred-ball pool table triggered a chain reaction of neutrons splitting nuclei, which creates more free-range neutrons, splitting more nuclei. When the water boiled away, the natural ore reactors shut down . . . until more seeped back in, and they started up all over again.

In real life, though, you are not imperiled by Brazil nuts, bananas, microwave ovens, smoke detectors, going to the dentist, working as a flight attendant, sleeping with a lover, visiting Gabon, or living in Denver. Even with these daily accumulations, a scientific majority believes that you would need over 10,000 mrem to get any increased cancer risk, and 200,000 for radiation sickness—so only your smoking nephew is in trouble. Though tobacco companies knew as early as 1959 that cigarettes were rife with polonium-210, they kept that fact quiet for over four decades, even after discovering that acid-washing could remove 99 percent of the problem—but deciding that it wasn't worth the cost—resulting in 138 deaths per 1,000 smokers per twenty-five years. And for those concerned about Denver, even though elevated Colorado gets between two and three times the cosmic-shower radiation of New Jersey, cancer rates in New Jersey are higher than they are in Colorado.

Irrational. Confusing. Conflicted. These are hallmarks of the whole of nuclear history. What was once an era heralded by Curie, Einstein, and Oppenheimer became degraded to *Dr. Strangelove*, $50 billion wasted on Reagan's failed Star Wars, the 1979 partial meltdown at Three Mile Island, the 1986 explosion at Chernobyl, and the 2011 crisis at Fukushima. We are now living in the twilight of the Atomic Age, the end of both nuclear arsenals and nuclear power, yet, simultaneously, radiation has become so ubiquitous in contemporary life that it is nearly invisible, at once everywhere and unnoticed. It is the source of medical diagnostics from X-rays to PET scans and barium tracers, as well as a significant weapon in vanquishing certain cancers; it powers submarines and aircraft carriers; provides 20 percent of America's domestic energy (and 80 percent of France's); is used by anthropologists and forensic scientists to date biological remains, and by farmers to destroy bacteria; it remains a Pentagon mainstay and a weapon held dear by a club of developing nations who see it both as a route to global prestige . . . and the ultimate in defense against that same Pentagon.

At its root, radioactivity is wholly irrational to the human mind, appearing to exist somewhere between the quick and the dead. Uranium, thorium, and their ilk aren't biological, yet they have half-lives, the amount of time it takes for a radioactive element to lose half of its radiant force, which is followed by a loss of a quarter, then an eighth, then a sixteenth, then a thirty-second, and on and on . . . mathematically immortal. That seemingly "dead" rocks can send out powerful rays without external stimulus is counterintuitive, reminiscent of quicksilver, the half-liquid, half-metal state of room-temperature mercury that so transfixed Isaac Newton, he died, poisoned by it. This fundamental condition is nothing less than magical—matter spontaneously converting itself into energy; atoms flying apart all of their own volition; a process disturbing to our common sense of the world.

Radiation's powers rise from a simple condition: fat atoms. Uranium, thorium, and their radiant cousins are built from atoms so morbidly obese that they burst the laws of attraction—the fundamental building block of matter—and spit out a little piece of themselves, radiating subatomic alpha particles (two neutrons and two protons); beta particles (electrons); and energy waves of gamma rays (similar to X-rays). Imagine a giant blob of Ping-Pong balls, held together with rubber bands, but there are slightly more balls than the bands have the strength to cohere—fat atoms. Years may pass, but sooner or later, the rubber ties that don't bind enough will falter, and the blob will spit out a ball. The changes in atomic structure caused by the spit induces alchemy, a transformation into an isotope, or into a different element altogether.

When uranium ore is left atop photographic paper, it leaves the image of a rock veined in energy, of matter seeming to pulse with life. Hold silvery plutonium in your hand and it feels warm as a puppy . . . a big enough lump will boil its own water. The meltdown of runaway nuclear reactors, meanwhile, results in what physicist Robert Socolow describes as "afterheat, the fire that you can't put out, the generation of heat from fission fragments now and weeks from now and months from now." A fire that, unchecked, becomes eternal. And say what you will about the less than pleasant qualities of nuclear weapons . . . their detonations are rapturously beautiful.

Formed by stars that exploded into the gas and dust of supernovas, radiance is the main source of heat within the earth, and its force propels the tectonic shift of the continents. Its invisible rays trigger biological damage, birth defects, tumors, and cellular mayhem. Hiroshima; Godzilla; Dr. Strangelove; Nagasaki; Bikini; Spider-Man. What other history combines unimaginable horrors with genetic monstrosities, Armageddon fantasies, Hollywood tentpole grandees, and a revolution in swimwear? No wonder it's rare today for someone not to be at least a little bit radiophobic, alarmed by this omnipresent, invisible, mythic force. Yet the same rays that cause cancer can be used to cure cancer—drink the poison, or die—and the development of the most hideous weapon in the history of humankind has wholly eroded that same humankind's ability to wage global war against itself. Every time another country with erratic political leaders—Pakistan, North Korea, Iran—develops the ability to manufacture nuclear weapons, the world responds with grave fear. Yet, two of the greatest mass murderers in human history—Stalin and Mao—were nuclear armed and never used their atomic weapons. Sixty-five years and counting, and still the only country to ever drop the Bomb is the United States of America. The deterrent benefits have led more than one expert on this history, after detailed analysis, to propose awarding the atomic bomb with the Nobel Peace Prize.

Before this research, you and I probably had similar thoughts about atomic energy (an eternal, potential menace) and nuclear weapons (a moral and mortal hazard). Then I found out that Chernobyl has become something of a human-free Eden, that the survivors of Hiroshima and Nagasaki are in much better health than any of us could ever have imagined, that except for the radiating blanket of fallout, nothing can be accomplished with atomic weapons that can't be done with conventional explosives, and that Marie Curie was one hell of a broad. The very term *atomic bomb* originated with science fiction writer H. G. Wells and was taken up by the physicists of Los Alamos as something of a joke. Since everything in the material world is composed of atoms, not just nuclear weapons are atomic—all weapons are atomic, this book is atomic, and you are atomic. But when it comes to radioactivity in the modern world, this "atomic everything" paradox makes a piquant kind of sense. For example, the 2011 disaster at Japan's Fukushima Daiichi nuclear power plant was triggered by a power failure when the emergency backup batteries, stored foolishly in the basement, were destroyed by a tsunami. That oceanic flood originated with an earthquake, which was the result of crashing tectonic plates, which moved from the pressure and heat created by radiation rising from the earth's iron core. In the end, the Fukushima nuclear disaster was triggered by organic atomic forces . . . so in today's world, the "joke" has reverberated back on Los Alamos. Nuclear in power, in medicine, and in weaponry has become so pervasive that it might as well be "atomic," and the story of its birth, of nuclear's startling rise and slow-motion collapse, of the men and women who changed our lives in ways they could never imagine, from Curie to Oppenheimer, Teller to Reagan, and "duck and cover" to Fukushima, defies belief.

2

The Astonished Owner of a New and Mysterious Power

THE cataract of discovery that inaugurated the Atomic Age was a fifty-year revolution that transformed our scientific comprehension of matter, energy, and the essential ingredients of all that we know of the material world. Nearly every one of these great leaps forward was made, astonishingly enough, by an academic nonentity.

On the late afternoon of November 8, 1895, at the University of Würzburg, a fifty-year-old scientist who had been expelled from the Utrecht Technical School (and had never received a diploma) was investigating the electrostatic properties of various glass vacuum tubes, fitted with metal posts at each end. At the turn of the century, physicists were obsessed with electricity; their laboratory's stature was determined by battery power and the size of their sparks, with an appearance that has been re-created on a more epic scale in the movies of *Frankenstein*. Everything in a scientist's lab was handmade in this, the "sealing wax and string" era, as red Bank of England wax was liberally used to seal up leaking vacuum apparatuses, and delicate blown-glass tubes were held up with strings.

The Tesla coil of 1891 provided electrical investigators with their first generator of lightning-quality bolts, but the most popular turn-of-the-century sparker was the Rühmkorff—a widely admired London Rühmkorff had a 280-mile-long coil that could throw forty-two-inch jolts. These induction coils were powered by sulfuric and nitric acid batteries with zinc anodes that had to be cleaned with mercury, a combination that produced a constant gust of unsavory odors. By 1895, physicists were attaching each end of a vacuum tube to these coils, and trying to understand why throwing the switch would make the insides glow in blues and greens, a philosophic conundrum as these were "vacuum" tubes with presumably nothing inside them (though the mechanical methods of producing vacuums at the time rarely achieved perfect zero emptiness). The more advanced of these men and women thought that the revelation of the source of these glows might reveal the mysteries of electricity . . . and they were right. English chemist William Crookes insisted that the evanescences within cathode tubes, these cathode rays, must be a new form of "radiant matter" in a "fourth state"—neither solid, liquid, nor gas. French physicist Jean Perrin theorized that they were "corpuscles" carrying a negative charge . . . eventually known as electrons.

Wilhelm Röntgen (*RUNT-gun*)—described by a *McClure's* magazine profile as "a tall, slender, and loose-limbed man, whose whole appearance bespeaks enthusiasm and energy"—was director of the Physical Science Institute at the University of Würzburg and lived with his wife, Anna Bertha, upstairs from his two-room office, "a laboratory which, though in all ways modest, is destined to be enduringly historical. There was a wide table shelf running along the farther side, in front of the two windows, which were high, and gave plenty of light. In the centre was a stove; on the left, a small cabinet, whose shelves held the small objects which the professor had been using. There was a table in the left-hand corner; and another small table . . . was near the stove, and a Rühmkorff coil was on the right. The lesson of the laboratory was eloquent. Compared, for instance, with the elaborate, expensive, and complete apparatus of, say, the University of London, or of any of the great American universities, it was bare and unassuming to a degree." Today the site of this lab is easy for medical-imaging enthusiasts to find, as it's directly behind the Würzburg bus station.

Röntgen used a Raps pump to vacuum out the pear-shaped Hittorf-Crookes and zeppelin-like Lenard tubes, which he then connected to a Rühmkorff that could throw sparks of four to six inches. On November 8, 1895, he covered a Lenard with black cardboard, drew the curtains to completely darken the room and ensure that the cardboard jacket was light-tight, and flipped the current. He then did the same with a different style of tube, a Crookes, but this time, though the cardboard still kept all light within, he noticed an odd, green glow coming from a lab bench, about a meter away. He turned the current off . . . the glow from the bench faded . . . then clicked the switch back . . . and once again the glow resumed. Lighting a match, he went over and found a piece of cardboard coated in barium platinocyanide—a standard fluorescent screen called a *Leuchtschirm*—and realized that, somehow, invisible rays from the cathode tube had to be passing through the black cardboard sheath and igniting this distant screen: "A yellowish-green light spread all over its surface in clouds, waves, and flashes. The yellow-green luminescence, all the stranger and stronger in the darkness, trembled, wavered, and floated over the paper, in rhythm with the snapping of the discharge. Through the metal plate, the paper, myself, and the tin box, the invisible rays were flying, with an effect strange, interesting, and uncanny."

It was mysterious, and alarming. Wilhelm immediately began a series of investigatory experiments to learn as much as possible about these rays. He kept using thicker and thicker objects to try to block the emanations, from paper to cardboard to books, then experimented with sheets of metal. Only a disk of lead would wholly interfere; otherwise, the *Leuchtschirm* continued to luminesce. At one point he photographed his door, which produced a strange effect in the plate, and he couldn't understand this result. He took the door apart, and the answer was plain: lead paint.

Once while he was waving a lead disk between the tube and the screen, his hand fell before the stream. On the *Leuchtschirm*, within the vague, dark outline of the shadow of his skin, the bones of his fingers could plainly be seen. He was so stunned that he decided to tell absolutely no one about this: "When at first I made the startling discovery of the penetrating rays, it was such an extraordinary astonishing phenomenon that I had to convince myself repeatedly by doing the same experiment over and over again to make absolutely certain that the rays actually existed. . . . I was torn between doubt and hope, and did not want to have any other thoughts interfere with my experiments. . . . I was as if in a state of shock."

For the next two months, Röntgen spent every possible moment exploring this discovery, photographing the effects of the rays passing through wood, metal, books, and flesh, spending so much time at the lab that his wife became upset. When her husband then described what he'd found, Anna Bertha thought Wilhelm had lost his mind. On December 22, 1895, he asked her to come downstairs with him and had her rest her hand atop a cassette holding a photographic plate. He showered her with rays for fifteen minutes, then asked her to wait. He returned with the developed plate: a photograph of the bones in her hands and the rings on her fingers, with her flesh in soft outline around the whole. He was so pleased with what he had discovered. She was horrified, and like so many others in that era who would, for the first time, view what will remain, she cried out, "I have seen my death!"

Even after constant efforts in the lab, Röntgen remained so mystified by his rays that he could only name them X . . . the unknown. Almost immediately after he began to report his findings, others started working with cathode-ray screens and published follow-up reports in scientific, medical, and electrical journals, which were in turn almost immediately taken up by the popular press. *McClure's*: "Exactly what kind of a force Professor Röntgen has discovered he does not know. As will be seen below, he declines to call it a new kind of light, or a new form of electricity. He has given it the name of the X rays. Others speak of it as the Röntgen rays. Thus far its results only, and not its essence, are known. In the terminology of science it is generally called 'a new mode of motion,' or, in other words, a new force. As to whether it is or not actually a force new to science, or one of the known forces masquerading under strange conditions, weighty authorities are already arguing. More than one eminent scientist has already affected to see in it a key to the great mystery of the law of gravity. All who have expressed themselves in print have admitted, with more or less frankness, that, in view of Röntgen's discovery, science must forth-with revise, possibly to a revolutionary degree, the long accepted theories concerning the phenomena of light and sound."

"Röntgen ray" articles appeared on January 5 in Vienna's *Wiener Press*; January 7, in Frankfurt's *Frankfurter* and Berlin's *Vossiche*; January 11, London's *Saturday Review*; January 13, Paris's *Le Matin*; January 16, *New York Times*. In a journalist's game of "telephone," each would rewrite the previous item with an ever-growing collapse in accuracy, which continually enraged Röntgen. As the public then became obsessed with the discovery, the era's newspapers fed the hunger by publishing thousands of haunting photographs illuminating the shadowy flesh and lacy, geometric skeletons of mice, chickens, puppies, and birds. Journalist Cleveland Moffett described one example of what were called *shadow photographs*: "A more remarkable picture is one taken in the same way, but with a somewhat longer exposure—of a rabbit laid upon the ebonite plate, and so successfully pierced with the Röntgen rays that not only the bones of the body show plainly, but also the six grains of shot with which the animal was killed. The bones of the fore legs show with beautiful distinctness inside the shadowy flesh, while a closer inspection makes visible the ribs, the cartilages of the ear, and a lighter region in the centre of the body, which marks the location of the heart." Inside of a year, over fifty X-ray books and a thousand articles were released, and at London's Crystal Palace, lucky visitors could have their change purses Röntgen-rayed as a souvenir. The fad was versified in *Photography* magazine:

The Roentgen Rays, the Roentgen Rays.
What is this craze?
The town's ablaze
With the new phase
Of X-ray's ways.
I'm full of daze,
Shock and amaze;
For nowadays
I hear they'll gaze
Thro' cloak and gown—and even stays.
These naughty, naughty Roentgen Rays.

Kaiser Wilhelm asked his nation's most famous scientist to give him a private royal lecture in the Star Chamber on January 13, 1896, after which Röntgen was decorated with the Prussian Order of the Crown. One newspaper summed up the revolution: "Civilized man found himself the astonished owner of a new and mysterious power," and for this power, Röntgen would be awarded, in 1901, the first Nobel Prize in Physics. At the same time, his embarrassment at the public's hands continued, with X-rays taken up by spiritualists, Christians, somnambulists, and the temperance movement, with claims as well that they could erase the mustaches of women and transmit anatomical drawings directly into the brains of medical students. One man announced a secret alchemy technique of x-raying ordinary metals into gold; another claimed to have photographed souls.

Cartoons in British *Punch* and American *Life* portrayed high-society swells reduced to skeletons, while eavesdropping maids no longer had to stoop with their ears to keyholes, as they could now see through doors. The state of New Jersey thought it necessary to debate a bill forbidding X-ray glasses in theaters, while a London manufacturer produced shielding undergarments for ladies who didn't want to be seen naked on the streets by hordes of X-ray-spectacle-equipped voyeurs.

Miracles in our time always seem to combine blessing with menace. Röntgen had used a zinc box and a lead plate to focus his beams, to protect the photographic plates in his lab from being accidentally exposed. This procedure coincidentally protected Röntgen himself. Others were not so fortunate. In 1896 Columbia's H. D. Hawks demonstrated Röntgen rays at Bloomingdale's department store in New York City. He noticed a dryness on his skin, which became something like a sunburn, and then scales appeared. Over the following months, his fingernails stopped growing, the hair on the side of his head fell out, and he had trouble seeing. His eyelashes and eyebrows fell out, and the skin became extremely painful.

The February 1896 *Electrical World* announced that Thomas Edison suffered from "Röntgenmania." Following the public euphoria, the great inventor ordered his employees to investigate any and all X-ray possibilities for seventy hours nonstop, keeping them awake and working by hiring a man to aggressively play the accordion. By May, the New Jersey lab offered a fluoroscope demonstration at New York's Electrical Exhibit so that the public could appraise its own bones. The exhibit was run by glassblower Clarence Madison Dally, who then spent a number of years helping to develop an Edison X-ray lightbulb. After eight years of work, Dally's hair fell out and his skin started erupting in lesions that wouldn't heal. Edison canceled the bulb, but Dally continued working with Röntgen rays. Burns on his hands became cancerous; both of his arms were amputated to save his life. It didn't work, and he died in 1898 at the age of thirty-nine, becoming the first human known to be killed by X-rays. His death stopped Edison's Röntgenmania for good; the wizard of Menlo Park never worked with radiation again.

Blessing, with menace. X-rays started being used for medical diagnosis eight weeks after Röntgen announced his discovery. A student at Hahnemann Medical College in Chicago, Emil Grubbe, stuck his hand in an X-ray machine and noticed that, after a while, the skin from that hand was falling off. Showing this to one of his professors, he convinced him to try the rays on a breast-cancer patient named Rose Lee, diagnosed as hopeless. With the rays, Lee improved; the cancer shrank and seemed to remit. Radiotherapy was born. In February 1896, Grubbe founded the first radiation-therapy facility in Chicago—he didn't graduate from medical school until 1898—and by 1929, the rays had so damaged his left hand with cancer that it had to be amputated. In 1960, he died of squamous cell carcinoma.

By 1959, Germany's Röntgen Society announced that 359 people had died of X-ray overexposure. The mixed blessing produced by science, and its disturbing qualities, triggered mixed feelings about the discoverer. Wilhelm Röntgen, with his unruly beard and hair, wild and untamed, would become the world's image of a mad genius.

When Röntgen sent preprints of his article announcing X-rays out to fellow scientists at the end of 1895, his discovery was the dramatic breakthrough in an investigation of the mysterious relationship between matter and energy that had been building long before he'd ever charged a tube. One of the recipients was French mathematician Henri Poincaré, who shared Röntgen's X-ray photographs with the fellow members of Paris's Académie des Sciences on January 20, 1896. In that audience was Antoine Henri Becquerel, who, inspired by the fact that the X-rays seemed to emanate from the area of the vacuum tube that glowed, immediately began experimenting with fluorescent materials and their emissions.

Henri's grandfather, Antoine César Becquerel, was a Parisian celebrity for having discovered the use of electrolytes to refine metal and was one of the first graduates of the École Polytechnique, which became so central to the military, scientific, and engineering cultures of France that anyone wanting a career in those fields needed to be a *polytechnicien*. After Antoine César was told in 1815 that he was terminally ill and that death would arrive shortly, he resigned from the army, became a physics professor at Paris's Musée d'Histoire Naturelle, developed a keen interest in electricity, electrochemistry, and fluorescence, became the museum's director, and lived to the age of ninety. After himself graduating from the Polytechnique, son Alexandre-Edmond worked at his father's museum and taught at his father's school. The Becquerels would assemble at their institution a profound collection of minerals that could absorb, and then radiate, light. They were of two kinds: those that could glow after the light was turned off phosphoresced; and those that only glowed with the light on luminesced. Antoine César told his son, "I will never be satisfied with explanations they give why some chemicals and minerals shine in the dark. Fluorescence is a deep mystery and nature will not give up the secret easily." By 1896 and the age of X-rays, grandson Antoine Henri had the Musée chair and taught at the Polytechnique, following in his father's and grandfather's shoes with fluorescence and phosphorescence; he also typified his era by sporting a commandingly luxuriant and astonishingly manicured barrage of facial hair.

After winning his doctorate investigating crystal phosphorescence, Henri decided to take the Becquerels' dynastic expertise into a new direction inspired by Röntgen, with this one experiment: "One wraps a Lumière photographic plate with a bromide emulsion [photography then being done on glass panes coated with a warmed colloidal suspension of potassium bromide and silver nitrate] in two sheets of very thick black paper, such that the plate does not become clouded upon being exposed to the sun for a day. One places on the sheet of paper, on the outside, a slab of the phosphorescent substance [Becquerel used uranyl potassium sulfate—a uranium salt], and one exposes the whole to the sun for several hours. When one then develops the photographic plate, one recognizes that the silhouette of the phosphorescent substance appears in black on the negative. If one places between the phosphorescent substance and the paper a piece of money or a metal screen pierced with a cut-out design, one sees the image of these objects appear on the negative. . . . One must conclude from these experiments that the phosphorescent substance in question emits rays which pass through the opaque paper." If Becquerel had been able to conduct his research with modern photographic paper, he would have been even more flabbergasted, for the results are the shadow of a rock veined in energy; matter seeming to pulse with an inner life.

Henri was trying to determine if uranium captures sunlight and emits it later. One day was so overcast that he decided it wouldn't be productive, so he put away that day's plate and his rocks in a drawer. By accident, he developed that pane as well and was stunned to find that "there is an emission of rays without apparent cause. The sun has been excluded." Even after leaving his ore in the dark for many months, it still inscribed its uranic form into the photographic gelatin, and no other element he tried matched this feat. When he discovered that uranium's emanations could penetrate aluminum, copper, and even platinum, he believed he'd discovered another form of Röntgen rays.

Becquerel's uranium results were in equal measure mystifying and alarming because all the other known incidents of phosphorescence and luminescence began with an external source of illumination. Instead, the rays of uranium emitted all on their own accord. Unlike what everyone had known so far about the boundaries of the material world, Becquerel's accident revealed matter creating energy through its own volition: "Its luminosity came from within." He had discovered light that comes from a stone, and called it *les rayons uraniques*—"uranic rays."

Despite this earthshaking revelation, scientists at the time continued studying Röntgen rays instead of uranics—which, after all, did not dramatically reveal skeletons—but that lack of broad interest appealed to a Sorbonne graduate student looking for a topic for a doctorate in 1897, as this meant there was no lengthy history of journal scholarship to research, and the properties of uranic rays could immediately be investigated firsthand in the lab. This academic "shortcut" led to six years of backbreaking toil and a discovery that would revolutionize the science of physics, as well as the stature of women in the world.

When they were teenagers together in Warsaw, Bronya Skłodowska (*Squaw-DOFF-ska*) made a pact with her littlest sister, Manya. If Bronya was accepted to medical school in Paris, Manya would work two years to support her; then if Manya was accepted to the university, she would be supported in turn. As Polish women were forbidden from anything approaching higher education in the czarist Russian colony of that time, though, they first attended the Floating University—"we agreed among ourselves to give evening courses, each one teaching what he knew best," as Manya described it. This illicit underground educational collective got its name from the fact that its classes met in changing locations, the better to evade the eyes of imperial authorities and local snitches, and its students' lofty goal went far beyond mere self-improvement. They hoped their grassroots educational movement would raise the likelihood of eventual Polish liberation, and many followed the "positivist" philosophy of Auguste Comte, which promoted a scientific method for understanding both human affairs and the universe. Even to this day in the Polish tongue, a *positivist* is a pragmatist, accepting of people as they are and the world as it is.

From the example of the Skłodowskas, the Floating University was exemplary in educating its brave students. Bronya was indeed accepted to medical school at the Sorbonne in Paris, and Manya, as promised, began work as a nanny, first in Kraków with a family whom she unreservedly despised, telling her cousin Henrietta on December 10, 1885, "My existence has been that of a prisoner [working for] a family of lawyers [who] when there is company speak a chimney-sweeper's kind of French. . . . I shouldn't like my worst enemy to live in such a hell. . . . They are sunk in the darkest stupidity. . . . I learned to know the human race a little better by being there. I learned that the characters described in novels really do exist, and that one must not enter into contact with people who have been corrupted by wealth."

Manya was then hired by the Zorawskis, the managers of a sugar-beet plantation north of Warsaw. The family house was stucco, with a pleasure garden, a croquet lawn, forty horses, and sixty cows, adjacent to the immense brick sugar-beet factory, which processed the bounty of two hundred acres, continually arriving in a stream of oxen carts, and discharged into the river a "dark, sticky scum." Her charges on the estate included Bronka, eighteen; Andzia, ten; Stas, three; and Maryshna, six months. Manya wrote Henrietta that "Stas is very funny. His *nyanya* told him God was everywhere. And he, with his little face agonized, asked: 'Is he going to catch me? Will he bite me?' . . . I ought to think myself very lucky." Manya and Bronka, with the parents' assent, spent two hours a day teaching the farm's eighteen peasant children how to read and write in Polish—an effort considered such a crime by Russian authorities it was punished with hard labor at a concentration camp. The Zorawskis paid Manya a good salary of five hundred rubles a year, and she would stay with them for four years. This job would change Manya's future course in the world, but it also included "moments which I shall certainly count among the most cruel of my life."

The cruelty arrived when firstborn son Kazimierz (Casimir), studying mathematics and agricultural engineering at Warsaw University, returned home for the holidays. He quickly fell in love with the nineteen-year-old Manya, and she fell back, tail over teacup. In time, he told his parents they wanted to get married, and the young couple expected a happy consent. Everyone adored Manya. But, they were wrong. The parents believed their brilliant son was destined to marry above, not below, his station and forbade an engagement with this penniless nobody. Casimir agreed to his parents' wishes, which made him seem weak to Manya, but she still loved him, and she couldn't afford to quit a job that paid so well.

Then, for a number of years, Casimir wavered, telling Manya he loved her and had to have her as his bride and, alternately, that he had to accede to his parents' wishes. She wrote Henrietta on April 4, 1887, "If [men] don't want to marry impecunious young girls, let them go to the devil! Nobody is asking them anything. But why do they offend by troubling the peace of an innocent creature?" On November 25, 1888: "I have fallen into black melancholy. . . . My existence strangely resembles that of one of those slugs which haunt the dirty water of our river. . . . I was barely 18 when I came here, and what I have not been through! . . . I feel everything very violently, with a physical violence, and then I give myself a shaking, the vigor of my nature conquers, and it seems to me that I'm coming out of my nightmare. . . . First principle: never to let one's self be beaten down by persons or by events."

The tormented young woman tried drowning herself in her studies, reading ferociously after dinner every night—sociology, literature, history, even an advanced math course she completed, by mail, with help from her father. Repeatedly her interests and her talents were sparked by physics and chemistry, so much so that she convinced one of the factory chemists to give her lessons. Then the Zorawskis learned that Casimir and Manya were still illicitly seeing each other. She was fired. Completely heartbroken, Manya returned to Warsaw, lived with her father, worked at a few more governess jobs, and, through her cousin Joseph, got laboratory experience at the clandestine Museum of Industry and Agriculture, where Joseph illegally educated a generation of Polish scientists.

Then in March 1890, Manya received a letter from Bronya. The elder sister announced she was engaged to a very different Casimir—a man who would be deported to Siberia if he ever returned to the Russian empire, as he was believed to be one of the conspirators behind Czar Alexander II's assassination—and that her studies were complete. Now it was Manya's turn, and Bronya invited her littlest sister to come to Paris to live with the new couple and be supported financially, as promised so long ago. But now, Manya wavered, for she was still so much in love. Finally in the fall of 1891, Casimir Zorawski wrote to say that their relationship was categorically finished. Manya left Poland, for Paris.

But this is not the end of that story. After growing up to become a well-regarded mathematician in Poland, the adult Casimir Zorawski would frequently be seen gazing up at Warsaw's monumental statue of the nation's great heroine, Marie Curie, the "penniless nobody" he had lost forever.

Born November 7, 1867, in the province of Vistula Land, a Poland brutally ruled by a vengeful Moscow, Marja Skłodowska was the baby of her family and like all Polish husbands, wives, pets, children, and cherished possessions, she and her brothers and sisters all had nicknames. Zofia was Zosia; Bronisława, Bronya; Helen, Hela; Joseph, Jozio; and the youngest, Marja, went by Manya, Manyusya, and Anciupeccio. The birth of her fifth and last child led mother Bronya to resign her position as head of a Warsaw school, where the family had resided in complimentary housing; she now worked from home, as a cobbler. Then, she became ill, the beginnings of a family cataclysm. Bronya had been taking care of her husband's younger brother, sickened with tuberculosis. The brother passed, and soon enough Bronya herself was infected, having likely contracted the disease from her good intentions.

Three years before Manya's birth, Polish nationalists waged revolutionary assaults against the Russian colonial authorities and were defeated. Tens of thousands were interned in Siberian slave-labor camps; hundreds of thousands fled to live in exile, and the rulers began a program of "Russification." Manya's daughter, Eve: "For the children, the dreadful nature of Czarist occupation was in the Russian-appointed head of the gymnasium, Ivanov. They were taught that Poland was a province and their language a dialect, and forced to recite their Catholic prayers in Russian." Manya: "Constantly held in suspicion and spied upon, the children knew that a single conversation in Polish, or an imprudent word, might seriously harm, not only themselves, but also their families."

Manya's grandparents had been comfortably well-to-do, and the furniture in the family's study revealed this prosperous ancestry: a desk of French mahogany, Restoration armchairs in red velvet, a green malachite clock, a table inlaid with a marble checkerboard; the portrait of a bishop; and a collection of scientific apparatuses, including an oak-mounted barometer and a gold-leaf electroscope. But Manya's parents' sympathies with the Polish liberation movement destroyed their earnings as a state-employed teacher and school administrator, and that combined with poor investments erased their inheritance.

Eve described what happened next: "While his wife was being treated on the Riviera in Nice, his brother-in-law had lost 30,000 rubles of the family's money in a steam mill; the father was forced to take in boarders. The proper family was unraveled into chaos and cacophony; Manya had to study with her thumbs in her ears. But she became so absorbed in whatever she turned her mind to that the family made a joke of making a tremendous noise around her and watching as she continued to read and pay them no attentions." The girl would have this power of concentration for the rest of her life, as Manya herself said, "Weak as I am, in order not to let my mind fly away on every wind that blows, yielding to the slightest breath it encounters, it would be necessary either to have everything motionless around me, or else, speeding on like a humming top, in movement itself to be rendered impervious to external things. . . . One must make of life a dream, and of that dream a reality."

Mother Bronya then returned from Nice, her condition unimproved. Eve: "One of the boarders infected Bronya and Zosia with typhus, killing the elder sister. Their mother, too weak to leave the house, watched from the windows as the cortege took her first born away forever." Manya was eight when her beautiful sister Zosia died, but this was not the end of the family's sorrow. The children prayed every night for their mother's health to be restored, but on May 9, 1878, she died as well. Manya was ten, and "would often sit in some corner and cry bitterly. Her tears could not be stopped by anybody," sister Hela said, while Manya remembered, "For many years we all felt weighing on us the loss of the one who had been the soul of the house." The family was forced to sell their country home, but at least some of their friends and relatives still had estates, where they were invited to spend the summers, tramping the woods for mushrooms and whortleberries, playing dress-up as traditional peasants, riding horse-drawn sleighs, and waltzing at night to fiddlers on the lawn. Manya: "We sleep sometimes at night and sometimes by day, we dance, and we run to such follies that sometimes we deserve to be locked up in an asylum for the insane."

Traveling as economically as possible, Manya Skłodowska left Warsaw for Paris carrying not only enough food and reading for the trip, but also a folding chair and a blanket, as German fourth-class travel did not include seats. She then joined Bronya and her husband at their apartment on the rue d'Allemagne at the city's edge in smelling distance of la Villette, the abattoir. Twice a day the small, young foreign woman climbed the spiral staircase of a horse-drawn double-decker omnibus for the hour-long commute to and from the Sorbonne. The top section, open to the elements and known as the imperial, held the cheapest seats, with the best views. It was impossible, but she was in Paris, taking classes at the finest university in all of continental Europe. She filled out her registration card not as Marja or Manya, but as Marie Skłodowska, half-French, and half-foreign, with a surname no one but a Pole could pronounce—regardless, everyone who knew her forever called her Manya. Marie's educational background and laboratory experience were dramatically scantier than that of her fellow classmates, and she struggled to keep up with schooling and with the mysteries of the French tongue. Stubborn, shy, dressed in the gray-wool-and-pomegranate-linen dress of a poor immigrant scraping through life, Marie was nevertheless a striking woman, with ash-blond hair and porcelain complexion. She was so pretty that one of her friends told a group of boys to leave them alone or risk a beating from her *parapluie*.

Bronya and Casimir turned their apartment into *petite Pologne*, with every night a gathering of homesick Varsovians in exile, enjoying wine, cake, vodka, tea, and a piano player by the name of Ignaz Paderewski, all in heated conversation over science and politics—two future presidents of Poland were among the regulars. It was a glorious reminder of home, and it was also too much of a distraction for an overburdened university student. After six months, Marie moved into a sixth-floor *chambre de bonne*—a one-room garret—at 3 rue Flatters off boulevard de Port-Royal in the *quartier latin*, paying twenty-five francs a month, with no hot water as she couldn't afford heat. Sometimes, she fainted from hunger. Her first real adult home had a folding iron bed, a white wooden table, a big brown trunk (which could be sat on in an emergency), an oil lamp, a stove, an alcohol-fed oven, a washing tub, one fork, one knife, one spoon, one cup, one cooking pan, two plates, and one teakettle with three glasses for herself and her only guests, the relatives. She knew how to sew, but never made herself any clothes, only repairing her Polish outfits again and again. On nights when it was too cold to sleep, she put everything she owned—clothes, coats, towels—on top of her coverlet, then put the single chair on top of all of it to weigh everything down and perhaps give at least the illusion of warmth. "My situation was not exceptional; it was the familiar experience of many of the Polish students whom I knew," she wrote later. "I carried the little coal I used up six flights. . . . I prepared my meals with the aid of an alcohol lamp. . . . It gave me a precious sense of liberty and independence. . . . [This was] one of the best memories of my life."

After all her troubles as a university student, undereducated because she was a woman and foreign-born, Marie finished first in her master's-degree physics course in the summer of 1893 and second in math the following year. Then a miraculous stroke of luck came her way: the same friend who'd defended her on the streets with an umbrella got her awarded the Alexandrovitch scholarship—six hundred rubles—enough for over a year of living expenses. She was saved. The following year, she was awarded her master's in mathematics, and by then, she spoke perfect French with a bare whisper of Polish flavor. In a few years' time, she would have her first paid assignment, and out of those fees she would repay, for the first time in its history, the scholarship, so another impoverished student could be given help when it was most needed.

Before completing the math degree, she was commissioned by the Society for the Encouragement of National Industry to do a study aligning the magnetic properties of different steels to their chemical compositions. She needed to find a lab where she could do this work, and a friend she'd met while working for the Zorawskis knew someone who might have a room, a teacher at the École Supérieure de Physique et de Chimie Industrielles de la Ville (the city's industrial engineering and chemistry school). Additionally, this friend of a friend of a friend had a number of remarkable similarities to Marie. She was investigating the magnetic properties of steel, and this city schoolteacher had discovered a remarkable interaction of heat and magnetism; Marie was educated outside any state system through the Floating University and her program of self-education while working as a governess, while Pierre had been homeschooled by his parents, his brother, and a tutor before qualifying to enter the Sorbonne. They were both workaholics, with his family nearly as well educated and as financially precarious as the Skłodowskas. And they were both outsiders in the French scientific community, who expected their members to have a Polytechnique education like the Becquerels.

Marie: "Pierre's intellectual capacities were not those which would permit the rapid assimilation of a prescribed course of studies. His dreamer's spirit would not submit itself to the ordering of the intellectual effort imposed by the school. . . . He grew up in all freedom, developing his taste for natural science through his excursions into the country, where he collected plants and animals for his father." Pierre: "I did not regret my nights passed in the woods, and my solitary days. If I had the time I would let myself recount my musings. I would describe my delicious valley, filled with the perfume of aromatic plants, the beautiful mass of foliage, so fresh and so humid, that hung over the Bievre, the fairy palace with its colonnade of hops, the stony hills, red with heather. . . . We must eat, sleep, be idle, have sex, love, touch the sweetest things in life and yet not succumb to them."

Pierre Curie had fallen in love as a young man, but then the girl died, and a lack of income forced him to put off work toward his doctorate indefinitely. Instead, he became a poorly paid laboratory instructor at the city school; at the age of thirty-five, he was still living with his parents. Working with elder brother Jacques, Pierre studied crystals—quartz, tourmaline, topaz, sugar—and found that, when they were compressed along the axis of symmetry, they produced a charge—piezoelectricity (from the Greek, "to squeeze"). For the precise measurements needed for this work, the physicist brothers created a highly sensitive instrument that combined tiny weights, microscopic meter readers, and pneumatic dampeners—the Curie scale. Then, heating various materials to 1,400°C (over 2,500°F), they discovered a link between heat and magnetism. Today the temperature that a given element loses its magnetism, the Curie point, is used in studying plate tectonics, treating hypothermia, measuring the caffeine in beverages, and understanding extraterrestrial magnetic fields, while piezoelectricity is found in mechanisms propelling the droplets of ink-jet printers, regulating time in quartz watches, controlling the shrill wail of smoke detectors, turning the adjustable lenses of autofocus cameras, acting as the pickups of electric guitars, giving a spark to electric cigarette lighters, reducing vibrations within tennis rackets, and sending out high-frequency audio to monitor the heartbeats of fetuses.

Lord Kelvin was so taken with the Curie brothers' work on electric quartz that he arranged for a number of visits with Pierre in his lab the first week of October 1893, the same period that the impoverished teacher was meeting the great love of his life. Marie Curie: "As I entered the room, Pierre Curie was standing in the recess of a French window opening on a balcony.

He seemed to me very young, though he was at that time thirty-five years old. I was struck by the open expression of his face and by the slight suggestion of detachment in his whole attitude. His speech, rather slow and deliberate, his simplicity, and his smile, at once grave and youthful, inspired confidence. . . . There was, between his conceptions and mine, despite the difference between our native countries, a surprising kinship, no doubt attributable to a certain likeness in the moral atmosphere in which we were both raised by our families. . . . Soon he caught the habit of speaking to me of his dream of an existence consecrated entirely to scientific research, and he asked me to share that life. It was not, however, easy for me to make such a decision, for it meant separation from my country and my family, and the renouncement of certain social projects that were dear to me. Having grown up in an atmosphere of patriotism kept alive by the oppression of Poland, I wished, like many other young people of my country, to contribute my effort toward the conservation of our national spirit."

Marie was the first woman Pierre had encountered in fifteen years who was both attractive physically and shared his great passion for science. She felt likewise; besides his professional achievements, Marie "noticed the grave and gentle expression of his face, as well as a certain abandon in his attitude, suggesting the dreamer absorbed in his reflections." But then he asked if she planned to remain in France permanently, and she said, "Certainly not. . . . I shall be a teacher in Poland; I shall try to be useful. Poles have no right to abandon their country."

After a few months passed, Marie made plans for a trip to Warsaw, for a vacation with her family, and to apply to graduate school in her native country. Pierre suddenly insisted, "Promise me that you will come back! If you stay in Poland you can't possibly continue your studies. You have no right to abandon science." Marie later said that she felt what Pierre really meant by this was "You have no right to abandon me." But she could never, in turn, imagine abandoning Poland, or marrying a man who wasn't Polish, and only allowed Pierre to consider themselves as friends.

While Marie was away, a torrent of letters arrived from Pierre in his childlike writing, signed "your very devoted friend" and begging her to return: "We promised each other (isn't it true?) to have, for each other, at least a great affection. As long as you do not change your mind! For there are no promises which hold: these are things that do not admit of compulsion." Pierre "had a touching desire to know all that was dear to me," Marie said. He even learned a bit of her difficult native language and, when she finally returned in October, made a remarkably abject offer: he would move to Poland and find some kind of position, if only she would marry him. Pierre Curie: "It would, nevertheless, be a beautiful thing in which I hardly dare believe, to pass through life together hypnotized in our dreams: your dream for your country; our dream for humanity; our dream for science. Of all these dreams, I believe the last, alone, is legitimate. I mean to say by this that we are powerless to change the social order. Even if this were not true we should not know what to do. And in working without understanding we should never be sure that we were not doing more harm than good, by retarding some inevitable evolution. From the point of view of science, on the contrary, we can pretend to accomplish something. The territory here is more solid and obvious, and however small it is, it is truly in our possession."

After Kraków University rejected her application because she was a woman, Marie decided to compromise, writing a friend, "It is a sorrow to me to have to stay forever in Paris, but what am I to do? Fate has made us deeply attached to each other and we cannot endure the idea of separating." In time, she would change her mind, falling deeper and deeper in love: "I have the best husband one could dream of; I could never have imagined finding one like him. He is a true gift of heaven, and the more we live together the more we love each other." Pierre: "I think of you who fill my life, and I long for new powers. It seems to me that in concentrating my mind exclusively upon you, as I am doing, that I should succeed in seeing you, and in following what you are doing; and that I should be able to make you feel that I am altogether yours at this moment—but the image does not come."

Their wedding was in every way untraditional. The couple needed no lawyers since their only possessions were two bicycles bought the day before with wedding money from a cousin. They would have no white dress or tails, no gold rings, no formal breakfast, and no religious ceremony, as Pierre was a freethinker. The bride's attire, a navy wool suit and blue-on-blue-striped blouse, was paid for by her brother-in-law's mother and sewn by Mme. Glet according to Marie's requirements that it be "practical and dark, so that I can put it on afterwards to go to the laboratory."

The couple rode, together, atop the omnibus across boulevard Saint-Michel to the Gare du Luxembourg for the train to Pierre's hometown of Sceaux, where his parents still lived. They were married at city hall, with a reception in the garden of the Curie home. Taking their *vélos* on the train, they honeymooned in Brittany. The two would in time discover they both enjoyed long bike rides and overseas travel, and over the next eleven years, Marie remembered, "My husband and I were so closely united by our affection and our common work that we passed nearly all of our time together."

Having refused Dr. Curie's marital gift of furniture, their drawing room in Paris had a wooden table with two chairs, one for each of them, and none for any guests. Photographs depict the Curies as a remarkably severe couple: Pierre with his gaunt face, his bristling salt-and-pepper Vandyke, and his military brush cut; and his wife, Marie, tough as cancer. But clearly, they were as meant for each other as any man and woman in history. Though he had the significant scientific background, it was the unstoppable, indefatigable force known as Marie Curie, with her seemingly infinite reserves of energy and ambition, that drove the couple professionally. With her encouragement (and very likely nudging), by 1895 Pierre had won the doctorate he'd long deserved and was promoted to a full professorship at the city school. In addition to the two master's degrees she held by the time of her marriage, Marie passed the French state exam to teach science to women, while continuing to experiment on magnetics and steel. The director at Pierre's school gave her a lab to use, and she convinced French metalworks companies to donate materials, a trinity of corporate, government, and academic funding she would juggle for the rest of her professional life, and which would become a standard for modern practice in the era of big science.

In the summer of 1897, when they would be separated by Pierre's work and Marie's difficult first pregnancy, they would write back and forth in Polish, he poetically, beginning each with "my dear little child whom I love"; she, in language plain enough that he might understand it. Then in her eighth month, during a bicycle trip when she said all was fine, they were forced to rush back to Paris where Marie's father-in-law oversaw the birth of his granddaughter, Irène (*ee-REN*), on September 12. Though impossible to imagine for the first twenty years of her life, Irène and her husband would in time achieve a professional stature nearly as prominent as that of her parents.

Soon after giving birth, Marie decided on her doctoral topic: Henri Becquerel's rays, which she picked since "the subject seemed to us very attractive and all the more so because the question was entirely new and nothing yet had been written upon it." With this pragmatic notion, Marie Curie had found a subject to study for the rest of her life; a partner to study it with; and a temple where she suffered and was redeemed.

Before Becquerel, German pharmacist Martin Klaproth named the element uranium in the spring of 1789 after the recently discovered Uranus. It was used to stain glass in the Roman empire, the bodies of American Indians, and the glazed pottery of Depression-era America (eating from Fiestaware orange-red plates produced before 1942 is hazardous, though the maker has argued, "In truth, the red glaze emitted far less radiation than some other consumer products").

Marie arranged to get a ton of ore donated from Bohemia's St. Joachim's Valley, where a mine in the 1500s produced 2 million silver coins called *joachimsthalers* (or *thalers*, which became the English word *dollars*). Only interested in silver, the Bohemians ignored the various yellow, orange, and green ores they called "bad luck tar rock," or *Pechblende* (English, *pitchblende*). Not realizing what bad luck this tar rock really was, though, the miners would, two decades later, choke up blood for about six weeks and then die from an unspecified "mountain sickness."

With Pierre's help, Marie built an ionization chamber out of wooden crates discarded by their grocer. Inside were two metal plates, with the element to be tested resting on the lower plate and one of Pierre and Jacques's delicate instruments on the upper. By charging the lower with a battery, Marie could determine if the element electrified the air—as Becquerel had noted—through a current detected by the instrument. Besides its being one of the great love stories of the twentieth century, then, Marie Curie's great professional luck in meeting Pierre was that he had coinvented the piezo electrometer.

What were these uranic rays, this invisible power somehow generated by inorganic minerals? Marie confirmed Becquerel's assertion that the rays' force was not affected by wetting, drying, heating, illuminating, compressing, or pulverizing; that nothing but the amount of uranium itself determined the amount of voltage emanated. But Marie and Pierre could not understand how uranium's rays were birthed. They first theorized that a special feature within uranium absorbed cosmic rays from space, then slowly released them. To test this, German schoolteachers Julius Ester and Hans Friedrich Geitel buried radioactive materials beneath 300 meters of Harz mountain rock as well as at the bottom of an 850-meter mine shaft, for forty-eight hours. Neither had an effect on their emanations, and Marie, along with Ester and Geitel, then went beyond Becquerel to theorize that the effulgence must arise not from chemistry (from the interaction of uranium with other elements), but solely from within the element's very atoms. By their showing all the ways in which its power was unaffected, radiance by default had to be an atomic property of the element uranium. For the rest of his life, Pierre Curie remained convinced that the process was an energy transfer, similar to thermodynamics, and spent many of his last years trying to apply theories of heat to radium and polonium. But if atoms were constantly losing their energy through a thermodynamic-like process, they would eventually either implode or explode, and his experiments to counter Marie's atomic assertion were all failures.

Out of everything Madame Curie would discover, as science this was the simplest, most significant, and most revolutionary. She had pointed to the first physical evidence that enormous energy lay within the very essence of matter. It was revolutionary because, as she noted, "from this point of view, the atom of radium would be in a process of evolution, and we should be forced to abandon the theory of the invariability of atoms, which is at the foundation of modern chemistry." The fundamental law of thermodynamics, which forbade the creation of energy from nothing, had been undone.

On February 17, 1898, Marie's piezo electrometer measured torbernite (or chalcolite) having twice the radiance of uranium, while pitchblende ore was four times as vibrant. This only made sense if some other, even more powerful, radiating element, still unknown, lay within these compounds, and clearly, Becquerel was mistaken in calling them uranic rays. She tested and recalibrated her instruments and still had the same results, working constantly to explain this mystery. Her speed came from a fear of being trumped, as she knew full well that if Becquerel (who was overseeing her doctorate) had not told the Académie des Sciences of his own findings the very day after he made them, the discoverer of Becquerel rays would instead have been Silvanus Thompson, who announced his identical discovery one day later.

Pierre was so fascinated by his wife's conclusions that on March 18 he dropped his work with crystals and joined her efforts. "Neither of us could foresee that in beginning this work we were to enter the path of a new science which we should follow for all our future," she later said. Eventually the Curies worked with seven tons of pitchblende from Bohemia, black ore suffused with pine needles, in a "laboratory" that was essentially a hut that the municipal medical school used for its students to dissect human corpses. But now it was in such disrepair, especially the leaking roof, that it wasn't even fit for cadavers. (The Institut Curie is now located on the same rue Lhomond as the shed, adjacent to rue Pierre et Marie Curie.) The hut's glass roof made summers roasting, winters debilitating, and rain an imminent presence; the stove used for heat was too weak to be useful; the only ventilation was the opening of a window and a door, meaning that processes involving fumes, which were innumerable, were conducted in the courtyard . . . with any rainstorms forcing the scientists to scurry their equipment back into the leaky shed, where they worked at remarkable physical labors from 1898 to 1902 . . . four toilsome years. Marie:

> The life of a great scientist in his laboratory is not, as many may think, a peaceful idyll. More often it is a bitter battle with things, with one's surroundings, and above all with oneself. . . . Between the days of fecund productivity are inserted days of uncertainty when nothing seems to succeed, and when even matter itself seems hostile; and it is then that one must hold out against discouragement. I had to work with as much as 20 kg of material at a time so that the hangar was filled with great vessels full of precipitate and of liquids. It was exhausting work to move the containers about, to transfer the liquids, and to stir for hours at a time, with an iron bar, the boiling material in the cast-iron basin. . . . I extracted from the mineral the radium-bearing barium, and this, in the state of chloride, I submitted to a fractional crystallization. . . . And yet it was in this miserable old shed that the best and happiest years of our life were spent, entirely consecrated to work. I sometimes passed the whole day stirring a mass in ebullition, with an iron rod nearly as big as myself.

The couple carried on between them the labors of a large chemical plant. Even though that winter was especially harsh, their work had to be done out of doors due to the fires and fumes. The first step was to melt the crude ore in a large, oblong tank until it was boiling like lava. Then acids were poured in to dissolve out the salts. The next stage was to melt down the residue in separate cauldrons, fired up twenty-four hours a day, with either Pierre or Marie present throughout. The reduced ore had to be filtered again and again to remove all other elements, and then evaporated in small bowls . . . revealing crystals. Marie: "We lived in our single preoccupation as if in a dream. We're very happy in spite of the difficult conditions under which we work. We passed our days at the laboratory, often eating a simple student's lunch there. A great tranquility reigned in our poor shabby hangar; occasionally, while observing an operation, we would walk up and down talking about work, present and future. When we were cold, a cup of hot tea, drunk beside the stove, cheered us."

On April 14, they ground up one hundred grams of pitchblende to prepare it for crystallization, knowing full well that they were searching through an agglomeration of thirty or so elements arrayed in multiple compounds, yet having no idea that the elements they wanted were so rare that seven tons of ore would have to be processed to extract one gram. With advice from their school's chemists, they heated, distilled, pulverized, and precipitated with ammonium, until Marie's samples registered 300 times as radiant as uranium's, and Pierre's 350 times. Each time they thought they were done, however, the spectroscope refused to produce clear lines revealing a new element. Inside of a month, they were able to isolate two concentrations of ore radiant enough to publish findings. In their report of July 1898, "On a New Radio-Active Element Contained in Pitchblende," they announced the discovery of a new member of the periodic table named for the home where Marie couldn't live, yet couldn't say farewell to: polonium. The same paper coined a new term for the emanation of Becquerel rays—"radio-active"—and called matter that emanated "radio-elements."

The more Marie learned about uranium and its emanations, the more in love she fell. Manya Skłodowska may have renounced religion with the death of her mother and sister, but she seemed a penitent in the arms of the Lord when it came to her approach to science: monastic, devoted, chaste, she lived her life in what Pasteur had called "the temples of the future": laboratories. This would be especially true after Bronya and Casimir decided to leave France and open a tuberculosis sanitarium in their beloved Zakopane, Poland. Marie was brokenhearted, writing to Bronya on December 2, 1898, "You can't imagine what a hole you have made in my life. . . . I have lost everything I clung to in Paris except my husband and child. It seems to me that Paris no longer exists, aside from our lodging and the school where we work." Yet in that period, she would also say, "Life is not easy for any of us. But what of that? We must have perseverance and above all confidence with ourselves. We must believe that we are gifted for something, and that this thing, at whatever cost, must be attained."

After three months of vacation in Auvergne, the Curies returned to work in November and made rapid progress, a barium concentrate producing results nine hundred times as strong as uranium's. One of the school's chemists could finally see their second element through the spectroscope, and around December 20 they named it: radium. After four years, forty tons of chemicals, and four hundred tons of water, on March 28, 1902, they produced one-tenth of a gram of radium chloride.

In time, English chemist Frederick Soddy would work with New Zealand physicist Ernest Rutherford to discover the secret of uranic rays, the remarkable ability of radioactive elements to, through the spontaneous loss of subatomic particles, change into other elements, producing an emanation of alpha, beta, or gamma rays over the course of what they called a half-life. Subatomically bloated, these elements are forced to constantly shed neutrons or electrons until they achieve a stable, nonradioactive form and are at nucleic peace. It was, to Rutherford and Soddy's great dismay, the transmutation that alchemists had pursued for centuries . . . dismay, as alchemy had been a laughable topic for generations. But half-lives themselves are pretty funny, when they aren't being cosmically grand, such as what happens to the most common form of uranium over its many lives as it ejects subatomic particles and alchemizes into various elements and isotopes:

> Uranium-238 has a ½ life of 4½ billion years, after which it turns into
> Thorium-234, with a ½ life of 24 days, after which it turns into
> Protactinium-234, with a ½ life of 1.16 minutes, after which it turns into
> Uranium-234, with a ½ life of 245,500 years, after which it turns into
> Thorium-230, with a ½ life of 75,380 years, after which it turns into
> Radium-226, with a ½ life of 1,620 years, after which it turns into
> Radon-222, with a ½ life of 3.8 days, after which it turns into
> Polonium-218, with a ½ life of 3 minutes, after which it turns into
> Lead-214, with a ½ life of 26.8 minutes, after which it turns into
> Bismuth-214, with a ½ life of 20 minutes, after which it turns into
> Polonium-214, with a ½ life of 0.164 microseconds, after which it turns into
> Lead-210, with a ½ life of 22.3 years, after which it turns into
> Bismuth-210, with a ½ life of 5 days, after which it turns into
> Polonium-210, with a ½ life of 138 days, after which it turns into
> Lead-206, which is stable, not radioactive, and has no ½ life.

While Pierre investigated radium's signature properties (including that it generated enough continuous heat to melt its own weight in ice in under sixty minutes—the first clue to nuclear power), Marie experimented with the industrial-chemistry recipes needed to isolate her new elements. They tried finding an atomic weight by measuring unrefined against refined samples, but couldn't, and from this they knew the element was in tiny amounts and very, very powerful. Three years later they would discover it was less than one-millionth of 1 percent, and this was only the start of its magic. Marie: "The chloride and bromide, freshly prepared and free from water, emit a light which resembles that of a glow-worm. . . . A glass vessel containing radium spontaneously charges itself with electricity. If the glass has a weak spot, for example, if it is scratched by a file, an electric spark is produced at that point, the vessel crumbles like a Leiden jar when overcharged, and the electric shock of the rupture is felt by the fingers holding the glass." Marie would then note the remarkable property that Irène would investigate and that in time would revolutionize both medical diagnosis and treatment:

"Radium has the power of communicating its radioactivity to surrounding bodies. When a solution of a radium salt is placed in a closed vessel, the radioactivity in part leaves the solution and distributes itself through the vessel, the walls of which become radioactive and luminous."

At that moment, there was no greater scientific achievement than adding new elements to the periodic table. The Curies had discovered two, publishing their proofs in nine months. Also, both elements brilliantly luminesced, radium with an aquatic shimmer reminiscent of absinthe. When the couple pressed glowing radium against their eyelids, they saw fireworks and meteors flashing across the retinas.

The other scientist investigating radium was a German organic chemist employed by a quinine factory, Friedrich Giesel, who said the blue light it produced was so powerful it could be employed as a night-light for reading. He advised the Curies to try bromide salts instead of chlorides during crystallization, and was able to deflect the path of Becquerel waves with a magnet, proving that they were, in fact, a form of matter. He also revealed that, when he fired the alluring radium with his Bunsen burner, it didn't ignite with a green flame, like barium, but with a blaze that was the color of Christmas cherries. Marie's beloved radium, then, had a sapphire light, but a carmine flame.

A little house at boulevard Kellermann 108 was now home to Pierre, Marie, the four-year-old Irène, and Pierre's widowed father. One evening at nine o'clock, after her daughter was put to bed, Marie turned to Pierre and asked if they could go back to the shed, to their radium. They went to look, making sure to not turn on the lights. There, in shelves and on tables, the aquatic glow of their babies shone in the night: "Sometimes we returned in the evening after dinner for another survey of our domain. Our precious products, for which we had no shelter, were arranged on tables and boards; from all sides we could see their slightly luminous silhouettes, and these gleamings, which seemed suspended in the darkness, stirred us with ever new emotion and enchantment. . . . The glowing tubes looked like faint, fairy lights." She loved the radiance so much that she would wait a few minutes before turning on the lights in her lab after arriving on dark, wintry mornings, to enjoy her shimmering vials. Other visitors noticed that, even after the samples were removed, the walls themselves continued to glow.

Only a heavy blanket of lead could contain the powerful rays of the Curies' greatest discovery. Radium produced light, heat, and helium; it ionized the air and excited photographic plates; it tinted glass a delicate purple and dissolved paper into ash; and it could infect other substances with its emanations. Diamonds when treated would phosphoresce brilliantly; imitations, poorly, if at all. Sir William Crookes (of the Crookes tube that had originally served Röntgen) prepared for the Royal Society a 1903 demonstration of radioactivity: "Viewed through a magnifying glass, the sensitive [zinc sulfide] screen is seen to be the object of a veritable bombardment by particles of infinite minuteness, which, themselves invisible, make known their arrival on the screen by flashes of light, just as a shell coming from the blue announces itself by an explosion." Marie called the process a "cataclysm of atomic transformation," and she tried to explain the magic through science: "The sensitive plate, the gas which is ionized, the fluorescent screen, are in reality receivers, into another kind of energy, chemical energy, ionic energy . . . luminous energy . . . and once more we are forced to recognize how limited is our direct perception of the world which surrounds us, and how numerous and varied may be the phenomena which we pass without a suspicion of their existence until the day when a fortunate hazard reveals them. . . . If we consider these radiations in their entirety—the ultra-violet, the luminous, the infra-red, and the electromagnetic—we find that the radiations we see constitute but an insignificant fraction of those that exist in space. But it is human nature to believe that the phenomena we know are the only ones that exist, and whenever some chance discovery extends the limits of our knowledge we are filled with amazement. We cannot become accustomed to the idea that we live in a world that is revealed to us only in a restricted portion of its manifestations."

During the age of the séance, this was a resonant notion, and radium, with its magical properties, appeared as an element of the otherworld. Electric lights, radio, telegraph, spiritualism—all unseen forces that were both magical to the human mind and, in their way, threatening. At that time, instead of being contrasted with science, spiritualism embraced it. The miracle of the telegraph bridged the divide of geography; séances bridged the divide of death. The Curies, along with Pierre's brilliant student Paul Langevin, attended a number of séances with medium Eusapia Palladino, and during one evening, Pierre weighed her and discovered she'd gained six kilos (thirteen pounds). Other French spiritualists were communicating with the spirit world through planchettes, small tablets cut in the shape of a heart, with two legs and a pencil. Two men in Maryland combined the planchette with an alphabet board; another named the new contraption by merging the French and German words for "yes." As the Great War unleashed a torrent of horror, grieving mothers and wives bought thousands of Ouija sets to communicate with their departed.

Two years before Manya's arrival in France, Paris celebrated an 1889 World's Fair by erecting Eiffel's tower, and now that tower was the centerpiece of 1900's Universal Exposition, which inspired Henry Adams to believe the age of the Virgin had been eclipsed by the age of the Dynamo. Electricity had begun to replace gaslights in Paris in 1891, but at this fair, it powered the first metro line and the *trottoir roulant*, a moving sidewalk of two tracks and two speeds carrying visitors—the women in the enormous, billowing dresses of that era, and the men in towering silk top hats framing topiary-like facial hair—across 277 acres. In the evenings, electrically powered fountains were lit by electrically powered lights, and the wonders of the age were displayed in the Palaces of Electricity, Civil Engineering, Transportation, Machinery, Textiles, Mining, and Metallurgy. Max Nordau, the Hungarian cofounder of the World Zionist Organization, became so alarmed by the new powers invested in science and electricity that he warned of a horrifying future where everyone would "read a dozen square yards of newspaper daily . . . be constantly called to the telephone [his era's version of e-mail] [and] think simultaneously of five continents of the earth." If Nordau had included "stare constantly at a blinking screen instead of living in the material world," he would have been a prophet with a Nostradamus-like following, yet he seems to have been nearly alone with these trepidations, for everyone else in his era believed that scientific progress would solve all problems, fix all economies, end all war, and create a civilized, Edenic planet. Louis Pasteur referred to laboratories as temples of humanity, and a sensation running for three decades in both France and Italy was Luigi Manzotti's 1881 *Excelsior* ballet, which chronicled the triumph of the Enlightenment over Darkness, ending with love, brotherhood, progress, and science. This fantasy ended in 1914, and as historian Barbara Tuchman noted, "A phenomenon of such extended malignancy as the Great War does not come out of a Golden Age." The Universal Exposition's two largest exhibits were, after all, Schneider-Creusot's immense cannon, and Vickers-Maxim's remarkable array of automatic machine guns.

Pablo Picasso's favorite exhibit at the fair was American dancer Loie Fuller, famed for her billowing phosphorescent veils, which she used as screens for projectors of color-shifting light. The effect was so dramatic it would appear in his revolutionary painting *Les Demoiselles d'Avignon*, and Miss Fuller would ask the Curies for advice on creating "butterfly wings of radium." After they helped her, she danced for them privately at their home and introduced them to sculptor Auguste Rodin; the four became regular friends and perhaps the only two people in the world the Curies saw regularly who weren't scientists or blood relatives. Their closest friends remained the next-door neighbors at boulevard Kellermann, Jean and Henriette Perrin; he was a physics professor at the Sorbonne who verified Einstein's explanation of Brownian motion, correctly estimated the size of water molecules and atoms, and established cathode rays as negatively charged particles—electrons.

Pierre presented his and Marie's scientific findings to France's Academy of Sciences on March 16, 1903, and the Swedish Academy of Sciences then awarded them and Becquerel the Nobel Prize. Behind the scenes, four members of the French Académie had recommended that Becquerel and Pierre alone share the Nobel, leaving out Marie's work entirely. But one of her champions was a Danish mathematician who had great influence with the Swedish academy, and his strenuous efforts repelled that slight.

The Nobel, which had only begun two years before, did not have the global prestige then that it has now, but awarding it to an obscure husband-and-wife team in that tabloid and suffragette era would change that dramatically. The Nobel made the Curies famous, and the Curies, in turn, made the Nobel significant.

In Pierre and Marie's miracle year of 1903, few foretold what was to come. One who did was English chemist Frederick Soddy, announcing that matter must now be thought of "not only as mass, but also as a store of energy . . . [and] the planet on which we live rather as a storehouse stuffed with explosives, inconceivably more powerful than any we know of, and possibly only awaiting a suitable detonator to cause the earth to revert to chaos." In 1904, Soddy told Canada's military leaders that whoever unlocked the power within the atom "would possess a weapon by which he could destroy the earth if he chose."

In the wake of the Nobel, the worldwide press created story after story about this unknown pair, their exciting discovery, their tremendous love for each other and for their work together, and whether Marie was equal, superior, or inferior to Pierre in scientific acumen and achievement. Only Röntgen could appreciate this kind of global attention, as the public enthusiasm for radium would now wholly eclipse that for X-rays. *Cosmopolitan* magazine called the metal "life, energy, immortal warmth . . . dust from the master's workshop," and the cover of the *Chicago Daily Tribune* of June 21, 1903, summed up the moment:

Radium Greatest Find of History

May Upset Vibratory Theory of Light and Questions Conservation of Energy.

New York Engineer First to Make Photographs by Rays from New Substance.

Discovery of Stupendous Import.

Radium, $2,721,555 a Pound.

Blessing, with menace. German chemist Friedrich Giesel, working with a dentist, first reported on radioactivity's biological effects in 1900, and Pierre, following up, was thrilled to find radium could enflame his skin "with a lesion resembling a burn that developed progressively and required several months to heal," as Marie reported. "Henri Becquerel had by accident a similar burn as a result of carrying in his vest pocket a glass tube containing radium salt. He came to tell us of this evil effect of radium, exclaiming in a manner at once delighted and annoyed: 'I love it, but I owe it a grudge.'" Pierre: "The inflammation of the extremities of the fingers lasted about 15 days and finished when the skin dropped off, but the painful sensation did not disappear for two months." An American journalist reported, "Pierre Curie pulled up his sleeve and showed me a forearm scarred and reddened from fresh healed sores." Then it occurred to Pierre that radium's emanations could kill malignant tumors, and a new medical specialty was born: curietherapy.

Just as Röntgen rays had been used against diphtheria, tuberculosis, and other germs, as well as certain cancers, the Curies' radium, when administered with care, could kill tumor cells without killing healthy tissue. When X-rays were used for treatment, radiologists adjusted the settings by putting their arm in the path of the rays; if they got a small sunburn, the machine was ready. This seat-of-the-pants method resulted in many of these technicians' dying of leukemia. Curietherapy was immediately adopted as a replacement since it required no difficult equipment. In one famously successful technique, a bare sip of radium was stuck to the tip of a glass needle or goose quill, then injected into the center of a tumor. Its alpha rays were of such short range that they only damaged the high-turnover cells of malignancy nearby. This would be called brachytherapy and is still used today, primarily for cervical and uterine cancers. Other techniques used radium and mesothorium in sealed glass vials, resting on skin cancers or inserted into body cavities for mouth, throat, and digestive cancers.

Some noticed that when animals were exposed to radium's emanations, they were miraculously invigorated, and beyond its use in oncology, an entirely new industry arose in the 1910s and 1920s purveying radium-infused bath salts, bread, chocolates, Radithor bottled water, suppositories, and "liquid sunshine" tonics. Revigator tanks lined with a radium skin stored drinking water overnight to produce a homemade health infusion. Austria's St. Joachimsthal, the source of the Curie's ore, became a health spa featuring the Radium Palace Hotel (which served Radium Beer), attracting twenty-five hundred customers a year—one of them the young J. Robert Oppenheimer. The price of radium rose to three thousand times the price of gold, as now it cured "anemia, arteriosclerosis, arthritis, asthenia, diabetes, epilepsy, general debility, gastric neurosis, heart disease, high blood pressure, hyperthyroid, hysteria, infection, kidney troubles, muscular atrophy, neuralgia, neurasthenia, neuritis, obesity, prostatitis, rheumatism, senility, and sexual decline." A 1929 pharmacy in Europe offered eighty radioactive products, including lotions, cigarettes, chocolates, bath salts, pillows, suppositories, condoms, and a face cream called Tho-Radia. Radium and X-rays were so magical and so admired they became brand names for many things that contained neither, including enamel, butter, cigars, playing cards, laundry starch, matches, hand cleanser, headache tablets, furniture polish, oil soap, liniment, and stain remover. A popular curio was the spinthariscope (after the Greek for "spark"), which held tiny bits of radium against a screen inside a brass viewer, like a kaleidoscope; the element's emanations would strike the screen, creating a microscopic fireworks. The toy was based on the scintillation method used in physics, a zinc oxide plate under a microscope that flashed each time it was struck by an alpha particle, allowing the electrical shadows of atoms to be viewed with human eyes. Interned as an enemy alien at an abandoned Berlin racetrack during World War I, British chemist James Chadwick—discoverer of the neutron—was able to secretly continue his research using a thorium-infused German toothpaste.

There was even a movement to replace electric lights (which had, after all, just replaced gas lamps) with radium's glow. The first electrics were so harsh and glaring that Robert Louis Stevenson, for one, pled for a return to gas, complaining, "Such a light as this should shine only on murders and public crime, or along the corridors of lunatic asylums, a horror to heighten horror. [Electric illumination will] never allow us to dream the dreams that the light of the living or the lab conjured up." The resonantly named Undark Paint combined radium with arsenic, manganese, thallium, uranium, copper, lead, and zinc sulfide to be used as a replacement, offering the "color and tone [of] soft moonlight." The Curies never profited from the fad, for they had published their exact method for extracting radium without (in the name of free and open scientific inquiry) patenting any of the details. Anyone could produce their own radio-element without any royalties paid to its discoverers or their school, and without the Curies having any say in its use.

As the Curies achieved greater and greater fame, with Marie becoming the most illustrious and renowned woman in the world, the couple seemed to turn more and more to each other. Neither's personality changed a whit. At dinner with President Loubet at the Élysée Palace, a woman asked Marie if she would like to be presented to the king of Greece, and Marie said, "I don't see the utility of it." When the woman appeared shocked, Marie suddenly realized whom she was talking to: the first lady of France. "But—but—naturally, I shall do whatever you please. Just as you please," she stammered. On January 22, 1904, Pierre described their novel situation to a colleague:

> I have wanted to write to you for a long time; excuse me if I have not done so. The cause is the stupid life which I lead at present. You have seen this sudden infatuation for radium, which has resulted for us in all the advantages of a moment of popularity.
>
> We have been pursued by journalists and photographers from all countries of the world; they have gone even so far as to report the conversation between my daughter and her nurse, and to describe the black-and-white cat that lives with us. . . . Further, we have had a great many appeals for money. . . . Finally, the collectors of autographs, snobs, society people, and even at times, scientists, have come to see us—in our magnificent and tranquil quarters in the laboratory—and every evening there has been a voluminous correspondence to send off. With such a state of things I feel myself invaded by a kind of stupor. And yet all this turmoil will not perhaps have been in vain, if it results in my getting a chair and a laboratory.

For reasons that remain unclear to this day, Pierre never trusted the Curies' most important scientific colleague, Henri Becquerel, even after Becquerel helped them professionally and financially and never seemed to bear them ill will for eclipsing his professional stature with his own discovery. When in 1902, Pierre applied for membership in the Académie des Sciences and lost, he was certain Becquerel was responsible. The same year, he lost his second attempt at a Sorbonne chair and, when informed he was up for the Legion of Honor, refused it preemptively. Many in the local scientific community began thinking of the Curies not as unappreciated outsiders struggling to live a life of honor and science, as Marie and Pierre saw themselves, but as antisocial ingrates.

In the middle of the Curies' great triumph, Pierre sickened and never improved. He was wracked by pain so debilitating it made him weak, and Marie joined him in sleepless nights, watching in fear as her husband groaned in agony. The only diagnosis physicians could offer was "rheumatism," which the Curies attempted to treat through a diet free of red meat and red wine. Pierre whispered to Marie, "It's pretty hard, this life that we have chosen."

Blessing, with menace. The radium boom lasted for two decades, then collapsed. Like X-rays, radium's emanations were discovered to be a miracle with two faces. Though exposure did indeed make people feel alive and energetic, it would be learned that the sensation came not from renewed health but from the body's producing a torrent of red blood cells to defend itself. When Pennsylvania steel titan and international playboy Eben Byers died in 1932 after drinking Radithor every day for four years, the *Wall Street Journal* headlined, "The Radium Water Worked Fine Until His Jaw Came Off." In the early 1920s, U.S. Radium in West Orange, New Jersey, hired eight hundred women to apply Undark Paint to watches and airplane instrument dials. Not knowing of any risk, the women used the radioactive ink as fingernail polish and moistened their work bristles with their lips for a sharper point. Nearly a decade later, they became horribly anemic, and their teeth and jaws fell out; a reporter noted, "One girl fainted at the sight of her own reflection; her body glowed as if lit from within." In 1925, five of them sued U.S. Radium, and with the public support of *New York World* editor Walter Lippmann, they won their case in 1928. By then, twenty-four of the eight hundred were already dead.

Manya Skłodowska Curie became the first woman in French history to be awarded a doctorate, in June of 1903. Sister Bronya, now practicing medicine in Poland, returned to celebrate. She insisted Marie buy a new dress for the occasion, and just as she had for her wedding, she got one that would work equally well as lab wear. Ernest Rutherford, the discoverer of the classical model of the atom (with electrons orbiting nuclei much as the planets revolve around the sun), visited from Canada and was astonished by the Curies' lab in the cadaver hut, as well as by the celebratory garden party at Paul Langevin's that evening, illuminated by radium vials—"The luminosity was brilliant in the darkness and it was a splendid finale to an unforgettable day"—and the sight of Pierre's deeply swollen, burnt hands. When earlier that month the Curies had been the guests of London's Royal Academy, Pierre was so ill and his hands so damaged that he needed help in buttoning his clothes, and during the middle of the Friday Evening Discourse before Lord Kelvin, he fumbled and spilled some of the precious radium.

Pierre's health now rapidly deteriorated, and he frequently woke up in the middle of the night, unable to stop moaning from an untreatable pain. He said he was "neither very well, nor very ill. . . . I am easily fatigued, and I have left only a very feeble capacity for work." His wife was also suffering; the couple's fingertips were permanently damaged. Marie had lost twenty pounds, and in August she miscarried, telling Bronya that she "had grown so accustomed to the idea of the child that I am absolutely desperate and cannot be consoled."

In the summer of 1904, Pierre's "rheumatism" was so severe that he had to cancel going to Stockholm to give their Nobel lecture. Marie learned she was pregnant once again and, worrying over the loss of her miscarriage, abandoned her work and joined her sick husband and daughter, along with *frère* Jacques and his family, in a farmhouse in Saint-Rémy, just outside Paris. The couple's second child, daughter Ève, was born on December 6, 1904. Marie: "I have been frequently questioned, especially by women, how I could reconcile family life with a scientific career. Well, it has not been easy."

Their life of penury, at least, was over. Baron Edmond de Rothschild became a substantial donor, and industrialist Armet de Lisle started a company to produce radium and gave the Curies a laboratory on the factory grounds. They could now afford a servant to cook and clean, but the woman never heard a word of praise. Finally she could take it no longer and directly asked Pierre what he thought of her roast. In a near parody of the absent-minded professor, he replied: "Did I eat a beefsteak? It's quite possible."

Marie confided to sister Helen, visiting over the summer of 1905, that Pierre's attacks of severe back pain were now more frequent, and more severe. She worried that he had contracted something the doctors couldn't diagnose, and that he would never recover.

On April 14, 1906, Pierre took the train from Paris to spend a day with his family at the Saint-Rémy farmhouse. The two parents sat together in a meadow, watching Irène, who was obsessed with butterflies. Though Marie had "a little heartache" about Pierre's continuing medical troubles, she later wrote, "We were happy. [I] had this feeling I had had recently a lot, that nothing was going to trouble us."

On the sixteenth, Pierre returned to Paris, and she joined him two days later.

On the nineteenth, he left his lab at around 10:00 a.m. to travel across town for a meeting of the Association of the Professors of the Science Faculties, of which he was vice president, at the Hôtel des Sociétés des Savants on rue Danton. Paul Langevin was there, as was Joseph Kowalski, the friend in common who had introduced Pierre to Marie twelve years before. Pierre was in such a good spirit that he invited everyone at the luncheon to his house that night.

He then headed out to Gauthier-Villars, publisher of the leading Parisian science journal, *Comptes rendus*. There, he would read proofs and, after, would do a little research in the library of the Institut de France.

He arrived at the publisher to discover that the offices were closed, due to a strike. So, he headed off to the library. A block away, he approached the Pont Neuf and the rue Dauphine and began to cross. It was raining, and foggy. Onetime milkman Louis Manin was driving two Percheron stallions hauling a thirty-foot wagon loaded with six thousand kilos of military uniforms and supplies. Manin started to rein his animals to allow a tram to pass, but the tram conductor signed for him to proceed. As he did so, Manin passed a carriage, and right behind it, Pierre appeared, rushing by foot across the street. He bumped into one of Manin's horses, tripped, and grabbed it to keep from falling to the pavement. Both horses reared. Pierre fell. Manin jerked left to keep the horses from trampling M. Curie and succeeded in keeping the hooves and his wagon's front wheels from striking the physicist. Daughter Ève: "His body passed between the feet of the horses without even being touched, and then between the two front wheels of the wagon. A miracle was possible. But the enormous mass, dragged on by its weight of six tons, continued for several yards more. The left back wheel encountered a feeble obstacle which it crushed in passing: a human head. The cranium was shattered and a red, viscous matter trickled in all directions in the mud: the brain of Pierre Curie."

Parisians surrounded the cart and began threatening the driver; the police arrived to protect him from a growing and enraged mob. Officials decided the tragedy was a result of bad weather, visibility, and the victim's not paying attention to what he was doing. Famously absentminded and distracted, Pierre Curie was in a weakened state from his poor health and likely did not see the wagon from under his large umbrella. Manin was acquitted of blame.

As no ambulance could reach through the mob, the officers carried the dying man on a stretcher to a pharmacy, where a druggist reported there was nothing anyone could do. The police then carried him to the station in the Hôtel des Monnaies, where a doctor pronounced him dead, and went to the

Sorbonne, where the dean of the science faculty, Paul Appell, was informed. Appell then went with Curie friend and neighbor Jean Perrin to boulevard Kellermann to tell Marie. But, she and Irène had gone to spend the afternoon in the countryside, at Fontenay-aux-Roses. The two scientists did not want to reveal the terrible news to the elderly Dr. Curie, but after learning who they were, he took one look at their drawn faces and announced, "My son is dead."

When the thirty-eight-year-old Marie returned that evening at seven, Pierre's father told her what had happened: "I enter the room. Someone says: 'He is dead.' Can one comprehend such words? Pierre is dead, he who I had seen leave looking fine this morning, he who I expected to press in my arms this evening. I will only see him dead and it's over forever. I repeat your name again and always 'Pierre, Pierre, Pierre, my Pierre,' alas that doesn't make him come back, he is gone forever, leaving me nothing but desolation and despair."

The notably tough Marie Curie became unmoored at this loss. "Crushed by the blow, I did not feel able to face the future [as] an incurably and wretchedly lonely person." She felt she couldn't go on, either as a person or as a scientist. Her *Mourning Journal* is a testimony to the horror of grief, loss, death:

> They brought you in and placed you on the bed. . . . I kissed you and you were still supple and almost warm. . . . Pierre, my Pierre, you are there, calm as a poor wounded man resting in his sleep, his head bandaged. Your face is sweet, as if you dream. Your lips, which I used to call hungry, are livid and colorless. . . . Your little graying beard; one can barely see your hair, because the wound begins there, and on the right one can see the bone sticking out from under the forehead. Oh! How you were hurt, how you bled, your clothes were inundated with blood. What a terrible shock your poor head, that I had caressed so often, taking it in my hands, endured. And I still kiss your eyelids which you close so often that I could kiss them, offering me your head with the familiar movement which I remember today, which I will see fade more and more in my memory. . . .
>
> We put you in the coffin Saturday morning, I held your head. . . . Then some flowers in the casket and the little picture of me . . . that you loved so much. . . . It was the picture of the one you chose as your companion, of the one who had the happiness to please you so much that you didn't hesitate to make her the offer of sharing your life, even when you had only seen her a few times. And you had said

to me many times it was the only time in your life when you acted without any hesitation, because you were absolutely convinced that it was right. . . .

Your coffin was closed and I could see you no more. . . . I was alone with the coffin and I put my head against it. . . . I spoke to you. I told you that I loved you in that I had always loved you with all my heart. . . . I promised that I would never give another the place that you occupied in my life and that I would try to live as you would have wanted me to live. And it seemed to me that from this cold contact of my forehead with the casket something came to me, something like a calm and an intuition that I would yet find the courage to live. Was this an illusion or was it an accumulation of energy coming from you and condensing in the closed casket which thus came to me as an act of charity on your part? . . . I got up after having slept rather well, relatively calm. That was barely a quarter of an hour ago, and now I want to howl again—like a savage beast. . . .

In the street I walk as if hypnotized, without attending to anything. I shall not kill myself. I have not even the desire for suicide. But among all these vehicles is there not one to make me share the fate of my beloved? . . . I feel very much that all my ability to live is dead in me, and I have nothing left but the duty to raise my children and also the will to continue the work I have agreed to.

Bronya had arrived to comfort her grief-wracked sister for two months, and now it was time for her to return to Zakopane. Marie asked her to come with her to the bedroom, where, in the middle of a hot June, the fireplace roared. The widow locked the door behind them and took a bulky package from her armoire. Retrieving a strong set of shears from the mantel, she asked her sister to sit beside her, before the fire. Then she cut open the parcel to reveal the bloody, mud-drenched clothes Pierre had been wearing when he was killed, which Marie had been saving all this time. She began cutting the material into pieces and throwing it onto the flames but, finding pieces of her husband's body on part of a coat, she burst into tears and began kissing it. Bronya took the scissors from her and continued cutting, and burning, until everything was consumed. Marie then asked, "Tell me, how am I going to manage to live. I know that I must, but how shall I do it? How can I do it?" She fell into a spasm of sobbing, and Bronya tried to comfort her.

Marie could no longer live where she and Pierre had spent the whole of their married life and went to look for another home, deciding finally to move to 6 rue du Chemin de Fer, in Sceaux, close by her husband's tomb. When Dr. Curie died in 1910, she had Pierre disinterred, putting her father-in-law's coffin first, then Pierre, with room for her on top.

Marie Curie never forgave France for what she considered its rude treatment of her husband in failing to give him either the honors or the laboratory facilities he merited. Following his lead, she, too, refused the Legion of Honor and devoted the rest of her life to erecting a laboratory in Paris worthy of Pierre's memory. In 1909, the Pasteur Institute offered to build a Curie lab, but this would have meant Marie's resignation from the Sorbonne. Suddenly the university rose to action, collaborating with the Pasteur to build the Radium Institute, with one lab for Marie, and another for Claude Regaud, who researched to perfect curietherapy.

Its street was named rue Pierre Curie.

Four years after the accident that took Pierre's life, in the spring of 1910, Marguerite Borel, the novelist daughter of Sorbonne chair Paul Appell, commented, "Everybody said Marie Curie is dead to the world. She is a scientist walled in behind her grief." But, after years of widowhood, Marie began to resurrect. She stopped wearing all black and physically appeared to regain decades of youth.

The secret was as old as time, and Paris. She was in love.

For a century after first becoming famous, the public would hold of Marie Curie an image of a brain without a heart; a scientist, but not a wife or mother; a hero of women's rights as iconic as George Washington . . . and as a figurehead, just as lacking in humanity. Even though her greatest work was achieved when she was in her thirties, she is remembered as an asexual, emotionless old woman . . . but there are clear reasons for these misperceptions. At the height of her fame, a journalist asked for details of her childhood, of her psychology, of her emotions, and Mme. Curie refused, explaining, "In science we must be interested in things, not in persons." She told another reporter, "There is no connection between my scientific work and the facts of private life." Albert Einstein wrote, "Madame Curie is very intelligent but has the soul of a herring, which means that she is poor when it comes to the art of joy and pain." Additionally, though he admired her immensely, her main method of expressing emotion, he said, was in griping. But a great reason for Marie Curie's denuded public reputation was that her *Mourning Journal* was unknown for decades, as was her heartbreaking love affair with Pierre's student Paul Langevin, a wildly handsome and extravagantly brilliant scientist famous for his magnetic theories, his quartz oscillators, and his termagant of a wife.

Their very public affair caused such a scandal that the Curie descendants would suppress its details for the next forty years.

Paul Langevin first met the Curies as a seventeen-year-old municipal school student under teacher Pierre in 1888 and was, in effect, a protégé of both husband and wife. When Pierre left for the Sorbonne in 1904, Paul was hired to replace him at the city school; he taught alongside Marie at Sèvres, and when she replaced Pierre at the Sorbonne in 1906, he was given her post. Paul said, at Pierre's funeral, "The hour when we knew we could meet him, when he loved to talk about his science, the walk that we often took with him, these bring back his memory day after day, evoke his kindly and pensive face, his luminous eyes, his beautifully expressive head, shaped by twenty-five years spent in the laboratory, by a life of unrelenting work, of complete simplicity, at once thoughtful and industrious, by his continual concern with moral beauty, by an elegance of mind which produced in him the habit of believing nothing, of doing nothing, of saying nothing, of accepting nothing, in his thought or in his actions which was not perfectly clear and which he did not entirely understand." Of Langevin, Einstein was equally fulsome: "In his scientific thinking Langevin possessed an extraordinary vivacity and clarity . . . it seems to be certain that he would have developed the special theory of relativity if it had not been done elsewhere."

Paul was brilliant, passionate about science, and good-looking. It was a wonderful match for the brokenhearted widow, but Langevin was also married, to Emma Jeanne Desfosses, a harridan who never tired of warmongering in the name of love. In their first year of marital unbliss, Desfosses's mother and sister took letters from the newlywed husband's pockets that described his troubled marriage so that Jeanne would have evidence in case of divorce. The following year, he appeared at the lab covered in bruises—during a fracas, the three women had thrown an iron chair at him. But M. Langevin was no bystander in this eternal drama, for when Jeanne stormed out after one fight and threatened to end the marriage, Paul begged her to return . . . a scenario that would be repeated endlessly over the years to come. Sorbonne physicist Jean Perrin and his wife, Henriette, were close with both the Curies and the Langevins, and after one violent spat Henriette recounted, "Often, during meals, M. Langevin, cruelly wounded by the words of his wife, left the table. The meal continued. . . . I was very sad to see the unhappiness of a friend that I liked with all my heart. . . . He said to me, 'I don't know who I can lean on. I have only my children and they are very small.' "

In the spring of 1910, Marie, having heard Jeanne's tales of how her husband woefully mistreated her, criticized Paul to his face for his vile behavior. Langevin replied that she only knew half of the story; that in fact just the other day, Jeanne had cracked a bottle on his head. Paul now found someone to lean on, regularly confiding in Marie about his terrible domestic conundrum—then suddenly, everything between them changed. Marie wrote, "I spent last evening and night thinking of you and the hours we had together. I hold the delicious memory. Still I see your eyes, kind and tender, and your warm smile and I can only dream of the moment that I find again the sweetness of your presence." He replied, "I am trembling with impatience at the thought of seeing you return at last, and of telling you how much I missed you. I kiss you tenderly awaiting tomorrow."

On July 15, 1910, Paul and Marie secretly rented an apartment together, at 5 rue du Banquier. They called it *chez nous*. Almost immediately, one of Jeanne's servants fished a love letter from Paul to Marie out of the postal box and gave it to the wife. During their next brawl, she warned him, "You are going to see quite a scandal in the newspapers," and asked their son, eleven years old, if he wanted to grow up to be like his father and cheat on his wife with a mistress. Marie told the Perrins that her and Paul's "great friendship angered Mme. Langevin [and] that she had declared to her husband that she was going to get rid of this obstacle." Paul explained to his lover, "That means that she would kill you." Marie: "As long as I know you are near her my nights are atrocious. I cannot sleep. With the greatest difficulty I fall asleep at two or three o'clock and awake with the sensation of fever. I cannot even work. . . . I must be attached to you by very strong cords to make up my mind to preserve these cords at the risk of my position and my life."

The following week, Jean Perrin "was astounded to see Mme. Curie run to me as I was entering the house. She had been waiting for me for several hours. . . . She said that she had been insulted in the street in crude terms by Mme. Langevin and by her sister, Mme. Bourgeois, and that this woman had threatened her [and demanded that Marie] leave France. . . . I think I will never forget the emotion I felt seeing the distress to which this illustrious woman had been reduced . . . wandering like a beast being tracked." The next day, Perrin went to try to talk some sense into Jeanne, but she "shouted threats for everyone to hear, that if Mme. Curie didn't leave in eight days she would kill her."

Perrin arranged for a meeting of Jeanne, Paul, and Jeanne's brother-in-law, Henri Bourgeois, an editor at *Le Petit Journal*. They agreed that, in exchange for Paul's no longer seeing Marie either personally or professionally, Jeanne would end her campaign. However, that the terminally aggrieved wife had a journalist for a relative and an ally remained a serious threat, for in belle epoque Paris, the tabloids had a loud public voice. Historian Barbara Tuchman: "Variegated, virulent, turbulent, literary, inventive, personal, conscienceless and often vicious, the daily newspapers of Paris were the liveliest and the most important elements in public life [and] represented every conceivable shade of opinion, calling themselves Republican, Conservative, Catholic, Socialist, Nationalist, Bonapartist, Legitimist, Independent, Absolutely Independent, Conservative Catholic, Conservative Monarchist, Republican Liberal, Republican Socialist, Republican Independent, Republican Progressivist, Republican Radical Socialist. Some were morning, some were evening, some had illustrated supplements. At four to six pages, they covered, besides the usual political and foreign affairs, news of the *haut monde*, of *le turf*, of fashion, of theater and opera, concerts and art, the salons and the Academy. . . . The press was daily wine, meat and bread to Paris. Major careers and a thousand minor ones were made in journalism. Everyone from Academicians to starving Anarchists made a supplementary living from it." Additionally, in this society at this time, married men were assumed to squire mistresses. But those mistresses were supposed to be socially invisible, not the most famous Frenchwoman in the world.

After returning from the International Congress of Radiology and Electricity in September of 1910—underwritten by sodium carbonate magnate Ernest Solvay and attended in Brussels by Albert Einstein, quantum discoverer Max Planck, Marie's neighbor Jean Perrin, as well as Paul Langevin—Marie vacationed with her children at L'Arcouëst on the Breton coast, a spot so overrun with Sorbonne professors it was nicknamed Port Science. There, she wrote Paul an echo of what Pierre had written to her so many years before: "I spent yesterday evening and night thinking of you, of the hours that we have spent together and of which I have kept a delicious memory. I still see your good and tender eyes, your charming smile, and I think only of the moment when I will find again all the sweetness of your presence. . . . It would be so good to gain the freedom to see each other as much as our various occupations permit, to work together, to walk or to travel together. . . . What couldn't come out of this feeling, instinctive and so spontaneous and so compatible with our intellectual needs, to which it seems so admirably adapted? I believe that we could derive everything from it: good work in common, a good solid friendship, courage for life and even beautiful children of love in the most beautiful meaning of the word." They renewed their affair and returned to *chez nous*.

On October 31, 1910, one of the Immortals, chemist Désiré Gernez, died, meaning an Académie Française chair was in contention. Marie was the only French laureate who was not a member, even though she had been elected to equivalent organizations in Sweden, the Netherlands, Czechoslovakia, Poland, the United States, and Russia. If she won, she'd be the first woman in the Institut de France's 215-year history; her competition was Édouard Branly, inventor of the wireless coherer, an element of the telegraph. The French tabloids turned the contest into breathless headlines, with French nationalists and the Catholic Church supporting Branly. At the January 4, 1911, meeting, the vote to admit women as academy members failed, 85 to 60, upholding "immutable tradition." During the ensuing science-branch vote, astronomer Henri Deslandres explained how it was so "very difficult to judge the works of Mme. Curie and to separate her research from the inspired work of M. Curie." Branly won, and for the next eleven years, Curie's research could not be published in the most globally read French science journal, *Comptes rendus*.

In the spring of 1911, Paul and Marie opened the door of *chez nous* to discover that someone had broken in and stolen their love letters. Tabloid editor/brother-in-law Henri Bourgeois informed Marie that Jeanne had hired the thief, and that he would now use the letters to scandalize the world and destroy Curie's reputation. Jean Perrin advised Marie to leave town until things calmed down, and she did, attending a scientific conference in Genoa where she poured out her troubles to another attendee, Sorbonne science dean Paul Appell's daughter, Marguerite Borel, who recalled "under the austere scientist, the tender and lively woman, capable of walking through fire for those she loves."

On November 4, *Le Journal*'s front page heralded, "A Story of Love: Madame Curie and Professor Langevin. . . . The fires of radium which beam so mysteriously . . . have just lit a fire in the heart of one of the scientists who studies their action so devotedly; and the wife and the children of this scientist are in tears." The following day, Bourgeois's paper, *Le Petit Journal*, also had a front-page report, and the two papers began to duel over plots and sources. The Curie-Langevin affair became known as "the greatest sensation in Paris since the theft of the *Mona Lisa*."

Three days later, Reuters announced that the first woman laureate now would be the first man or woman to have two Nobels. The 1903 award had been in physics, for the discovery of radioactivity; now she would receive the 1911 chemistry prize for discovering radium and plutonium. But when the Scandinavians learned of the French hubbub, there was an effort to disinvite her, with laureate Svante Arrhenius writing Marie, "I beg you to stay in France; no one can calculate what might happen here." Albert Einstein told her, "I am convinced that you [should] continue to hold this riffraff in contempt. . . . If the rabble continues to be occupied with you, simply stop reading that drivel. Leave it to the vipers it was fabricated for."

To end the war of love, Jeanne demanded full custody of the four children and a thousand francs in monthly support. Paul refused. She went to court, charging him with "consorting with a concubine in the marital dwelling," including as evidence the stolen love letters. On November 23, alongside ten pages of the Curie-Langevin epistles, *L'Oeuvre* explained that since Marie was Polish, this affair proved "France in the grip of the bunch of dirty foreigners, who pillage it, soil it and dishonor it."

On December 10, 1911, Marie Curie, accompanied by Bronya and Irène, attended the Nobel ceremonies in Stockholm. The scandal in France had no effect on the Swedish ceremony; in fact, King Gustaf himself would be accused, years later, of carrying on a sordid love affair with a married man—something not even thousands of kronor could keep from being made public. Meanwhile, *L'Action Française* and *L'Intransigeant* daily attacked Mme. Curie on their front pages, calling her work "overrated" and explaining her alien perfidy: "There is a mother, a French mother who . . . wants only to keep her children. . . . She has above all the eternal force of the truth on her side. She will triumph." By the end of December, *L'Oeuvre* claimed that Marie's middle name was Salome and that "her father is in fact a converted Jew." Marie and Irène returned home to Sceaux to find a mob outside their house screaming, "Down with the foreigner, the husband-stealer!" Marguerite Borel took the Curies in, which angered her father, the Sorbonne's science dean. He told her that the university was planning to suggest that Marie leave France.

On December 29 at the age of forty-four, Marie was rushed by ambulance to a hospital bed, where she spent nearly a month recovering from a kidney infection. Everyone in Paris assumed that she was pregnant with Paul's baby. She had an operation in March, and instead of returning to Sceaux, where mobs still gathered, she rented an apartment in Paris. Now she was ashamed at what she had done to the name of her husband and began officially calling herself Madame Skłodowska. She insisted that, when Irène wrote her letters, she use that name, and not Mme. Curie (privately, both daughters called her Mé). She did no lab work for fourteen months.

Eventually, five armed duels were conducted over the affair, and Paul and Jeanne settled out of court, she getting the full custody and eight hundred francs in support—nearly everything she'd asked for before publishing the letters. Marie and Paul's romance, though, did not survive. By 1914, Paul and Jeanne were reunited, and soon after, with her assent, he took another mistress.

At the start of the press frenzy, Marie was the most famous woman in the world, a living symbol of the great heights to which women could now aspire, regardless of the inane voting of the French academy. Afterward, she was a home-wrecking, foreign Jewess whose saintly French husband had killed himself when he learned of her many adulteries. *L'affaire Curie* riveted the French press for longer than the actual romance had lasted. Daughter Ève's biography, *Madame Curie*, published in 1937, three years after her mother's death, was in many ways an effort to rein in the Langevin scandal in history and keep it from tainting her mother's legacy. Marie's heirs then arranged that the Curie-Langevin letters and the Perrin testimony would be hidden for four decades in the archives of the Paris city school that had been Pierre and Marie's professional home.

In time, Marie's granddaughter Hélène and Paul's grandson Michel would fall in love and get married. But most crucially, Paul Langevin would be an agent for Marie's last great triumph, the shocking professional ascension of her daughter Irène.

As the German front line marched toward Paris in the autumn of 1914, the city evacuated, leaving it deserted to the poor, the struggling working class, and Marie Curie, who stayed behind to safeguard her just-built laboratory and its precious cache of radium. She then realized what she could really do for the war effort was bring X-ray technology to wounded soldiers on the front lines, and she tried to equip Red Cross hospitals with bare-bones radiology departments, but so many of them didn't even have electricity that the effort made little progress. She then saw that an automobile with a generator running from its motor; a supply of glass vacuum cathode tubes; a radioscopic screen; photographic plates and chemicals; an armature to position the tube over the needed target in the body; a table for the patient to lie on; black curtains to create the necessary darkness; and an operator's lead apron would be invaluable in saving lives—a mobile X-ray lab; a *voiture radiologique*, which in the field became known as *petite Curies*.

As soon as possible, Irène returned from their Port Science summer home on the Breton seaside—the Curies now had an Île Saint-Louis apartment, a seaside house of white plaster in l'Arcouëst, and another house on the Mediterranean in Cavalaire—to take a course in nursing, get her diploma, and truly become her mother's daughter. On November 1, 1914, Marie, Irène, a mechanic, and a driver rode a *petite Curie* to the army hospital in Creil, the first of thirty trips the women would make to the front. With seven hundred thousand francs donated by the French charity Patronage de blessés, Marie established two hundred radiology clinics and outfitted eighteen mobile labs. In 1915 when Irène turned eighteen, she began teaching women how to be X-ray technicians and traveled the country to solve problems with the various labs and outposts, all while qualifying for her Sorbonne *certificats* in math, physics, and chemistry.

At war's end, Poland became an independent republic for the first time in 123 years.

In May of 1920, Marie met her greatest and most useful fan, Marie "Missy" Meloney, the editor of a wildly popular American women's magazine, the *Delineator*, and the Oprah Winfrey of her day. Missy's opinion was that Curie was "the Greatest Woman in the World," who deserved immense financial support from the United States. In fact, because of her long-standing relationships with industry and her glossy fame that attracted well-to-do benefactors, few scientists in the world—and perhaps not a single one—were better supplied in every way than Marie Curie. But after Marie told Missy that America had nearly fifty times as much radium as the one gram its discoverer possessed, Missy began a "Marie Curie Radium Campaign." Giving her readers story after story on Marie's poverty and her likelihood of curing cancer—both wildly exaggerated—in eight months Missy raised $100,000 to buy Marie a gram of radium, which the editor finagled to be given to the scientist, at the White House, by Warren G. Harding. In his ceremonial speech on May 20, 1921, the president of the United States echoed the recently passed Nineteenth Amendment to the Constitution, giving women the vote—which they would not have in France for another quarter of a century: "As a nation whose womanhood has been exalted to fullest participation in citizenship, we are proud to honor in you a woman whose work has earned universal acclaim and attested woman's equality in every intellectual and spiritual activity."

The *New York Times* trumpeted her arrival with a front-page story, "Mme. Curie Plans to End All Cancers," and noted that her Carnegie Hall appearance before thirty-five hundred was "the largest meeting of American college women ever held in this country." Throughout the trip, Marie suffered dizziness, nausea, and anemia—the same health troubles that had so weakened Pierre—and had to cancel a number of appearances, sending Ève and Irène in her stead. But her appearances were electrifying. Dressed all in black, her health damaged into frailty by occupational radiance, yet still commanding that historic staunch presence, Marie Curie had what one audience member described as the moral force of a Buddhist monk. The public would further understand this extraordinary woman as explained by Hollywood, through Greer Garson:

Pierre: "What if there is a kind of matter in the world we never even dreamed of? What would that mean?"
Marie: "What if there exists a matter which is not inert, but alive, dynamic?"
Pierre: "Marie, that would mean our whole conception of the nature of matter would have to be changed."

Marie went to Warsaw for another great triumph, the inauguration of a Polish institute under her name, to be managed by Bronya. At the front of its plaza stood an immense effigy of Manya Skłodowska Curie, the same statue that would regularly draw the contemplative gaze of one Casimir Zorawski. Then in 1925, Paul Langevin recommended that Marie hire one of his recent graduates, Jean-Frédéric Joliot, as a junior lab assistant. Like Pierre Curie, Fred did not have an elevated academic provenance, but he had graduated first in his engineering class at Pierre and Paul's alma mater, the Municipal School of Industrial Physics and Chemistry. Just as Paul had been a student of Pierre's, so Fred was a student of Paul's, and like his mentor, Fred was brilliant, charming, and appallingly handsome. Soon after his appointment, Fred and the rather plain Irène, who'd just won her doctorate for studying polonium's alpha rays, fell in love. "I rediscovered in [Pierre Curie's] daughter the same purity, his good sense, his humility," Fred crooned. Marie, though, was far from won over, writing brother Józef that the Joliots were "well-respected but they are industrialists."

On October 9, 1926, Irène, age twenty-nine, married Fred, twenty-six. That romantic quickness unnerved Marie, who insisted the groom sign a prenuptial agreement, which included that, whatever happened in their marital future, Irène would retain the Curie radium. A bit later, Marie in a letter secretly revealed why she was still not a fan of her first son: "I miss Irène a lot. We were so close for such a long time. Of course, we often see each other, but it's not the same."

Fred, though, refused to let his mother-in-law consider herself abandoned. The newlyweds ate dinner with Mé four times a week until she accepted him all out as a member of the family, finally admitting to Jean Perrin, "That young man is a ball of fire." Irène: "My mother and my husband often debated with such ardor, answering back and forth so rapidly, that I couldn't get a word in and was obliged to insist on having a say when I wanted to express an opinion."

In time, Madame's early fears proved wholly unfounded. Fred and Irène had a long and happy marriage, while their partnership as physicists would accomplish so much that they would emerge from her parents' burdensome professional shadow. At first, the couple struggled with the financial sacrifice of pure research; a number of times, Fred thought he should leave the Institut for a better-paying job in private industry. But in 1928, they began publishing jointly, and just as Pierre and Marie were better together than apart, Fred and Irène were unbeatable, striking scientific pay dirt again and again.

In 1931, the Joliot-Curies showed that when beryllium, a lightweight metal, was bombarded with alpha particles from polonium, it gave off powerful rays that could make protons burst at high speed from the atomic nuclei in paraffin wax. They concluded that the rays were a new type of gamma ray, the most powerful form of particle radiation then known, and called them recoil protons. Reading their articles in Rome, physicist Ettore Majorana said to his colleagues, "What fools. They have discovered the neutral proton and they do not even recognize it," and when British chemist James Chadwick repeated their experiment, he, too, realized that the rays included a new kind of subatomic particle, which he called the neutron. He later won a Nobel Prize for this insight, instead of the Joliot-Curies. Fred and Irène then completed another experiment with another mistake about odd results. This time, they ceded the discovery of the positron to C. D. Anderson.

Though increasingly ill from radiation poisoning, Marie lived long enough to see her children then make the great discovery of their careers in 1934. At the time, Fred had become taken with his newest apparatus, a Wilson cloud chamber, a magnificent little vitrine housing a dense fog of supersaturated vapor that revealed the movements of particles as trails in the mist. Alpha and beta particles produced distinctive vapor shapes, as did electrons. "An infinitely tiny particle projected in this enclosed region can trace its own path thanks to the succession of drops of condensation. Isn't it the most beautiful experiment in the world?" Fred wrote to a colleague.

On January 15, 1934, the Joliot-Curies tried striking the light end of the periodic table with polonium emissions in their cloud chamber. Something strange happened; when elements such as aluminum and boron were bombarded, they continued to exude rays after the bombardment stopped.

Aluminum would stay radioactive for three minutes all on its own. Was it trapping some of the alphas, shooting out a neutron, and alchemizing into a radioactive isotope?

After bombarding aluminum and then boron with alpha particles, chemistry revealed isotopes of radioactive phosphorus in the first and radioactive nitrogen in the second—the first time human beings had created radioactivity. Fred "began to run and jump around in that vast basement." "With the neutron we were too late. With the positron we were too late. Now we are in time," he joyously told a student.

Fred Joliot: "I will never forget the expression of intense joy which overtook [Mé's] face when Irène and I showed her the first [man-made] radioactive element in a little glass tube. I can see her still taking this little tube of the radioelement, already quite weak, in her radium-damaged fingers. To verify what we were telling her, she brought the Geiger-Muller counter up close to it and she could hear the numerous clicks. . . . This was without a doubt the last great satisfaction of her life." Marie told Irène that this achievement meant "we have returned to the glorious days of the old laboratory!"

Marie's fingers were now crusted in radiation-triggered, wartlike ulcers, and she tried surgery to repair the radiation-caused cataracts in both eyes. But her health continued to leak away. As with Pierre before her, doctors still had no idea what was wrong and suggested tuberculosis—the disease that had killed her mother and given sister Bronya a life's work. At three thirty on a perfect afternoon in the Paris spring of May 1934, she came down with a fever at the Institut and went home. She never came back. Doctors found tubercular-like lesions on her lungs and again suggested she rest at the Sancellemoz sanitarium in Passy, Savoy. Waiting for the train at the Gare du Nord, she collapsed, and Ève had to help her into the cabin. But doctors at Sancellemoz found no tuberculosis. Ève: "Then began the harrowing struggle which goes by the name of 'an easy death'—in which the body which refuses to perish asserts itself in wild determination." On July 4, 1934, at dawn, "when the full light of a glorious morning had filled the room," Marie Curie died of what the Sancellemoz director said "was a plastic pernicious anemia of rapid, feverish development. The bone marrow did not react, probably because it had been injured by a long accumulation of radiations."

Originally, Marie was buried next to Pierre and her father-in-law, as she had always planned, in the little cemetery of Sceaux. Six decades later, in 1995, the French government exhumed their bodies, and that is how a little girl from Poland named Manya became the first (and only) woman interred in the French Panthéon. Langevin rests there as well. Today her notebooks and cookbook are still so radioactive they are kept in a lead-lined safe at the Bibliothèque Nationale, and scholars need protective clothing to work with them.

Radium and its effluvium, radon gas, would be used in medical therapies for years to come—replaced only because of Fred and Irène's discovery—while polonium would be used as a trigger for nuclear weapons (including Hiroshima's), and to alleviate static cling. In 2006, ex–KGB agent Alexander Litvinenko was dosed with Marie's beloved polonium, and before dying, he accused Vladimir Putin of orchestrating the assassination.

Irène and Fred won the 1935 Nobel Prize in Chemistry for the discovery of artificial radiation. Marie had been the first woman laureate, and now her own daughter was the second. After Hungarian George de Hevesy won the chemistry Nobel in 1943 for inventing the tracer technique (in which radioactive isotopes can follow chemical changes and physiological processes), that process, combined with Joliot-Curie's artificial radioactivity, became as important for modern science as the microscope.

Her daughter and son-in-law's breakthrough meant that the backbreaking and wildly expensive labors of isolating radium from its ore—Marie Curie's great triumph—was no longer required. Nuclear medicine could create its own irradiated materials at any time . . . and so could nuclear physics. This discovery dramatically changed the course of modern medicine, as well as forged a path for Enrico Fermi, Leo Szilard, and "Germany's Madame Curie," Lise Meitner, to ignite the Atomic Age.

3

Rome: November 10, 1938

THERE is a still in the night—but not this night. The palatial apartment on via L. Magalotti, in the neighborhood of Il Duce's own Villa Borghese, was enrobed in marble, from the entryway's Carrara to the bathrooms' obscure sea-green—a cool, aquatic trance—and echoed with dozens of clinking, chiming, tocking mechanical clocks, nearly drowned out by the ruckus of the family's cook making supper, which in turn was a murmur of distant thuds compared to the excited voices of the two children, Nella, eight, and her three-year-old brother, Giulio, who played and squabbled as voraciously as the wolf-bred children of Rome's nascence. Nella's father had at one point given her the nickname *bestiolina*—"little animal"—and she would prove for many years deserving of this honor.

The only quiet came from the parents in the living room, reading the paper while listening to an immense mahogany-and-bakelite radio. The two were suffering silently, both tense and in conflict. Laura, thirty-one, dark and elfin, was reminiscent of Audrey Hepburn; her husband, Enrico, thirty-seven, was black-haired, muscular, and easy to like, with slate-blue eyes as soulful as any basset hound's, and a reputation as a genial workaholic. One of his colleagues called him "completely self-confident, but wholly without conceit," and onetime pupil and lifetime friend Emilio Segrè said Enrico was "a steamroller that moved slowly but knew no obstacles."

Enrico was a titan of Roman society whose membership in the Royal Academy of Italy came with both a princely salary and the title Your Excellency. Many decades later, astronomer Carl Sagan summed up his legacy: "There is a Fermi Sea, a Fermi Energy, a Fermi Paradox, Fermi Statistics . . . a Fermi class of elementary particles, a Fermi Constant, a Fermi

Surface, a Fermi Mechanism (for the acceleration of cosmic rays), a Fermi Age (neutron diffusion), a Fermi unit of distance (which is roughly the size of a nucleon), two Fermi Golden Rules, a Fermi Prize, a Fermi Institute, a Fermi High School, a Fermi National Laboratory, and a chemical element named after Fermi. . . . It's hard to think of another physicist of the twentieth century who's had so many things named after him—and this surely is an indication of the respect and affection with which he is thought of in the community of physicists, and in a larger community as well." Meanwhile, another lifelong friend, Franco Rasetti, called him "a very very common man, in fact, he was common as an old shoe."

Like his older siblings, Maria and Giulio, Enrico Fermi had spent over two years of his infancy with a rural wet nurse, the European practice even for the middle class of that era. Brother Giulio then became Enrico's idol and partner in crime, the two kids together investigating mysteries of science and engineering, building motors and drafting detailed technical drawings, especially of the era's cutting-edge technology: aeroplanes. Ardent as both colleagues and competitors, the two brothers had no other friends.

Then, at the age of fifteen, while being operated on for a throat sore, Giulio suffocated from the anesthesia. Mother Ida became inconsolable, abandoning her other two children to withdraw into a sobbing mantle of grief. Each day, the boy Enrico forced himself to pass by the hospital where his older brother and only friend had died, until the crippling pain somewhat abated. He then withdrew himself, into a life of books. The family's apartment near the train station was in a district hurriedly built to accommodate Rome's turn-of-the-century doubling in population and would, in time, become famous for its vile urbanity. The quarters were unheated, so in the winters the young Enrico Fermi would sit on his hands to keep them warm and turn the pages with the tip of his tongue.

The years went by with both mother and son hidden from the world, until Enrico finally found another friend, who turned out to be one of his brother's classmates, Enrico Persico. The boys often took long walks through the city and browsed for math and science books in the flea market of the Campo de' Fiori, the Renaissance square where scientist Giordano Bruno was burned at the stake in 1600. One of their scavenges was a two-volume survey of mathematical physics, written in 1840, which taught the two Henrys the Newtonian equations of planets, waves, and tides, the language of numbers with which physics describes everything from the flow of water across a bed of stones, and the gravity that binds planets to their homely stars, to the evidence of things unseen. Enrico Fermi was so enthralled he hadn't noticed until he finished both volumes that they were written in Latin.

Three years later, when Fermi was ready for university, a family friend insisted he learn German to read scientific publications without waiting for them to be translated into Italian or French, as well as attend the Scuola Normale Superiore in Pisa instead of the University of Rome, to get him away from his mother's unabated and suffocating melancholia. For all her faults, though, Enrico inherited from Ida a remarkable trait: the idea that, if you needed something, you could just learn how to make it yourself. The schoolteacher mother made her own pressure cooker, and her physicist son made his own lab apparatus.

To win a spot at Pisa, which included free room and board, he needed to prepare an essay; Fermi's was on the vibration of strings. After reading it, the examiner, a University of Rome geometry professor, asked Enrico to come in and see him, for he had never seen anything in his professional life quite like Enrico's essay. At the meeting, the professor told the student that he was extraordinary, sure to become famous, which, along with the full scholarship to Pisa, changed Fermi's life. For the first time since the loss of his brother, he felt that what he was doing was right, and that he was good at it.

The Scuola Normale Superiore's director of physics was so overwhelmed by Fermi that the boy taught him Einstein's theories of relativity, and since Pisa did not yet teach quantum mechanics, Fermi had to learn it on his own. Pisa's days of greatness were in the Middle Ages, centered on its son Galileo, who'd been inspired by the physics of the pendulum from watching the great cathedral's swinging lamps. Like his Roman childhood home, Enrico's room was unheated, but instead of sitting on their hands, the students tried to keep warm with charcoal-burning ceramic braziers, *scaldini*.

A fast friend at Pisa was Franco Rasetti, a self-taught polymath deliberately studying physics because it was difficult—the boy wanted to prove to himself that he could accomplish anything. Besides hiking in the Alps near the Carrara marble quarries, the two loved pranks, with a twist. At that time in Italy, public urinals were built with pools of water. The nineteen-year-olds Fermi and Rasetti would sneak up behind a man using the facilities, loft a bit of metallic sodium into the water, then listen to the victim's cries of horror as the pool exploded into flames. Though the university may have lagged academically, it was a little Eden for young men falling in love with science. Fermi wrote of Pisa's three lucky physics students, "They were allowed to use the research laboratories at all times, received keys to the library and instrument cabinets, and were given permission to try any experiment they wished with the apparatus contained therein. [Enzo] Cararra and Rasetti, who in the previous year had come to recognize Fermi's immense superiority in the knowledge of mathematics and physics, henceforth regarded him as their natural leader, looking to him rather than to the professors for instruction and guidance."

In another echo of his mother's can-do attitude, he said, "Fermi, after much reading of the pertinent literature, decided that X-rays were the field that offered the best chance for original research [but] it soon appeared that the sealed tubes were not fit for research, and the experimenters decided to build their own tubes. The glass part was made to specifications by glassblower, while the physicist had to seal windows and electrodes. No diffusion pumps were available; hence the tubes were evacuated by means of rotary mercury pumps." As hard as it is to imagine, X-rays, vacuum tubes, and electrodes would be the foundations of the Atomic Age.

Emilio Segrè explained his friend's academic stance during this period: "Fermi was almost entirely self-taught; all that he knew he had learned from books or rediscovered by himself. He had found no mature scientist who could guide him, as he would have found at that time in Germany, Holland, or England, and did not personally know any older scientists with whom he could compare himself. He knew that he was better than those around him, but this he also knew meant little because these men were not in the forefront of active science. And he was in a hurry to get to the top."

The Italian Ministry of Education had one fellowship for postdoctoral study in the natural sciences, and Fermi won it in 1923, going to work under Max Born at a paradise of academic physics, Göttingen. Though his colleagues included the stellar Werner Heisenberg and Wolfgang Pauli, it seems that Fermi did not become a signature member of that extraordinary clique, even though his German was good enough. It may have been too big a pond for someone used to being a great star of the provinces; or maybe his tendency to be shy, proud, solitary, and aloof kept him from being welcomed and engaged. It was a profound opportunity wasted . . . yet, Enrico's life was soon to change dramatically . . . through the efforts of a modern-day Medici.

Orso Mario Corbino grew up in a small town on the east coast of Sicily where his family owned a handmade-macaroni factory with the product sold on the premises. Instead of pasta, though, Corbino worked in magneto-optics at the University of Rome; joined government committees to manage the nation's water resources; was made a senator of the kingdom in 1920; and the minister of national economics in 1923, even though he was not then and never became a Fascist. As the University of Rome's dean of physics, Corbino was determined to begin a world-class program, and he brought the twenty-six-year-old Fermi aboard with lifetime tenure—an achievement that most academics needed thirty-five years to acquire. Franco Rasetti transferred from Florence in 1927, Emilio Segrè and Edoardo Amaldi enrolled as students, and the group immediately made such an international splash in journal publishing that in the fall of that year, a physics conference in the lakeside resort of Como drew such international superstars as Rutherford, Planck, Bohr, and Heisenberg.

After decades as an academic backwater, Italian science was suddenly, all through Corbino's efforts, now at the global forefront. The Italians were honored with overseas invitations to learn state-of-the-art experimental technique, with Amaldi off to Leipzig, Segrè to Hamburg, and Rasetti voyaging first to California and then to Berlin to work with the acclaimed Lise Meitner. The University of Michigan invited Fermi to teach summer school in 1930; he loved Ann Arbor so much he returned every few years. Segrè: "Mechanical proficiency and practical gadgets in America counterbalanced to an extent the beauty of Italy. . . . We bought a car, the Flying Tortoise, which we drove back to New York, not without some mechanical difficulties along the way. These did not scare Fermi, who is a good mechanic. Once at a gas station he showed such expertise in repairing the automobile that the owner instantly offered him a job. And these were depression days."

Ragazzi Corbino—"Corbino's boys"—became so close that they developed their own accent. College friends Fermi and Rasetti, imitating each other, jointly evolved a deep speaking voice and a slow, modulated cadence that was in turn adopted by all of their professional colleagues. One of them was riding the train and started chatting with a seatmate, who quickly asked if he was a Roman physicist. The stranger said he recognized "your way of speaking."

Teaching quantum theory as Rome's first atomic physicist, Enrico was nicknamed the Pope, since it took such profound faith to believe that matter was energy, energy was matter, and that both were sometimes particles and at other times waves. Rasetti was the cardinal, and Segrè, known for his temper (and for breaking furniture as a result of that temper), was the basilisk, a mythical creature whose eyes were alight with fire. The department was housed in an 1880 monastery on via Panisperna 89a, with ocher walls, tile roofs, and a cupola. Gabriel Maria Giannini: "Everything around us was moldy with its eight hundred years, and we were young—bound together by youth and by Fermi's ageless thinking, which managed to find expression in spite of the sound of the church bells pouring in torrents from a Romanic tower next door." Emilio Segrè: "The location of the building in a small park on a hill near the central part of Rome was convenient and beautiful at the same time. The garden, landscaped with palm trees and bamboo thickets, with its prevailing silence . . . made the Institute a most peaceful and attractive center of study. I believe that everybody who ever worked there kept an affectionate regard for the old place, and had poetic feelings about it." One neighbor was G. C. Trabacchi, chief physicist of the Health Department, who shared his excellent collection of instruments and materials with untoward generosity, earning him the Fermi-engineered nickname Divine Providence. Trabacchi was especially providential when he loaned Fermi the radon-gas effluence of a gram of radium, which was at the time worth around $34,000. Hans Bethe (*Beta*), who won a Nobel for his fusion theory on the origin of starlight and who would become the chief of the Los Alamos theory group (inadvertently thrusting Edward Teller into a career combining Dr. Strangelove with Ronald Reagan), spent most of 1931 with the Fermi team:

> Fermi worked in the Institute of Physics, which was on a small hill in the middle of Rome, surrounded by a sea of traffic but very quiet on that little hill. There were trees, ponds, a nice garden, a fountain—really quite an oasis in the hectic traffic of Rome. Fermi was twenty-nine years old when I got there. He was a full professor since he published Fermi's Statistics at the age of twenty-five.
>
> I had studied with [Arnold Sommerfeld, one of the cofounders of quantum mechanics], and Sommerfeld's style was to solve problems exactly. You would sit down and write down the differential equation. And then you would solve it, and that would take quite a long time; and then you got an exact solution. And that was very appropriate for electrodynamics, which Sommerfeld was very good at, but it was not appropriate at all for nuclear physics, which very soon entered all of our lives.
>
> Fermi did it very differently. . . . He would sit down and say, "Now, well, let us think about that question." And then he would take the problem apart, and then he would use first principles of physics, and very soon by having analyzed the problems and understood the main features, very soon he would get the answer. It changed my scientific life. It would not have been the same without having been with Fermi; in fact I don't know whether I would have learned this easy approach to physics which Fermi practiced if I hadn't been there. . . .

> Fermi seemed to me at the time like the bright Italian sunshine. Clarity appeared wherever his mind took hold. . . . Depending on how we count, Fermi training led to ten, eleven, or twelve Nobel Prizes. I estimate the probability that an existing Nobel Prize winner in physics "gives birth" to another winner is less than 1/10. So if this is purely random, the probability of one winner giving birth to ten other winners would be one-tenth to the tenth power or one in 10 billion, which is essentially impossible.

Physics seemed to infuse Fermi's every waking moment, as American physicist Phil Morrison remembered from his time with Enrico: "I want to mention the 'Fermi Questions.' Fermi was the first physicist to my knowledge who enjoyed doing physics out loud walking through the hall. . . . As we walked, the sounds of our footsteps reflected off the high surface—wood, no acoustic treatment—and seemed to bounce throughout. And he said, 'How far do you think our footsteps can be heard in this building?' And then he began to tell me what the yield of sound would be from the impulse, how far that would go, how you have to worry about the wood conduction and the air passage. And pretty soon, by the end of the hall, he had [an answer]. It was a fast calculation. Sounded very reasonable. And when I tried to recalculate it, I got something like the same result—slowly and looking at the numbers over and over again. This was my idea of a Fermi Question: Turn every experience into a question. Can you analyze it? If not, you'll learn something. If you can, you'll also learn something."

At that moment in the nuclear science of the 1930s, there was a whole series of astonishing new questions to answer. Inspired by Fred Joliot and Irène Curie's revelations, those who'd taken an interest in radioelements began to focus on the atom's nucleus as the source of uranic powers. This was a difficult proposition, as physicist Amir Aczel noted, since "if an atom were the size of a bus, than the nucleus would be the dot on the letter *i* in a newspaper story read by a passenger on the bus." Hitting that dot on the letter *i* would make Enrico Fermi a laureate. His efforts began in 1934, when he combined the Joliot-Curie method of artificial irradiation with Chadwick's neutron discovery and his own theory of the weak force to imagine a tiny, uncharged particle that would not be waylaid on its path to crashing into an atom's nucleus. Neutrons fired in the right way, Fermi believed, should be able to excite radiation and produce isotopes—subatomic variations of elements—on just about any member in good standing of the periodic table. Hans Bethe: "Fermi organized a group to do this—of course, his old collaborators and friends, but they added d'Agostino, who was a chemist, and most importantly, Trabacchi, who was a biophysicist in charge of the biophysics in the Department of Health of the City of Rome. He had a very precious possession, namely one gram of radium. And radium produces all the time radon, a gas, which can easily be separated because it escapes from the radium, and then you can expose any sample you want to the alpha rays from radon." Emilio Segrè: "Radon plus beryllium sources were prepared by filling a small glass bowl with beryllium powder, evacuating the air, and replacing the air with radon. Rasetti was vacationing in Morocco so Fermi, Amaldi, and Segrè got to work. Fermi did a good part of the measurements and calculations; Amaldi did the electronics; and I secured the substances to be irradiated, the sources, and the necessary equipment."

It was a good thing these physicists were young and in shape, as this turned out to be a multistage investigation requiring a great deal of sprinting. Bombarding beryllium with radon produced neutrons, which Fermi and his team would in turn use to irradiate as many elements as they could get, to dramatically extend the Joliot-Curie findings of man-made radiance. Making neutrons from radon-charged beryllium, however, triggered their homemade version of Geiger counters that would be used to measure whether they'd succeeded in creating isotopes, making it look as though everything was already radioactive (Geigers weren't yet commercially available, so every scientist working on radiation crafted his or her own). To keep this from affecting their results, Fermi and his grad students would bombard the test element with their neutrons in one room, then run the irradiated subject to the other end of the hall to measure it with the counters. Bethe: "The experimenters had to run as fast as they could along the second-floor corridor from the exposure place to the counter. . . . I believe Fermi had the record of time of running from one place to another. There was a visit one time from a very dignified Spanish physicist, who wanted to see His Excellency Fermi, and he was shown a man in a very dirty lab coat, running like mad along the corridor."

They began at the beginning, with the periodic table's slot number one—hydrogen—and proceeded up the grid: oxygen, lithium, beryllium, boron, carbon. Nothing worked. Even with Trabacchi's precious seed as a source, they could find no induced radiance. Element after element failed, then failed again, then again and again . . . until they got to fluorine. From then on, the success rate was incredible: out of sixty elements tested, forty could be alchemized into radioactive isotopes. Joliot-Curie's quirk of happenstance had been turned by the Fermi team into a scientific procedure.

When Fermi's team bombarded uranium, their chemical tests showed its nucleus capturing the neutrons, spitting out photons of gamma radiation, and becoming heavier, turning into an isotope with an atomic number (in protons) of 93 and an atomic weight (in protons and neutrons) of 239—an element that had not yet been discovered. Would this be as epochal as the Curies' discovery of radium? The Nobel committee thought so, as did the Fascists. But Fermi wasn't absolutely sure since the chemistry needed for proof was inconsistent. Even so, on October 22, 1934, a professional's intuition would trigger the discovery of a fundamental ingredient in the birth of nuclear power. "One day, as I came to the laboratory, it occurred to me that I should examine the effect of placing a piece of lead before the incident neutrons," Fermi remembered. "Instead of my usual custom, I took great pains to have the piece of lead precisely machined. I was clearly dissatisfied with something: I tried every excuse to postpone putting the piece of lead in its place. When finally with some reluctance I was going to put it in its place, I said to myself: 'No, I do not want this piece of lead here; what I want is a piece of paraffin.' It was just like that with no advance warning, no conscious prior reasoning. I immediately took some odd piece of paraffin and placed it where the piece of lead was supposed to have been. About noon everyone was summoned to watch the miraculous effects of the filtration by paraffin. At first I thought the counter had gone wrong because such strong activities had not appeared before."

Hans Bethe: "Neutron research led to many surprises. It turned out that if you (as I remember it from the tales, since I wasn't there) put the sample on top of a wooden table, the radioactivity was stronger than if you put it on top of a marble table. Of course, everything in Rome was of marble, if it wasn't of wood. And so, I guess they got the idea that maybe different surroundings might make a difference, and so instead of using a lead box around the sample, they decided to use a paraffin box. And the paraffin box was tremendously effective. The radioactive count increased about 100-fold with most of the elements. That was a great surprise, of course. And Fermi, having discovered that in the morning, went to lunch, and over lunch he decided what was the reason for it. . . . The hydrogen, which was in paraffin and in wood, would slow down the neutrons." The slowing down made neutrons more likely to collide with neighboring nuclei, and more likely to sustain a chain reaction. Laura Fermi: "Physics was comprehensible, as long as atoms were small planetary systems and discoveries could be made in goldfish ponds . . . like the discovery of slow neutrons. . . . Back in the laboratory after their siesta, the group decided to test Fermi's theory using the most abundant hydrogenated substance at hand; and so they plunged neutron source and target in the goldfish pond at the back of the old physics building. Lo and behold! Fermi was right. Water too increased the radioactivity in the target by many times."

After the Fermi team announced that bombarding uranium produced short-lived transuranics—an array of isotopic variants that, to the inexperienced radiochemists of the time, appeared to be innumerable—the University of Rome experiments were taken up by the Joliot-Curie team in Paris, and by Lise Meitner and Otto Hahn in Berlin. Racing to uncover uranium's secrets, the three labs appeared to generate more and more transuranes, with ever more half-life decays. The method used by modern science, especially within a focused group such as this—sending details of experiments and results to each other, publishing findings as soon as possible, colluding and at the same time rapaciously competing to be first with a groundbreaking discovery—would in the web argot of the next century be called hive mind, a collective effort of human brainpower that would create far more than any one person or team could achieve alone. Scientists have been hive-minding, it turns out, since the Royal Academy began publishing during the Enlightenment.

By the end of 1935, however, the *ragazzi Corbino* were undone, with Rasetti at Columbia, Segrè at Palermo, Pontecorvo in France, and the atmosphere in Italy relentlessly gloomy as the country prepared for war in Ethiopia and the limitations brought by globally imposed sanctions. Only Amaldi and Fermi remained in the department of physics' garden monastery.

The phone call Enrico and Laura were waiting for that night of November 10, 1938, would affirm his decision to abandon their relatives, their friends, their heritage, the Eternal City, which Laura loved with such a passion, their extremely comfortable life, and the whole of their worldly possessions (including a lemon-yellow Peugeot Bébé convertible with celluloid windows and a hand crank for emergency start-ups, which were frequent). The family would immediately flee, as resident aliens, to the United States. Laura's most significant previous American experience had been in joining her husband when he taught summer school in Michigan, and regardless of the many charms of Ann Arbor, it was not Rome. During one visit coinciding with America's Prohibition, the university's chemistry department had to bury the alcohol it used for experiments to keep it from being stolen and drunk. Enrico, however, regularly discussed emigrating to the USA. Coming back from one semester accompanied by the Swiss physicist Felix Bloch, the two noted how superior the Burma Shave billboards in Michigan had been compared to Mussolini's Fascist exhortations along the Roman highways.

But that was one of the few jovial moments outside the lab. In 1936, a month after Hitler occupied the Rhineland, Enrico thought it prudent to supply each member of his family with a gas mask. He wasn't being fearful, just pragmatic. Nella Fermi: "For the most part, my father had very little to do with us when we were children, and I think it's too simple to say that he was too busy with his work and that he had no time for my brother and me. I think he was certainly absorbed in his work, but beyond that, he was a man of reason, and he was a physicist through and through. And he could not relate to us on an emotional level, so it wasn't until we were old enough (and I quote from him) 'to talk to' that he could approach us, and that he could approach us on his own level. With adult hindsight I am convinced that it wasn't that he lacked emotions but that he lacked the ability to express them."

Depending on the phone call, Laura and her children would immediately be deserting the culture and refinement of Europe, the magnificence of Rome, and a life of wealth and status, for some backwoods of hillbillies on the other side of the globe. A third-generation Italian cosmopolite, Sra. Fermi felt she could never fit in over there. Her English was rudimentary schoolgirl; her husband's came from reading Jack London novels. He loved everything about America. She thought otherwise.

Signore Fermi met Signorina Capon when he was twenty-two, and already so prominent as to hold a professorship. She was a mere sixteen. It was a Sunday in the spring of 1924, and a group of friends were taking the air in the countryside of suburban Rome, in a meadow adjoining the fork where the Aniene meets the Tiber. He was dressed in a black suit and black bowler, still in mourning over the death of his mother; yet, he decided that they should all play soccer.

Two years later, the Capons were planning to spend the summer in Chamonix, the French resort shaded by Mont Blanc. But Mussolini's new monetary policy kept them from being able to get any francs on the foreign exchange, and even Laura's father, an officer in the Italian navy, could not overcome this setback. Friends recommended the Dolomites instead, and they arrived to find many of Laura's school chums there for the season, including that acquaintance Enrico Fermi, who was now living with his father, Alberto, and sister, Maria, in Città Giardino, a new suburb reserved for civil servants—Alberto worked for the railroads and sang Verdi arias during his morning shave—not far from that meadow where Enrico and Laura first met.

On his arrival, Fermi immediately arranged for the group to make a series of hikes and climbs, always using his thumb to measure distances, both on maps and in real life. His great passion besides physics was mountain hiking, and this seemingly odd mix of scholar and athlete would be common among his peers. Niels Bohr was both a famed soccer player in his youth and a Ping-Pong champion as an adult, while Werner Heisenberg spent his lifetime downhill racing, at one point being clocked at an alarming fifty miles an hour. Physicist Valentine Telegdi: "Fermi was completely devoted to physics, and his whole existence centered around it. He appeared to have very few outside interests such as literature or the fine arts. He engaged in sports, e.g., in mountaineering and tennis, but one often got the impression that it was all for *mens sana in corpore sano*—i.e., to be in the best physical condition for doing physics; it must be added that in sports as well as in parlor games (which he occasionally organized in his home) he liked to win, being fiercely competitive [though he] was totally secure in his own physics talent and almost never displayed jealousy of another. The only exception, as one of his students recalls, was Einstein. More than once Fermi expressed annoyance at the attention Einstein received from the press." Laura Fermi: "One day that summer I asked Fermi to quiz me and see if I was well prepared for the approaching exam on the two-year physics course. We were at Ostia, and Fermi was sitting cross-legged on the sand, in his bathing-suit, which came up almost to his neck. As he quizzed me, his usual grin faded and his lips tightened. In the end he said: 'I am sorry, Miss Capon, but you don't understand a thing.' What an encouragement!"

Fermi told his friends that the woman of his dreams would be tall, athletic, blond, with ancestors from the countryside and no thoughts of religion—practically the opposite of Laura, who was descended from urban Romans for many generations, unathletic, and relentlessly brunette. Laura: "Fermi had always said he wanted to do something really exciting and outstanding. Either buy a car or get a wife. So when he bought a car I was a little disappointed, although I didn't have any real idea of getting married. But then he was more extravagant and got both a car and a wife. . . . I remember a sense of not even knowing whether he had asked me to marry him or whether he was posing a theoretical question of what would happen if I got married to somebody and he at the same time would get married to somebody else."

On July 19, 1928, they were wed, honeymooning in the Alps, hiking through the shadows of the Matterhorn, which Enrico thought was a perfect opportunity to turn Laura into a physicist . . . but when she refused to accept the mathematical proof that light was electromagnetic radiation of waves and particles, he gave up. Together, though, they wrote a physics textbook for Italian secondary schools, which brought the family income during their lean salad years. Laura: "The next winter was the coldest on record in Rome and we began talking of storm windows. Fermi pulled out his slide rule, calculated the effects of drafts on the inside temperatures, misplaced the decimal point, and we froze all winter."

Fermi had wanted to leave Italy ever since the government had passed the Manifesto della Razza on July 14, 1938—the Italian version of the Nazis' Law for the Restoration of the Professional Civil Service, which prevented Jews from government jobs and meant that, as European higher education was civil service, the Universities of Berlin and Frankfurt had, overnight, lost a third of their professors. Though Fermi and the children were Catholic, Laura was Jewish. As a result of the new decrees, Laura's father, practically of Roman nobility from his decades in the navy, was dismissed from active duty and placed on reserve. Even so, Laura was convinced that the Razza was a minor legal kerfuffle, a temporary annoyance. Italy's 1870 nationalist movement had freed Jews from the ghettos and given them full equality; they were now so few in number and so thoroughly assimilated that they were practically invisible. A third of them were Fascist Party members; Mussolini's own mistress was Jewish. Just after the law was announced, Laura overheard one man on the street ask another, "Now they are sending away the Jews. But, who are the Jews?" Mussolini received a telegram from a Sicilian mayor: "Re: Anti-Semitic Campaign. Send specimen so we can start campaign."

But Fermi clearly knew the history behind these laws. When astronomers confirmed Albert Einstein's theory of general relativity on November 7, 1919, Berlin's *Illustrierte Zeitung* transformed its entire front page into his photograph, calling his ideas "on a par with insights of Copernicus, Kepler, and Newton." But by February of 1930, students interrupted his lectures, with one screaming, "I'm going to cut the throat of that dirty Jew." On August 24, 1930, Nazi scientists Philipp Lenard and Johannes Stark held the first meeting of the Working Group of German Scientists for the Preservation of Pure Science at Berlin's Philharmonic Hall, attacking relativity as "Jewish physics" and Einstein as a plagiarist and charlatan. Einstein attended, watching from a private box, saying nothing. He was eventually compelled to renounce his German citizenship for a second time and leave for England, and then Princeton. When he arrived in New York harbor on October 17, 1933, he was smuggled ashore in a tugboat to ensure his safety. Three months later, he was spending a night at the White House with the Roosevelts. After the 1935 Nuremberg decrees, gangs regularly gathered outside his Berlin home to scream insults about "Jewish physics," and a magazine included him on a list of enemies of the state—with the notation "not yet hanged"—and a $5,000 bounty promised to his successful assassin.

Knowing all this, while the Fermis returned to vacation in the Dolomites that summer of 1938, Enrico had written to four American universities that had previously offered him posts, vaguely explaining that his earlier reasons for not accepting were no longer in effect. He then mailed these letters from four different towns to avoid suspicion and received five offers, accepting Columbia's. In his follow-up letter to the school's dean on September 4, he explained the precautions needed to leave Axis Rome and tried to help other *ragazzi Corbino* reach safe harbor:

> For reasons that you can easily understand however, I should like to leave Italy, without giving the feeling that this is due to political reasons. I could manage this much more easily if you could write me officially to teach at Columbia through the Italian Embassy in the U.S. Of course you need no mention, or stress, in this request, that it would be a permanent appointment.
>
> In order to get a non-quota visa for myself and my family, I should need besides an official letter from Columbia stating that I am appointed as professor and mentioning the salary. In case that you cannot write me through the Embassy, please send me only this second letter. And in any case please do not give unnecessary publicity to this matter until the situation in Italy is finally settled.
>
> I shall take the opportunity that I am writing to you from Belgium, in order to give to you some information about the situation of the Italian physicists that have lost their positions on account of racial reasons.
>
> They are Emilio Segrè, whom you already know. He is now at Berkeley and has, so far as I know, a small research fellowship for one year from the University of California. I don't think that I need to inform you about his scientific work.

> Bruno Rossi, formerly professor at the University of Padova (married with no children; age about 32). He is one of our best young physicists, his work on the cosmic radiation is probably known to you. He has lately acquired some experience on high tension work, since he had built in Padova a one million volt Cockroft Walton outfit, that was just now being tested.
>
> Giulio Racah, formerly professor at Pisa (not married; age about 30). He has a very extensive knowledge of theoretical physics. Has published many papers on atomic physics and quantum theory; in particular he has obtained independently and published only a few days after Heitler and Bethe equivalent results on the theory of the emission of high energy gamma rays from cosmic ray electrons colliding against nuclei.

Enrico then notified the Fascist government that he was planning a six-month visit to New York, and he would be accompanied by his family. He had to use all of his influence to keep his wife and children's Italian passports secure. The Americans, meanwhile, were so impressed by his stature that even the family maid was approved for a visa.

There was, however, an unresolved practical matter. At an industry conference in Copenhagen that fall, Niels Bohr had taken Fermi aside to reveal he was on the Nobel short list. Before the rise of Hitler, laureates were never informed in advance, but the Swedish Academy had then seen scientists living under dictatorships get harassed and attacked for the prize and wanted to make sure that Fermi wouldn't be embarrassed by it. He would in fact be embarrassed, but additionally if the Fermis returned to Italy with his Nobel winnings and then left the country, the family would only be allowed to take fifty dollars with them. Enrico decided that if he won, Laura and the children would accompany him to Stockholm for the ceremony, and then they would leave directly for a Southampton sailing to New York. Even considering the generous terms of the award, they would still be abandoning an extremely comfortable life in Rome. In May of 1938, for one example, baby Giulio and the housekeeper, out to get some fresh air in the park, had come across Il Duce taking the sun with Hitler.

The phone call from Sweden, then, would determine if Enrico Fermi was a laureate, and if he and his family were now to be refugees. Additionally, if he had to split the prize with one or more other physicists, the family would be starting their life over again under seriously reduced financial circumstances.

While they waited that evening, Italian radio described Germany's *Kristallnacht* of the day before and explained the new laws: Jewish children and teachers were barred from public schools; Jewish professionals such as doctors and lawyers could only have Jewish clients; "Aryans" could not work for Jews as servants; and all Jewish passports were withdrawn.

Laura's nonchalance about this state of affairs would prove to be misplaced. Though Mussolini consistently refused to hand over Italian Jews to the Germans, after his overthrow in July 1943, the Nazis occupied Italy and began murdering them, with over a thousand Romans sent to Auschwitz, including Laura's father. Decades later, daughter Nella Fermi tried to find out what had happened: "I think that my mother was having a lot of guilt about leaving her father behind [and going to America]. My grandmother had died of natural causes some years before we left, and there were 2 sisters and a brother who were still in Italy, so it wasn't as if she was abandoning him altogether to himself. I'm not sure when they learned about it. I know that at some point, my mother told me that they had heard . . . that he had been taken by the Nazis, but . . . I think that it might have been a way of protecting me, rather than the strict truth. . . . She said that he had died on the train. My aunt, my father's sister, was practically running an underground railroad in her basement, and she had gone over to persuade him to come and stay with her, and he had other friends and connections who were not Jewish who he could have stayed with. . . . It was really easy to hide in Rome . . . largely because the population was simply not behind it. [But] he thought that being a high-ranking naval officer . . . he was an admiral . . . that he had given his life to the service of his country, and he was a gentleman of the old school and was convinced that they would not bother him. . . . About three or four years ago, I talked to a man who had done some research into the subject, and he seems to have come up with some very conclusive evidence that my grandfather made it as far as Auschwitz. He was one of the first to go in the gas chambers."

The telephone finally rang. For his work with slow neutrons and his discovery of element 93, Enrico Fermi, at the age of thirty-seven, had won the Nobel Prize in Physics.

Even with such an immense honor, no one would have thought at that moment, least of all Fermi himself, that in a mere three years, he would join an elite group who would revolutionize the academic and scientific fabrics of the United States. Before the rise of Fascism, while European scientists, backed by their countries' military research budgets, were able to pursue fundamental scientific research, American science was wholly a backwater (though its engineers created the telephone and the lightbulb). What changed science in the United States for all time was the immigration of nuclear physicists displaced from Europe. By creating a safe haven for rejected genius, America transformed herself from an R&D Appalachia to the center of everything nuclear, and then of everything in science. In the years to come, joined by a phalanx of genius, Enrico Fermi would invent and perfect the engine that would create nuclear power, nuclear weapons, and the Atomic Age.

On December 6, the family took the forty-eight-hour train to Stockholm, and at the ceremony on the tenth, they learned that Enrico would not have to share the reward monies with another physicist. In October of 1933, they had dined with King Albert of Belgium and Marie Curie (Laura: "She would not take notice of insignificant wives like me"), and at the Nobel ceremonies, Laura danced with Crown Prince Gustavus Adolphus, so it is easy to understand her misgivings on leaving this life. But it quickly became apparent that the Academy and Bohr were right to notify Fermi in advance that he was a candidate, for, in the wake of the award, instead of pride at a hometown boy who made good, the Italian press was filled with criticisms of Enrico for not wearing a Fascist uniform, not giving the Fascist salute, and for shaking the king's hand, which was thought unmanly. He was attacked by the more extreme of the press for "having transformed the Physics Institute into a synagogue."

The Fermis went directly from Sweden to Southampton, sailing for Manhattan aboard Cunard's RMS *Franconia II*, a ship so luxurious she would ferry Winston Churchill to Yalta, and whose first-class smoking lounge was a detailed re-creation of a classic English hotel lobby, down to the aged-oak paneling and the brick inglenook fireplace.

One evening while crossing the Atlantic, Laura and the children were waiting for an elevator. Its doors opened to reveal a strange man in a red-and-white suit, who immediately invited them to a party he was having that night, where every child aboard would get presents! Giulio and Nella looked at their mother with a mix of shock and awe. Enrico liked to say, grandly, that they were off to establish the American branch of the Fermi family . . . but to Laura's mind, here was one more example of how far they had to go. Recently, Mrs. Fermi had been mortified to learn that one of her lifelong beliefs was wrong—she'd always assumed Abraham Lincoln was Jewish since, in Italy, only Jews were named Abraham. While her husband enjoyed the professional embrace that came with being a laureate, she would now have to explain to her new American children who, or what, Santa Claus was.

4

The Mysteries of Budapest

On January 2, 1939, the Fermis checked into the King's Crown next to Columbia University's main campus and immediately ran into another hotel resident in the lobby—Hungarian physicist Leo Szilard (*SIL-ard*). It was a happy coincidence as the two scientists had been writing to each other for the past three years. In the period when Fermi discovered slow neutrons, Szilard had patented a method for starting a nuclear chain reaction. Now, pure happenstance had brought them together . . . the two scientists who would, in time, jointly create the foundation of nuclear power and atomic weapons. With Leo, Enrico would achieve his greatest triumph; and with Leo, Enrico, the most genial man in the community of physics, would stop speaking for months at a time.

In Hungarian, *szilard* means "solid," yet Leo was anything but—he was an occupational gadfly, outrageously frank and forthright, to the point of being uncivil, undiplomatic, and obnoxious. One telling incident: In the 1940s, Szilard wrote his memoirs, *My Version of the Facts*. He showed the manuscript to Los Alamos Theoretical Division chief Hans Bethe, but explained he had no plans to publish. He just wanted God to know the facts. When Bethe asked, "Don't you think God already knows the facts?" Szilard replied, "But He may not know my version."

Leo's parents, Louis and Tekla Spitz, changed their name to Szilard two years after his birth on February 11, 1898, in Austria-Hungary to comply with a government campaign to unify its disparate ethnicities including the Jews. This simple attempt at social engineering resulted, because of birth order, in their children having different surnames: Szilard, Szego, Salgo, and Spitz. The family lived in a stucco art nouveau villa, kitted out with stained glass and wooden turrets in Budapest's Garden District. Leo's notorious personality revealed itself early. When his grandparents tried to teach the young boy to say "please," he refused, explaining, "It is beneath my dignity."

Fluent in German and French by the age of ten, Leo never learned to swim, or ride a bicycle. Drafted to fight in the Great War, he was stricken with Spanish flu and recuperating in a sanitarium when the armistice was signed. At that moment, Hungary had been headed up by a Jewish Communist, Bela Kun, for a disastrous 133 days. The peace treaty of 1920 carved up the nation, Kun was thrown out, and the right-wing Miklós Horthy became regent to the nation's now-powerless king. Horthy remained in charge until 1944, beginning a history of Hungarian anti-Semitism from on high, which exploded when the Nazis in time replaced him with the virulently Jew-hating Arrow Cross.

On July 24, 1919, in the wake of Horthy's rise, Szilard went to his neighborhood's Reform church and officially changed his religion from "Israelite" to "Calvinist." Two months later in September of 1919, a gang of students blocked the twenty-one-year-old Leo and his brother, Bela, on their way to engineering classes at Budapest's Technical University, screaming, "You can't study here! You're Jews!" When the Szilards tried to continue to their classes, the gang beat them up.

Leo immediately applied for an exit visa, which was difficult to get, as the Szilard brothers had joined various student political organizations and were considered "dangerous" by the new regime. But finally, Leo was released. With a suitcase full of books and practically his entire family's life savings hidden in the soles of his shoes, he rode a steamship up the Danube to Vienna. On board, a man asked why Leo was so sad, and he replied morbidly that he was "leaving his country, perhaps for good." The man explained that he himself was a Hungarian who'd lived in Canada for the past four decades, and that "as long as you live, you'll remember this as the happiest day of your life!"

What should then have been a one-day train ride from Vienna to Berlin took a week as the engine kept running out of coal, and Leo kept running out of food. In March 1920, Bela joined him, and the brothers shared a room with an equally impoverished landlady in the economic free fall of the Weimar Republic. They continued studying engineering at the Berlin Technical School, but since that institution offered no physics classes, Szilard decided to attend the Wednesday colloquium of the German Physical Society, held at Friedrich Wilhelm University, which counted among its regulars the titans Albert Einstein of relativity, Hans Geiger of the counter, Max Planck of the quantum, Lise Meitner (whom Einstein called "our Madame Curie"), Max von Laue of superconductivity, and Fritz Haber of man-made fertilizer. The Berlin colloquiums had a well-defined protocol: On its front bench sat five Nobel laureates who argued over ideas and experiments as the presenting lecturer's theory was refined, destroyed, or anointed. In the back sat students like Szilard. Almost imperceptibly, Leo moved from the back to front, until he was close enough to casually chat with the front-row immortals. Ordinary men might've been intimidated, but not Szilard, who told Max Planck when they first met, "I only want to know the facts of physics. I will make up the theories myself."

At both schools, Leo met other Hungarian exiles, including physicists Dennis Gabor (who would invent the hologram), Eugene Wigner, Edward Teller, and mathematician Janos von Neumann (*NOY-man*). These young expatriates together generated a slew of inventions, such as an electrically charged chair that would speed barbering by making the customers' hair stand on end, a "Bride-o-Mat" vending machine for love letters that the romantically inclined could not receive at home, and magnetic hosiery that would keep a woman's stockings up. The mix of odd-yet-useful would be a hallmark of Szilard's future, as the gadfly's most regular income would flow from his many patents.

It was the greatest time to be a physicist in the history of matter, and anyone who even casually studies this period always ends up pondering the same mystery: *Budapest*. Three of the great creators of modern science—the coinventor of the nuclear reactor (Szilard), the coinventor of the hydrogen or fusion bomb (Edward Teller), and the coinventor of the modern computer (von Neumann)—all grew up in the same town of Pest a few kilometers from one another and would all end up working together in the United States. There, they would be known alongside fellow physicist Eugene Wigner as the Hungarian Quartet. The Quartet more accurately should have included mathematician Theodore von Kármán and been a Quintet, but who doesn't yearn for a Beethoven allusion with their Atomic Age history?

When as a boy Eugene Wigner announced he wanted to be a physicist, and his father asked how many physicist jobs existed in Hungary, Gene exaggerated and said four . . . there were a mere three professorships in the entire country. Gene accepted his father's suggestion to study chemistry as well, for perhaps this would lead to a good job with a nearby tannery. Instead, Gene would end up in Tennessee, at one of the greatest manufacturing concerns in the history of humankind, as the patron saint of America's atomic secret.

Gene Wigner remembered Leo Szilard from their first meeting in 1921 as "a vivid man about five feet six inches tall. . . . His eyes were brown. His hair, like my own, was brown, poorly combed, and already receding." As Wigner got to know Szilard, he would expand on this: "[Frequently and suddenly] he appeared at your front door with several bold ideas, and not quite enough patience. Leo Szilard was always in a hurry. . . . If he saw the president of the United States meeting with the president of Soviet Russia, Leo would probably introduce himself and begin asking pointed questions."

The mysteries of Budapest. While working at the University of Chicago, Fermi brought up the question now known as the Fermi paradox: Why hadn't alien life-forms noticed a planet as beautiful as our earth? "They should have arrived here by now. So, where are they?" he asked. Szilard retorted, "They are among us. But they call themselves Hungarians." This was a twist on a famous assertion of von Kármán's, that the reason so many natives of Pest—which means "furnace"—became internationally acclaimed scientists was that they were Martians (he included Zsa Zsa Gabor in this extraterrestrial cotillion). Polish mathematician Stanislaw Ulam: "Johnny [von Neumann] used to say that it was a coincidence of some cultural factors which he could not make precise: an external pressure on the whole society of this part of Central Europe, a feeling of extreme insecurity in the individuals, and the necessity to produce the unusual or else face extinction." Additionally, there is an old Pest saying that if you enter a revolving door with a Hungarian behind you, by the time you come out, he'll be in front of you.

The mysteries of Budapest. After the Hapsburgs tried to conquer Hungary, failed, and needed to beg for military help from Russia, they had so much trouble governing their new province that they had to promote it to equality, creating an Austro-Hungarian empire under Franz Joseph I, with Vienna and Budapest as co-capitals. The government then pressed forward with immense liberalizations, which made the nation the second-most-popular destination for the 2 million Jews fleeing czarist Russian pogroms and anti-Jewish rioting in the Pale of Settlement (Poland, Ukraine, Belarus, and Lithuania). Budapest became nineteenth-century Europe's fastest-growing city, and within two decades the first generation of itinerant farmer and peddler émigrés begat descendants who created a vibrant, educated, cosmopolitan, professional, Yiddish-speaking class of doctors, lawyers, and storekeepers, becoming a fifth of Budapest's population. By the turn of the century, the city was a belle epoque jewel, with the largest parliament building in the world, the first European subway (horse-drawn, and then electric), the largest single-span bridge on earth, and a plethora of first-class hotels, electric trolleys, and plate-glass windows. Like Weimar Berlin—where many of these Martians would initially prosper—Buda and Pest were politically chaotic, with two revolutions in 1918–19, the incompetent Communist rule of Bela Kun, two years of White Terror focused against Communists and Jews, the loss of a war with Romania and Romanian occupation, and the ascent of the Hungarian Mussolini, Miklós Horthy (the grotesqueries of the Nazi and Soviet occupations have been memorialized in one of the world's most remarkable museums, the House of Terror). For decades, Berlin and Budapest were as alike and not as sisters: tough but warm; cosmopolitan but spiritual; perfectionist yet tolerant; and just a little too aware of their role in history. To this day, even after five decades of Soviet brutality and incompetence, the Hungarian capital easily competes with Paris and Vienna in its sophistication, beauty, and urban grace.

The men who would become America's Hungarian Quartet all came from Budapest's newly Jewish upper-middle class, but all of them save Teller had, like Szilard, nominally converted to Christianity. They did not follow Judaism as a religion or as an ethnic identity, but they would follow its culture and traditions. Though lifelong friends from college, they retained Magyar formalities, calling each other by surnames and often including the *ur*, meaning "master." That formality extended to their attire; as adults, both von Neumann and Szilard always wore suits and ties, even, most notoriously, when hiking in the deserts of New Mexico.

Each of the Quartet emigrated from Hungary to Germany, globally renowned for its science education and research—the teenaged Enrico Fermi's having been forced by his mentor to learn German was one sign of the country's prominence—with Teller going to Leipzig under Werner Heisenberg, von Kármán to Göttingen, and Szilard, Wigner, and von Neumann to Berlin (their fellow Austro-Hungarian in America Isidor Isaac Rabi emigrated to the Lower East Side of New York as an infant and became Columbia University's first Jewish professor). In Berlin, Leo became famous as "an intellectual bumblebee." His brain spewed a torrent of notions. Some friends believed that had he pursued all his ideas in the real world, today he'd be spoken of in the same breath as Thomas Edison. On Szilard, Rabi would say, "You didn't know what he was up to. He was always a bit mysterious." One chum nicknamed him "the inventor of all things," while Gene Wigner concluded, "I never met anybody more imaginative than Leo Szilard. No one had more independence of thought and opinion. You may value this statement better if you recall that I knew Albert Einstein as well." Szilard designed an electron microscope, a low-fat cheese (being a man constantly trying to lose weight), and the basis of the cyclotron. Leo Szilard: "I believe that many children are born with an inquisitive mind, the mind of a scientist, and I assume that I became a scientist because in some ways I remained a child."

For most of his life, Leo's friends would have terrible problems telling when he was serious and when he was kidding. Yet this childlike sense would be shared with Einstein, with whom Szilard would become extremely close. Each Wednesday, Einstein and his wife had students to their home for tea and pastries, and starting in 1920, Szilard attended religiously. Both men were notably shy, except with each other, and became so friendly that Einstein tried to talk Szilard into working as a patent clerk, since that's what Einstein did to earn a living while creating the greatest of his theories: "When I worked in the patent office, that was my best time of all." Both became enamored of Baruch Spinoza, the seventeenth-century Dutch lensmaker and philosopher, with Einstein explaining that he believed in "Spinoza's God who reveals himself in the harmony of all that exists, not in a God who concerns himself with the fate and actions of men."

As an adult, Szilard began almost every morning with breakfast and the newspapers in a café, then returned to his quarters to soak in a bathtub for up to three hours and think without distraction . . . suddenly leaping out, soaking wet, to jot down notes on a yellow legal pad. Scientists studying creativity have recently uncovered why this strategy produces results—when a problem is mulled again and again, frequently the inspired solution arises after a serious break from concentration, whether with a long walk or a good soak.

When Szilard saw a newspaper story that a family had died from a leak in their refrigerator's coolant, he and Einstein wanted to prevent that tragedy from ever happening again. They wondered if the body's method of circulation could be applied to refrigerators, and designed an electromagnetic pump that circulated liquid-metal coolant. This would never be as popular in home kitchens as General Electric's design of motor and coils, but it would become a fixture in the first generation of atomic breeder reactors.

Szilard's close friendship with Einstein would not lead to any jobs—mostly because Szilard couldn't settle on just what kind of job he should be doing—but it did result in letters of support for American visas, saying that Szilard was a dear colleague carrying out "work in which I myself have an interest." Szilard would eventually make a tidy income from certain of his twenty-nine German patents, including a mercury-vapor lamp, and the Szilard-Einstein refrigerator pump, which alone led to eight joint patents. Of him, Einstein said, "He tends to overestimate the role of rational thought in human life."

Beginning in 1923, Leo regularly dropped by for visits at the world's premier scientific research facility, the Kaiser Wilhelm Institute, where he interrupted the scientists in their labs to ask questions, often explaining why they were wrong, and suggesting entirely new avenues for them to pursue. Besides his lack of professional courtesy, Leo was widely disliked by fellow scientists for his vigor and speed in applying for patents, which he thought was his only course of guaranteeing an income, but which many found selfish and contrary to the hive-mind keystone of the modern scientific method. He was obnoxious, rude, haughty, and staggeringly offensive, yet so often right and inspiring that he was made a *Generaldirektor*, teaching nuclear physics and chemistry with Germany's Marie Curie, Lise Meitner. In 1925, he was promoted to the choice spot of being von Laue's assistant—Max von Laue had won his Nobel for showing how X-rays could be focused with crystals. Now Leo lived his dream life as a flaneur, soaking in his tub, smoking and drinking in cafés, then wandering the streets of the Eden that was Weimar Berlin, frequently with Dennis Gabor, to whom he proposed a method for uniting the bloodstreams of an old dog and a young dog to create a supercanine that would live forever.

Over Easter holiday in England in 1929, Leo's dinner companion was a man who would write over a hundred books from 1895 to 1946 and who was globally famous for what he called his "scientific romances"—today known as science fiction—H. G. Wells. In nearly all of his books, Wells showed how humanity could improve itself through the application of science, and in his spare time, he proselytized for free thought, free speech, and the end of war through world government, working to launch both the League of Nations and the United Nations. Wells also thought the human species would be greatly improved through a vast and calculated program of eugenics.

Three years after their dinner, Szilard read Wells's *The World Set Free*, in which, Leo reported, the author "proceeds to describe the liberation of atomic energy on a large scale for industrial purposes, and the development of atomic bombs, and a world war which was apparently fought by an alliance of England, France, and perhaps including America, against Germany and Austria [in which] the major cities of the world are all destroyed by atomic bombs." Just as rocket science had begun with Tsiolkovsky, Goddard, and von Braun inspired to their life's work from reading Jules Verne's *From the Earth to the Moon*, so nuclear power and weaponry would be triggered by Szilard's infatuation with this 1914 science-fiction novel.

Written in Switzerland at the dawn of World War I, *The World Set Free* describes a radium-like material, carolinum, which scientists use as the basis for what Wells called, for the first time, an atomic bomb. Wells had originally been inspired by Frederick Soddy's *Interpretation of Radium*, which predicted that such a bomb could be made to destroy the earth, but that the same amount of fuel could also generate enough power to light the lamps of London for a year. *The World Set Free* is not very good and has never been all that popular, but it is easily the most historically significant book H. G. Wells ever wrote because of the effect it had on one reader: Leo Szilard. Years after the Manhattan Project had completed its work, when Szilard was asked who the father of nuclear weapons was, he would always reply: H. G. Wells.

When Hitler became chancellor on January 30, 1933, Szilard told his Budapest relatives, "Hitler and his Nazis are going to take over Europe. Get out now." But he himself didn't, even though Wigner and von Neumann had left for Princeton three years before. Then on March 27, Reich minister for people's enlightenment and propaganda Joseph Goebbels announced a boycott of Jewish businesses and a limit on Jews entering universities and law and medical schools to their ratio of the population—1 percent. The next day, Einstein announced from Belgium he would not return to Germany until such anti-Semitic measures were ended, and two days later Szilard packed his things and took the train to Vienna. The following day when the anti-Jewish laws began being enforced in earnest, "non-Aryans" were refused exits at the Czech border and their valuables were confiscated. But Szilard had been blessed with perfect timing and had no trouble. He and Bela then rode to Switzerland to safeguard the family's savings and returned to Vienna to monitor the situation.

Through friends of friends, Leo then met London School of Economics director Sir William Beveridge, who agreed that a refugee effort for Germany's displaced Jewish scientists needed to be created, and who encouraged Szilard to come to London to "prod" Beveridge into it. On April 7, just four days before all Jews were removed from civil service, Szilard moved to Russell Square to organize the Academic Assistance Council, which rescued more than twenty-five hundred exiles over the next six years (and still exists today as the Society for the Protection of Science and Learning). As for Szilard himself, "practically everybody who came to England had a position, except me." His sole income? Refrigerator patents. While working so hard to help others professionally, he continued to dither about his own future, unable to decide whether to go to America, to India, or to remain in England. In a letter dated August 11, 1933, Szilard said, "I'm spending much money at present for traveling about and earn of course nothing and cannot possibly go on with this for very long. At the moment, however, I can be so useful that I cannot afford to retire into private life."

On September 13, Leo was walking the streets of London as he always did, in an absentminded haze, a man neither here, nor there, pondering Wells, Hitler, and especially Ernest Rutherford's pronouncement in the *Times* the day before that "anyone who looked for a source of power in the transformation of the atoms was talking moonshine." Nothing bothered Szilard more than hearing a scientist claim something to be impossible if that impossibility hadn't categorically been proven.

On Southampton Row in Bloomsbury, "as I was waiting for the light to change and as the light changed to green and I crossed the street, it suddenly occurred to me that if we could find an element which is split by neutrons and which would emit two neutrons when it absorbed one neutron, such an element, if assembled in sufficiently large mass, could sustain a nuclear chain reaction. I didn't see at the moment just how one would go about finding such an element, or what experiments would be needed, but the idea never left me. In certain circumstances it might become possible to set up a nuclear chain reaction, liberate energy on an industrial scale, and construct atomic bombs. The thought that this might be in fact possible became sort of an obsession with me."

On June 4, 1934, Szilard met with Rutherford about a possible slot at his Cavendish Laboratory. He wanted to conduct experiments proving or disproving the chain-reaction theory, and thinking it would help, he described what these experiments might be based on what Rutherford and his associates had achieved with alpha particles, instead of on the neutrons he himself envisioned. Rutherford immediately saw the limits of using alphas, and when Szilard then explained he'd taken out a patent on the whole notion, the Hungarian became the sole visitor that the New Zealand émigré physicist ever threw out of his Cambridge office. But the world is very, very small, and in time Rutherford would become president of Szilard's Academic Assistance Council.

Leo meanwhile remained so obsessed with the promise of a neutron-triggered chain reaction that he spurned all his friends and social life to soak in the tub, leaving his room only to eat. After running through calculation after calculation, he finally had an answer, winning a British patent on March 12, 1934, for a method of inducing a reaction with beryllium (it would turn out that an incorrect assessment of the element's atomic weight misled him on its nuclear potential), as well as uranium and thorium (the only two elements in nature that can in fact chain-react). He also applied for a patent to reduce libraries to images on a roll of film viewable by a "microbook," not knowing that German industrial giant Siemens had already patented their own "microfilm." But the only place in England where he was allowed to do research was at St. Bartholomew's Hospital, working with medical radium, where he and St. Bart's Thomas Chalmers discovered a new method of producing isotopes.

After donating his beryllium reactor patent to the British navy in the autumn of 1935, Leo decided to carry out an experiment completely on his own—the only time in his life this would happen—using gamma rays from radon gas to release slow neutrons from beryllium, which were corralled through a sixteen-inch tube of paraffin and then absorbed by sheets of cadmium or indium (a soft metal with a sheen like mercury, but made from zinc). He became so devoted to this investigation that everywhere he went he carried two black leather satchels, one for clothes and papers, the other for a Geiger counter, wax, metal foils, boxes, and tubes—the apparatuses of his experiment. He found on November 14, 1935, that "residual" neutrons, those not absorbed by the sheets, were affected very differently by cadmium and indium, and his results, published in *Nature*, won acclaim from Rutherford, Niels Bohr, Wigner, and Fermi. Finally, Szilard was being taken seriously by the nuclear community, so much so that Joliot-Curie offered him a position at their Radium Institute. His isotope-separation patents developed at St. Bart's would give him $14,000 that year, his first income since leaving Germany. But after the Nazis occupied the demilitarized zone in the Rhineland on March 7, 1936, and England did nothing to refute them, Szilard decided Hitler was unstoppable and that he had to flee the Continent.

On Christmas Eve 1937, he sailed on RMS *Franconia*, arriving in New York on January 2, 1938, and moving into the nine-story King's Crown. His inamorata, Dr. Trude Weiss—a woman with the stolid character of a Paleolithic Venus—had arrived a few weeks before and was working in the emergency room at Bellevue. Szilard quickly met Lewis Strauss, a Wall Street financier whose parents had recently died of cancer, and with whom he experimented on artificially irradiated cobalt. Strauss pulled strings to get Leo's brother through the immigration quota after Hitler annexed Austria with Bela in Vienna; then, when Bela arrived in the United States, he chose a surname spelling of Silard, so at least Americans would pronounce it correctly. Leo then worked with Sidney Barnes and the University of Rochester's cyclotron to see if indium would shed extra neutrons depending on what hit it: neutrons, protons, or electrons. "I don't remember him ever sitting down," Sidney said. "If I had anything to say, I just waited until he stopped for breath, and I'd get it in. I generally didn't say much, though."

Nothing worked. After five years, Szilard's chain-reaction experiment, which he was so certain of, had crashed into a dead end. But almost immediately after, when all hope had evaporated, in the lobby of the King's Crown Hotel, Leo Szilard met Enrico Fermi. The gravid hand of fate and coincidence set the course of their entwined lives for better, and for worse. Enrico and Leo would jointly work a miracle, and like all its predecessors, theirs would be decidedly two-faced.

On January 16, 1939, Laura and Enrico returned to the West Fifty-Seventh Street piers to welcome Niels Bohr, arriving on the SS *Drottningholm* to lecture at Princeton, where Einstein was in residence at the Institute for Advanced Study. While Einstein was pursuing unified field theory—the calculation that would unite everything, from electrons to planets, but as of this writing has not yet come to fruition—Bohr was developing quantum mechanics, which became so difficult to comprehend, and such a mixture of wave, particle, mass, energy, spin, momentum, and angle, that it seemed to approach the supernatural, even among those who studied it. Instead of the beloved planetary model, Bohr's atom was something that could not be visualized, a blur of matter and energy with electrons that could appear in one spot and then, instantly, in another. When, to take one example, Wolfgang Pauli—considered a genius on par with Einstein, with an acid wit that earned him the nickname Wrath of God—described a theory in a letter, Bohr replied with the view of his team from Copenhagen: "We are all agreed that your theory is crazy. The question, which divides us, is whether it is crazy enough to have a chance of being correct. My own feeling is that it is not crazy enough."

Bohr carried with him to America that January day an incredible secret—a signature theory of twentieth-century science—a secret he had promised not to divulge until the Austrian physicists who'd created it, working in exile, could polish and publish their work. But when he met with Einstein at Princeton, instead of quantum mechanics versus unified field theory, all they could talk about was this latest discovery. The Austrians had proved that part of what had earned Enrico his Nobel—the discovery of a new element—was entirely in error. Additionally, their theory's far-ranging implications so terrified Leo Szilard that he would work with Fermi to create nuclear power, and with Einstein to inaugurate the Manhattan Project—the birth of the atomic bomb.

This great revolution in physics that so captivated Bohr and Einstein began at the same time that the Fermis were crossing the Atlantic in December 1938, when chemists Otto Hahn and Fritz Strassmann at the Kaiser Wilhelm Institute in Berlin published an article in *Naturwissenschaften* revealing that, after they bombarded uranium with neutrons, instead of the next-larger element on the periodic table Fermi had claimed to find, all they ended up with was barium. When an element is irradiated and transforms into another element—as illustrated in the half-life chart of uranium devolving to lead—it moves one or two notches up the periodic table if it absorbs the neutrons, and one or two down if some of its own are knocked out. Iron, for example, moves one spot down to become manganese, but barium, at fifty-six, was dramatically far down the table from uranium, at ninety-two. It was strange, baffling, and annoying. Hahn and Strassmann were convinced something must be wrong, and so they recalibrated their instruments, asked for second opinions from others at KWI, repeated the experiment over and over, but still the results were the same: irradiated uranium produced lots of barium. Frustrated and not knowing what to do next, chemist Hahn turned to his ex-partner, the physicist Lise Meitner, who had recently escaped Nazi Germany and was living in Stockholm. He described the details and the results of their experiments in a letter and asked for her help.

Has there been in the history of science a less likely personage to revolutionize her field and, with that, the fundamentals of human knowledge than the forgotten Lise Meitner? Striking as Curie, with jet-black hair, dark-ringed, deep-set eyes, and skin pale as a Klimt, Lise rarely weighed more than 105 pounds and was such a lady of her Victorian era that, in nearly every photograph, her neck is fully covered by a blouse's collar. Meitner spent her first twenty-nine years in Vienna, then Europe's most cosmopolitan city . . . yet, also the one with the highest rate of suicide. Born on Kaiser Franz Josefstrasse 27 in the Vienna suburb of Leopoldstadt, Lise was so Viennese that, after fleeing the Nazis, she refused to accept Swedish citizenship until she was allowed to retain her Austrian passport as well. Like the Hungarian Quartet, she was descended from Russian Jews, her mother, Hedwig, having emigrated to Slovakia to escape the pogroms, and like most of the Hungarians' parents, the senior Meitners had nominally converted to Christianity, becoming Lutherans, though Lise herself was baptized as an Evangelical, and two of her sisters became Catholics. Just as all brainy boys and girls around the world have done since the dawn of time, the young Lise covered the crack at the bottom of her bedroom door so her parents wouldn't catch her staying up all night, reading.

At the age of four or five, Albert Einstein's father gave him a compass as a present, and when the boy turned it this way and that and saw the needle returning to its magnetic truth, he got so excited that chills ran through his body, because now he understood that "something deeply hidden had to be behind things." Lise had a similar epiphany as a child, becoming exuberant to understand "how a puddle with a bit of oil on it showed lovely colors." Becoming amazed "that there were such things to find out about our world," she pursued "more and more questions of that kind." She was a fan of Mme. Curie in an era when science, especially X-rays, was a great fad in Vienna, nearly as popular as music.

An Austrian girl's schooling at the time was almost always finished by the age of fourteen, unless she was wealthy enough for a Swiss university. Then, beginning in the late 1890s, the empire allowed women to pursue higher education, and all five of the Meitner daughters went to college. To make up for missing the gymnasium that boys attended that trained them to pass the *Matura* (a test required for university entrance), the women hired private tutors to acquire eight years of knowledge—history, literature, religion, philosophy, Greek, Latin, math, mineralogy, botany, zoology—in two. Photographs of Lise from this period show her looking physically exhausted. In July 1901, she passed her *Matura* and began studying at the University of Vienna in a physics department close to Sigmund Freud's office and so famous for its rotting stairs and decrepit ceiling beams that the Viennese joked about its students being would-be suicides. Within, however, were two great professors, Franz Exner, friend to Wilhelm Röntgen and the Curies, and the great atomist Ludwig Boltzmann. Until the end of her life, Meitner would remember Boltzmann's teaching as "the most beautiful and stimulating that I have ever heard. . . . He himself was so enthusiastic about everything he taught us that one left every lecture with the feeling that a completely new and wonderful world had been revealed."

As a student, Lise uncovered an error in an Italian mathematician's calculations. Her professor worked with her to trace the mistake and find the correct formula and told her that she should publish. Since he had helped her so much, though, she refused to take the full credit, which he thought foolish. This high-minded deprecation to the point of self-sabotage would be Meitner's Achilles' heel in her professional life. All the same, her achievements in a field where she was the only woman besides the Curies cannot be undervalued. In February 1906 at the age of twenty-seven, she received her PhD in physics, the second woman's doctorate in the school's five hundred years. Even with this remarkable achievement, during Meitner's first months at Berlin's Kaiser Wilhelm University, she was shy "bordering on fear of people."

She wanted to work with Heinrich Rubens, the Department of Experimental Physics chair, but instead, Rubens introduced her to Otto Hahn, a chemist working in radiation who needed to team up with an industrious physicist to reach the next level of his research. Fair-haired, chipmunk-cheeked, and always debonair with his hair carefully pomaded and his shirt collars crisp and celluloid, Hahn worked as *Privatdozent* at the university—a teaching post with the salary paid directly by student fees—but had a lab with Emil Fischer's institute where both Hahn and Meitner could work, and which included electroscopes for measuring alpha, beta, and gamma rays. Emil Fischer, however, did not allow women into his Chemistry Institute after fears that a Russian student's "exotic" hair would catch fire (though he never developed the same fear about his luxuriant beard). Hahn and Meitner had to convince Fischer to let them turn a carpenter's work area into a lab for themselves—a situation redolent of the Curies' cadaver hut—and they would collaborate as physicist and chemist over the next thirty years. Otto was patient, thorough, detail-oriented; Lise was mathematically adept and brilliant at thinking in broad strokes beyond the pale. His salt-of-the-earth personality made it easier for her to overcome that paralyzing shyness; in time, she called him Hänchen, "little rooster," and began all of her letters with "Dear Otto!" Meitner, however, was never allowed to put one foot into the institute itself. To use the bathroom, she was forced to walk to a nearby restaurant.

"For many years I never had a meal with Lise Meitner except on official occasions. Nor did we ever go for a walk together," Otto Hahn remembered. "Apart from the physics colloquia held at the university that we attended together, we met only at the carpenter's shop. There we generally worked until nearly eight in the evening, so that one or the other would have to go out to buy salami or cheese before the shops closed at that hour. We never ate our cold supper together there. Lise Meitner went home alone, and so did I. And yet we were really very close friends."

"My strongest and dearest remembrances are of Hahn's almost indestructible cheerfulness and serene disposition, his constant helpfulness and his joy in music," Meitner recalled. "We would frequently sing Brahms duets, particularly when the work went well." She also remembered, though, that when they walked the streets together, a majority of the other scientists at the institute would pointedly say, "Good day, Herr Hahn!"—and say nothing to her. Hahn's courage in working with a woman at this stage in history deserves commendation, especially as balance to his future ill treatment of this historic colleague.

Otto Hahn had worked under Ernest Rutherford at Canada's McGill University, where the New Zealander who'd split the atom had said the

Frankfurter had "a nose for discovering new elements," and in 1908, Hahn and Meitner began their historic breakthroughs as a team by finding a short-lived radioelement, actinium C, using a leaf electroscope. Before the time of Geiger and his counters, a leaf of gold or aluminum was attached to a rod and sealed in a glass orb. When charged by electricity, the leaf was repelled from the rod and stiffened. When radioactive materials were placed nearby, their radiance ionized the air inside the orb and the leaf relaxed back to its original position. The rate of relaxation revealed the quantity of radiation.

One day the postman arrived with a package, and Meitner decided to play a little joke. From the other side of the lab, she announced how happy she was to get something from Rutherford. The clerk looked at the address label and was flabbergasted to see that she was right. He had no idea that the leaves of her electroscopes were flailing in response to the radioactive materials throbbing within the box in his hands. Eventually, Meitner developed a reputation with the locals as a psychic, and when she and Rutherford finally met in person, he was shocked: "But I thought you were a *man*!"

At the university, though, in time she was fully accepted, because the school's most powerful man took her on as his protégé. European schools at this time did not directly oversee a curriculum or course of study; students could take classes (and pay for them) in any order or on any topic they wished, meaning a *Privatdozent*'s salary, such as Hahn's, was somewhat precarious. The success of an education depended on a student disciple's finding a professor mentor, and these were almost unheard of for women. But Lise Meitner would have a mentor, and he would be the wondrous Max Planck.

When heated, objects radiate heat and light, beginning with various reds, then orange-yellow, and finally white-blue. The math explaining the relationship of heat and color should be simple. But it eluded physicists for decades. On October 19, 1900, Max Planck conceived of a formula developed from stringent lab results . . . a formula that violated the fundamental laws of both electromagnetism and thermodynamics. Planck realized in going over a graph comparing temperature to color spectra that the numbers did not rise evenly, like a smooth graph, but in steps, like a staircase. He called the base unit of these energy steps *h*, the quantum. But at the time, no one including Planck thought this discovery would change the future of science—everyone imagined that *h* would eventually be explained and integrated into the classical physics of electromagnetism and thermodynamics. Then Albert Einstein read Planck's article, and "it was as if the ground had been pulled out from under one, with no firm foundation to see anywhere, upon which one could have built." Einstein used quanta to explain light,

Bohr used Planck's quanta to explain atoms, and Erwin Schrödinger theorized that these particles were not "corpuscles," as Einstein had described them, but condensed packets of waves, creating the illusion of a discrete object—a subatomic whitecap—reuniting quantum mechanics with classical physics.

With his Prussian heritage, his isosceles bush of a mustache, his enormous forehead, and his critical role in modern physics, Max Planck would seem born to intimidate humdrum mortals. Instead, he was as sweet as two Fermis, one of the most generous of scientists or academic leaders, even helping an obscure Swiss patent clerk by approving Einstein's "miracle year" papers for publication in *Annalen der Physik* . . . and the magnificent Planck did this even though he didn't believe in the signature article, which was Einstein's application of Planck's quantum theory to photons . . . the article that would years later win Einstein his Nobel. "As soon as we were in [Planck's] home he played tag at least as eagerly as we young students: he tried to catch us while we ran, really he did," Lise happily recalled. "In the summer we ran races in the garden, and Planck joined us with an almost childlike eagerness and pleasure. Planck once told us that [one colleague] was such a wonderful man that when he went into a room, the air in the room became better. Exactly the same could be said of Planck."

In 1912, when Planck asked Meitner to succeed Max von Laue as his assistant, it would mean her first paycheck since arriving in Berlin six years before. Until this salaried position, the thirty-four-year-old Meitner had been getting an allowance from her parents for a dozen years, though she did make a few pfennig translating English papers into German and contributing articles to *Naturwissenschaftliche Rundschau*, the most important science journal in Germany, which published articles by the world's greatest scientists. She said of Planck's gift, "It was the passport to scientific activity in the eyes of most scientists and a great help in overcoming many current prejudices against academic women."

But even the great Planck's support could only do so much. When Meitner moved with Hahn from Fischer's to the new Kaiser Wilhelm Institute chemistry department, built as one of the first true ivory towers in the forested suburb of Dahlem—a sylvan glade far from urban life, where scientists could think and dream without distraction, and which distance from reality would nearly cost Meitner her life—Hahn was made chief of the radioactivity lab, given the title of professor, and paid five thousand marks. Lise's title was "guest," and her salary was 0. The lab, however, was unusual in that Hahn and Meitner insisted their staff make strenuous efforts to keep the environment as clean as possible to reduce background radiation and achieve a purity of experiment. A side effect of this rigor was that Meitner and Hahn would be two of the few nuclear pioneers to die of old age . . . both at eighty-nine. The city of Hamburg meanwhile erected a monument to martyrs of nuclear science. By 1959, it would be inscribed with 360 names.

Nothing seems to be known of Meitner's romantic or sexual life. When asked why she never married, she said she never had the time. One of Lise's close friends in the 1910s was physicist James Franck. Together they played Brahms lieder, he on the violin, she on the piano, and when they were both in their eighties, he confessed that he had fallen in love with her. She said, "Late!"

Meitner was in fact all too busy breaking ground—truly, as Einstein had said, the Marie Curie of Germany. In 1913, she was promoted to a salaried slot, the Hahn-Meitner Laboratorium opened, and she received royalties for mesothorium, the isotope she and Hahn had discovered, which became so commonly used in medicine that it was known as "German radium." In the autumn of 1914, she worked as an X-ray technician for the Austrian army—efforts nearly identical to Marie Curie's on the other side—but Lise's fifteen hundred colleagues at KWI chemistry spent their war years researching poison gas, chemical weapons, and explosives under Fritz Haber. While Otto Hahn worked on phosgene, mustard gas, and chlorine—he would be applauded as a "Gas Pioneer"—Hans Geiger designed a gas mask. Watching the horrors of the Great War, Einstein believed that science and technology had become "like an axe in the hands of a pathological criminal." He was shocked by how eagerly German scientists helped with the killing and considered Haber one of those pathological criminals. Haber then personally went to field-test KWI's efforts. German soldiers at the front waited for the wind to turn, then opened cylinders of chlorine gas. Even though they then ran away as quickly as possible, during the war it was realized that poison gas killed as many on the offensive side as it did on the defense. Besides being impractical, gas was thought immoral, and its use was stopped—a thoroughly fine analogy in every way to what would happen, another world war later, with nuclear weapons.

When Lise returned to KWI in 1917, her salary was raised to match Hahn's, and she became head of her own physics department. With Hahn's assistance through letters, she discovered protactinium, and in an echo of her self-denigrating behavior from her university years, she published those results as a team, even putting Hahn's name first, though he wasn't even physically present in the lab when she made the find. Twenty years later, when both would be involved in the greatest discovery of their lifetimes, Otto Hahn would not return this favor. When in 1926 at the University of Berlin, Lise Meitner became the first female professor in the history of Germany, social misogyny remained in force; a lecture she gave on "cosmic physics" was written up in an academic review as being about "cosmetic physics."

When *La Ricerca Scientifica* published Fermi's irradiation findings in March and May of 1934, his discoveries reinvigorated the Joliot-Curies in Paris and Meitner-Hahn-Strassmann (Hahn's new assistant) in Berlin. The competing German, French, and Italian teams, all firing neutron cannons at uranium targets, produced one new element or isotope after the next, chemically detected yet short-lived creatures that were even atomically fatter than naturally occurring uranium, which they called transuranes. In Rome, after the group's chemist left to develop insecticides and Segrè departed for a job in Palermo, Fermi lost interest in transuranics, but Berlin and Paris battled, driving the science to new heights, churning out a torrent of experimental results, each side arriving at different interpretations for what was happening, and each claiming the other was inept. On January 20, 1938, Otto and Lise told Irène Curie that "she had committed a gross error" in claiming to have produced a three-and-a-half-hour transuranic thorium isotope and said that if she didn't release a "public retraction," they would humiliate her themselves.

But in fact they were all wrong. Their radioactive chemistry was misreading a fundamental process, and though elements atomically grander than uranium would in time be artificially engineered (and called plutonium, americium, et al.), the transuranes themselves were a mirage. Only one scientist seemed to grasp what was happening, German chemist Ida Noddack, who wrote to Fermi, "One can imagine that when heavy nuclei are bombarded with neutrons, these nuclei break apart into several large fragments, which are indeed isotopes of known elements but not neighbors of the irradiated elements." Noddack's revolutionary concept was, though, only a side note in her criticism that Fermi hadn't excluded such known elements as polonium in his chemical testing for transuranes. As she offered no proof of nuclei breaking apart into large fragments, no one in the physics community took this comment as revelatory. In fact, as Hungarian Quartet member Edward Teller explained, "Fermi was a very careful experimenter. He covered his uranium with a thin sheet of inert material to stop the normal alpha particles (without the extra energy) in which he was not interested. That sheet also stopped the fission products, which had a short range but extremely high energy-density. Had Fermi forgotten to cover this sample even once, fission would have been discovered years earlier."

While Meitner was making history as a woman physicist in the leafy Berlin suburb of Dahlem, outside that ivory tower in the economically collapsed Weimar Republic, "mystics, magicians and religious fanatics drew followers desperate for rescue. Each was called a *Heiland*, or savior. But in German, there is no plural word for 'savior.' There can be only one," as historian Nicole Rittenmeyer said. When on February 28, 1933, the Reichstag, Germany's Capitol, was attacked by arsonists, President von Hindenburg ordered the "Decree of the Reich President for the Protection of the People and the State," suspending private property, personal liberty, freedom of assembly, privacy, and press freedom, "until further notice." Chancellor Hitler dissolved parliament, dismissed the Weimar constitution, and sent eight hundred thousand Germans to prisons or concentration camps. Six weeks later, on April 7, 1933, the anti-Semitic "Law for the Restoration of the Career Civil Service" was passed. Over the next three years, sixteen hundred scholars were fired or resigned, a third being scientists, twenty being Nobel laureates, and quarter of Germany's physicists were exiled. Lise's nephew Otto Robert Frisch lost his Rockefeller fellowship to study with Fermi in Rome since, on his return to Germany, he would no longer be employed. Szilard's Academic Assistance Council got Frisch to Birkbeck in the UK, and after a short term he was able to join Niels Bohr in Copenhagen. Lise's next-door neighbor, James Franck, could continue working because of his service in the Great War, but he resigned on principle as did Erwin Schrödinger.

On May 16, the greatest scientist still living in Germany, Max Planck, tried to reason with Hitler. "I have nothing against the Jews. But the Jews are all Communists, and these are my enemies. My life is against them," Hitler insisted. Planck argued that many of the greats of German culture and many of its oldest families were Jews, and besides, "there are different sorts of Jews, some valuable for mankind and others worthless . . . distinctions must be made." "That's not right," Hitler immediately replied. "A Jew is a Jew; all Jews stick together like leeches. Whenever there is one Jew, all Jews of all sorts immediately gather." When Planck then said that expelling Jews wholesale would damage German science, Hitler became enraged: "Our national policies will not be revoked or modified, even for scientists. . . . If the dismissal of Jewish scientists means the annihilation of contemporary German science, then we shall do without science for a few years!"

Kaiser Wilhelm Institute employees were not civil servants and so not subject to this law. Lise Meitner was additionally exempt as a university professor since she wasn't German, but Austrian. Yet now, everything began to change. Her chief assistant became an ardent party member, while the new head of organic chemistry, Kurt Hess, was a "fanatic" Nazi. Even so, Lise spent the next years seemingly unconscious of the reality of the new Germany, perhaps because of her life of physics and music and nothing else, or her remove within the academic exile of Dahlem, or because of her joy in her professional achievements, which would be taken from her if she fled.

On May 10, 1933, University of Berlin professors and students, wearing their brown shirts, held a bonfire to burn twenty thousand books, notably those of Einstein, Kafka, Proust, Thomas Mann, and Helen Keller. Then, starting in the summer of 1933, one of the titans of world physics, Max von Laue, decided enough was enough. He campaigned to keep "Aryan physicist" Johannes Stark out of the Prussian academy; opened a physics conference with a paean to relativity following the history of Galileo and political opposition; wrote a spring 1934 *Naturwissenschaften* obituary for Haber praising him; was reprimanded by the Ministry of Culture, and welcomed it. But this kind of heroism was rare, and rarely effective.

On September 6, 1933, even though she was Austrian, Meitner was dismissed from the university and could no longer present findings at scientific conferences. When Hahn and Planck tried to reverse the decision, they found they were powerless, even after Hahn resigned in protest.

In November, Bohr got Lise a Rockefeller grant to spend a year with him in Copenhagen. She turned it down. Planck and Hahn convinced her she could stay; emigration was especially difficult for such a socially awkward woman; the world was still in an economic depression, and decently paying jobs were rare; and there was KWI physics, which she had created and run; her own department; her baby. In a letter to Gerta von Ubisch of July 1, 1947, she explained, "I built it from its very first little stone; it was, so to speak, my life's work, and it seemed so terribly hard to separate myself from it."

By the end of 1933, KWI chemistry was entirely staffed with party members and once again devoted to weapons R&D. Hahn's assistant Strassmann was offered a good job with an industrial firm, but had to first join the Nazi Party. When he refused, he was blacklisted and spent the 1930s trying to live on a quarter of his income. Regardless, he hid a Jewish friend in his apartment. *Naturwissenschaften* continued to accept Jewish contributions, becoming Meitner's sole outlet for the 1930s, but was so hounded by Nazi boycotts that its founder and editor of twenty-two years was dismissed.

On September 15, 1935, at the Nuremberg party rally, Hitler spoke on "the Jewish problem. . . . This law is an attempt to find a legislative solution. If it fails, it will be necessary to transfer the problem to the National Socialist Party for a final solution." The new laws: Jews were no longer German citizens, and sex and marriage between Jews and Germans were illegal. Historian Erich Ebermayer: "After these three laws were read, the halls rang with minute-long applause. It was the call of a wild animal, a beast that smells blood."

Planck, von Laue, and Heisenberg nominated Lise for the Nobel for 1936 and 1937 both because she deserved it, and because they thought it might save her. Then on March 12, 1938, Austria "united" with Germany (removing the exemption of her nationality), and the day after, KWI organic chemistry chief Kurt Hess publicly denounced Meitner, saying, "The Jewess endangers the institute." Four days after, KWI's treasurer told Hahn that Lise should resign since "nothing more could be done; perhaps she could continue working unofficially," and Otto began distancing himself from Lise so as not to be tainted by decades of collaboration with a Jew. The tension became so great that Hahn's mentally unstable wife, Edith, muttered over and over, "The great misfortune has happened! The great misfortune has happened!" Kaiser Wilhelm Society president Carl Bosch, however, insisted that the institute was under his management, not the government's, and that Lise was staying. The same month, Meitner received lecture offers from Paul Scherrer in Zurich and Niels Bohr in Copenhagen. From the University of Chicago, James Franck filed an affidavit ensuring that if she emigrated to America, she wouldn't become a public ward (the first document needed for US entry).

She just couldn't decide what to do. So, she did nothing. She turned them all down, even though, if she lost her job and then tried to leave Germany, she could be arrested and imprisoned. Finally, on May 9, she accepted Bohr's offer to join her nephew Otto, in Copenhagen. But the Danish consulate declared her Austrian passport no longer valid and refused entry.

Meitner and Bosch decided the answer was for him to formally request an *Auslandpass* (a passport allowing travel outside the Reich), which he submitted, filled with praise for Meitner's international reputation and worth to the KWI, on May 22. He quickly received an answer: "It is considered undesirable that renowned Jews should leave Germany for abroad to act there against the interests of Germany. . . . The Kaiser Wilhelm Society will certainly find a way for Prof. M. to stay in Germany after her retirement. . . . The Reichsführer SS and the Chief of German Police in the Reich Ministry of the Interior [Heinrich Himmler] in particular have advocated that position." Bosch wanted to appeal directly to Himmler, but was told that would not be effective.

Now, it was too late. Not even Max Planck could save Lise Meitner.

But Niels Bohr could. Bohr had made his career by noticing errors, omissions, or points unexplored in others' work, yet these seemingly trivial origins would make him a titan in the history of science. He had terrible difficulty thinking and writing at the same time, so he ended up having assistants—such historically significant physicists as Heisenberg, Pauli, Dirac, Gamow, Urey, and Bohr's own wife, Margrethe—take dictation and used their responses as something of a sounding board. He spoke in murmurs, coming up to his acolytes (who were legion), leaning in close, and whispering into their ears. Einstein said that Bohr "is like a sensitive child and walks about this world in a kind of hypnosis," but that his atomic theory "appeared to me like a miracle and appears as a miracle even today. This is the highest form of musicality in the sphere of thought." A true gentleman like Planck—whenever Bohr thought an idea was wrong, he told its originator that it was "very interesting"—Bohr had been so good at soccer as a young man that he'd won a spot on one of Denmark's best teams. Now, he was a Ping-Pong champion, and for his fiftieth birthday, the Danish public surprised him with 0.6 grams of radium. He was then awarded the Aeresbolig, the mansion of the Carlsberg brewery founder, which had been bequeathed to the state as the House of Honor. Edward Teller: "Bohr loved paradoxes and did his brilliant best to explain them, carefully emphasizing the contradictions. He also liked to talk about subjects that he did not understand, although he always made sense in an inspirational and ambiguous way. His sentences were long and convoluted. I remember his friend Paul Dirac once asked him, 'Were you never taught in school that before you begin a sentence you should have some plan as to how you're going to finish it?' Bohr turned to the rest of us and commented: 'Dirac may think that one should not start life until one has a plan about how to end it.' "

The minute Niels Bohr returned from a trip to Berlin that June, he launched a campaign across Scandinavia and Holland to rescue Lise Meitner from Hitler. In the Netherlands, Dirk Coster and Adriaan Fokker knew that lab space was easy but university jobs were hard, so they campaigned among their professional and industrial colleagues for funds to support Lise, then discovered they would need permission from both the Ministries of Justice and of Education for her immigration. They just didn't have enough time.

On June 14, Meitner learned that "technical and academic" citizens would not be allowed to leave Germany.

On June 16, Bosch received a letter from the Ministry of the Interior, categorically refusing Meitner a German passport.

Now she knew she had to get out. Otto Hahn asked Paul Rosbaud, the well-connected new editor of *Naturwissenschaften*, if he could get a passport forged for her. But then Adriaan Fokker, Dirk Coster, and W. J. de Haas achieved a miracle. By June 27, 1938, they had raised enough money to finance her for a year abroad, and Coster gave her the news. Then on July 11, the Netherlands approved her entry.

On July 12, Lise spent the night at Hahn's. Carrying only two valises of summer clothes, ten marks in cash, and a diamond ring Hahn had inherited from his mother and given her since "I wanted her to be provided for in case of an emergency," Paul Rosbaud drove her to the station. All the way, she begged him to take her back home.

The train took seven hours, but Coster and Fokker arranged for a Dutch border officer to overlook that her Austrian passport was invalid and she had no official entry visa, merely papers assuring a visa was on its way.

Otto Hahn: "The danger consisted in the SS's repeated passport control of trains crossing the frontier. People trying to leave Germany were always being arrested on the train and brought back. . . . We were shaking with fear whether she would get through or not."

Inexplicably, the SS let her pass. By six that evening she was safely in Groningen. Ardent Nazi chemist Kurt Hess learned she was escaping and informed the government, but some sympathetic members of the police force delayed processing his information until she was safely across the border. Wolfgang Pauli was so thrilled he cabled Coster, "You have made yourself as famous for the abduction of Lise Meitner as for [the discovery of] hafnium!"

In the meantime, Bohr had turned to Manne Siegbahn, then overseeing the construction of Sweden's Research Institute for Physics and the country's new cyclotron. After being repeatedly nagged by Bohr, Siegbahn finally agreed to offer Meitner both refuge and a one-year contract, even though he was in the building and not the staffing stage. In loyalty to Bohr, Lise accepted, turning down the Dutch, who were dispirited. But she had nothing, no money, no winter clothes for Scandinavian life, and, as it turned out, no entry visa for Sweden, as Siegbahn hadn't yet gotten permission.

On July 28, she flew to Copenhagen, terrified that the plane would be forced to land in Germany and she would be repatriated against her will. She stayed with the Bohrs at their seaside villa, Tisvalde, where one guest had asked Bohr about the horseshoe on the door, if he believed that it brought good luck. Bohr replied, "No, but I was told that they also bring luck to people who do not believe in them."

In August, Lise sailed for Sweden, stopping along the way at the coastal village of Kungälv to visit Eva von Bahr-Bergius, one of the few other internationally known women scientists and a longtime friend from Berlin. Eva told Lisa that she must officially retire from KWI and receive her pension now, when she needed it most.

Meitner finally arrived at the Physics Department of the Nobel Institute of the Royal Swedish Academy of Sciences, finding it spacious, empty, and completely disorganized, still under construction, with equipment always on order but never arriving, no assistants, her salary equivalent to what assistants would be paid if there were any, and all of her would-be professional colleagues talking in Swedish, which she could not understand.

The full force of exile now crushed her spirit and Lise collapsed in despair, sending out letter after letter to Hahn over that autumn of 1938: "Perhaps you cannot fully appreciate how unhappy it makes me to realize that you always think that I am unfair and embittered, and that you also say so to other people. If you think it over, it cannot be difficult to understand what it means to me that I have none of my scientific equipment. . . . I see no real purpose in my life at the moment and I am very lonely. . . . [Siegbahn] is not at all interested in nuclear physics and I rather doubt whether he likes to have an independent person beside him. . . . I can't do anything but live my life just as it is. . . . I often feel like a wound-up puppet that does certain things, gives a friendly smile, and has no real life in itself. By that you can judge how valuable my activity is."

November 7 was Lise's sixtieth birthday, and November 9–10, *Kristallnacht*: twelve hundred synagogues and prayer rooms destroyed; thirty thousand Jews sent to concentration camps; and the Nazis fining the Jews a billion marks for these damages. One man riding the train the next day noticed that few of his fellow passengers bothered looking out the windows, deliberately ignoring all the synagogues burning, burning, burning. Three hundred thousand German Jews applied for US visas; a third were granted. Storks migrating from Europe to Africa had notes taped to their legs: "Help us. The Nazis are killing us all."

On November 10, when Otto Hahn came to lecture in Copenhagen at Bohr's invitation, Lise Meitner took the nine-hour train across Scandinavia for a reunion. Both of their eyes filled with tears. Hahn and Strassmann had continued working in Meitner's laboratory with her equipment to uncover the mystery of the transuranes. Joined by her handsome, eager-eyed nephew, Otto Robert Frisch (who called her *Tante* Lise), Bohr and Meitner went over Hahn's mysterious findings. It was why he had come all the way to Denmark, even though his wife was suffering from an ever-growing mental illness. Both Meitner and Bohr told Hahn his findings of radium isomers after bombarding uranium with neutrons needed to be reexamined.

Lise returned to Stockholm and became even more depressed after learning that her brother-in-law, Otto Robert's father, Jutz, had been arrested along with thirty thousand other Jewish men and was now interned at Dachau. She tried to arrange for her sister Auguste to emigrate. Then she learned that Otto Hahn had referred to concentration-camp victims as "human undesirables" and begun joking that people assumed he was Jewish because of his name, because he was born in Frankfurt, worked at KWI, and had resigned from the University of Berlin in protest in 1933. The "joke" turned on Otto when he discovered his name included in "The Eternal Jew," a traveling exhibit of anti-Semitism, and he had to give institute officials affidavits confirming his "Aryan" lineage. Otto Hahn did, though, keep working to get Lise her due during this period and was able to unfreeze her bank account for her sister to use. But he had no luck with her pension. Then after months of delay, her possessions arrived in Stockholm severely vandalized, the Swedish shipper saying he'd never seen anything like it.

On November 20, Meitner wrote von Laue, "One always thinks life in this world cannot get much harder, but one is mistaken."

While the Fermis were in Sweden for the Nobel ceremonies, Laura recalled, "One day, while strolling in the streets of Stockholm we ran into a mousy little woman with a tense expression. She was Lise Meitner, then a refugee from German persecution. . . . Fermi and Meitner did not talk about physics that day in the street. To me the significance of that encounter lay in Meitner's tense, almost scared look, a look that I was to see time and time again on the face of other refugees." Who could imagine—least of all Lise Meitner herself—that this woman, in this predicament, would soon be as significant and as revolutionary a figure in the history of physics as Röntgen, Fermi, and both Curies?

In December, the Swedish sun rose at noon and set at three, and Lise's sense of hopelessness ballooned. She decided she couldn't spend the holiday brooding and alone and instead arranged to return to stay with Eva. Otto Robert was also invited, and wanting to be with family in the wake of his father's arrest and his mother's sorrow, he accepted. Before she left, however, Meitner received a letter from Hahn dated December 19, 1938 . . . perhaps the most important letter in the history of nuclear science: "Monday evening, in the lab. . . . It is now practically eleven o'clock at night. Strassmann will be coming back at 11:45 so that I can get home at long last. The thing is: there is something so ODD about the 'radium isotopes' that for the moment we don't want to tell anyone but you. . . . We are more and more coming to the awful conclusion that our Ra [radium] isotopes behave not like Ra, but like Ba [barium]. . . . Perhaps you can suggest some fantastic explanation." On the twenty-first, he then wrote, "How beautiful and exciting it would be just now if we could have worked together as before. We cannot suppress our results, even if they are perhaps physically absurd. You see, you will do a great deed if you can find a way out of this." Meitner responded on the twenty-first with "Your radium results are very startling. A reaction with slow neutrons that supposedly leads to barium! . . . At the moment the assumption of such a thoroughgoing breakup seems very difficult to me but in nuclear physics we have experienced so many surprises, that one cannot unconditionally say: It is impossible."

Otto Hahn: "Miss Meitner—Professor Meitner—had left our laboratory on July 1938 on account of these Hitler regime things and she had to go to Sweden. And Strassmann and myself, we had to work alone again and in the autumn of '38 we found strange results. . . . We could conclude that the substances could be really only radium because barium was prohibited by the physicists that we didn't dare to think it barium in those times. We always tried to explain what is wrong in our experiments, not to say we do have barium, but we always thought it can't be there and therefore we have to say, 'What is the nonsense we are doing?' So really, it is so, that we poor chemists—isn't it the same with you?—we are so afraid of these physics people."

It was now Christmas Day 1938. Just when she had reached the top of her field and her greatest research had begun, she had been exiled from both her longtime collaborator, her work, and the institute that had been her life's passion. Alongside Marie and Irène Curie, she had been the most famous and acclaimed woman in global physics. Now, exiled into Siegbahn's inhospitable refuge, she was essentially homeless, stateless, and impoverished. She felt, suddenly, old, useless, a failure, a woman without hope.

On Christmas Eve in Kungälv, after Lise and Otto Robert both "gagged" on the traditional dish of lutefisk, all Lise could talk about was Hahn's barium. It didn't make sense! Otto Frisch: "Was it a mistake? No, said Lise Meitner; Hahn was too good a chemist for that. But how could barium be formed from uranium? No larger fragments than protons or helium nuclei (alpha particles) had ever been chipped away from nuclei, and to chip off a large number, not nearly enough energy was available. Nor was it possible that the uranium nucleus could have been cleaved right across. A nucleus was not like a brittle solid that can be cleaved or broken. . . . Bohr had given good arguments that a nucleus was much more like a liquid drop."

Every time an experiment revealed new elements of the atom, a new model (or new analogy) was proposed. Was the atom's nucleus equivalent to a magnetic fog? A roiling plum pudding? Or a solid amalgamation of particles that, when bombarded, could be turned into chips? Wilfrid Wefelmeier was inspired to propose heavy nuclei stacked in a lump . . . a "nuclear sausage." Besides nuclear "mush," Bohr, in a 1936 *Nature* article, had proposed a "liquid droplet model" of the nucleus, amending and supporting his Russian student George Gamow, who'd published something similar in 1934.

The aunt and nephew went out that afternoon for a walk in the snow, Frisch on skis, and Meitner in boots. They sat on the logs of a fallen tree and discussed Bohr's raindrop model of the atom versus Rutherford's planetary version. Resting against a branch, Meitner took out paper and pencil and drew a Bohr nucleus. She drew it pulled apart and then split. She wrote the calculations of energy and mass that held the nucleus together, then started recalculating the numbers, assuming that the "surface tension" of the uranium nucleus was weaker than previously believed. Frisch: "Perhaps a drop could divide itself into two smaller drops in a more gradual manner, by first becoming elongated, then constricted, and finally being torn rather than broken in two? We knew that there were strong forces that would resist such a process, just as the surface tension of an ordinary liquid drop tends to resist its division into two smaller ones. But nuclei differed from ordinary drops in one important way: They were electrically charged, and that was known to counteract the surface tension."

If the nucleus was split, the resulting electric charge would repel the two pieces away from each other in a burst of energy, about 200 million electron volts (an electron volt is the energy of a single electron after being charged with a single volt). Meitner remembered what was called the packing-fraction formula and thought that the two split nuclei should be lighter than the original by one-fifth the mass of a proton. When taking into account Einstein's $E = mc^2$, her calculations produced . . . 200 million electron volts.

Frisch: "The charge of a uranium nucleus, we found, was indeed large enough to overcome the effect of the surface tension almost completely; so the uranium nucleus might indeed resemble a very wobbly unstable drop, ready to divide itself at the slightest provocation, such as the impact of a single neutron. But there was another problem. After separation, the two drops would be driven apart by their mutual electric repulsion and would acquire high speed and hence a very large energy, about 200 MeV in all; where could that energy come from? . . . [Lise] worked out that the two nuclei formed by the division of a uranium nucleus together would be lighter than the original uranium nucleus by about one-fifth the mass of a proton. Now, whenever mass disappears, energy is created, according to Einstein's formula $E = mc^2$, and one-fifth of a proton mass was just equivalent to 200 MeV. So here was the source for that energy; it all fitted!"

Meitner was flabbergasted, and Frisch, amazed. When she finally had a chance to later read their joint paper, she said, "These results, I realized, had opened up an entirely new scientific path, and I also realized how far we had gone astray in our earlier work!"

On December 28, 1938, Hahn wrote Lise, "What is the possibility that uranium 239 could split into one Ba and one Ma? One Ba 138 and one Ma 101 gives 239." On January 3, 1939, she replied, "Dear Otto! I am now almost certain that the two of you really do have a splitting to Ba [barium] and I find that to be a truly beautiful result, for which I must heartily congratulate you and Strassmann. . . . Both of you now have a beautiful, wide field of work ahead of you. And believe me, even though I stand here with very empty hands, I'm nevertheless happy for these wondrous findings."

Otto Frisch: "When I came back to Copenhagen I found Bohr just on the point of parting, of leaving for America, and I just managed to catch him for five minutes and tell him what we had done. And I hadn't spoken for half a minute when he struck his head with his fist and said, 'Oh, what idiots we have been that we haven't seen that before. Of course this is exactly as it must be.' And he added, 'This is very beautiful,' and, had we written a paper? So I said no, we were in the process of writing one."

On January 3, Otto Robert wrote Lise, "Dear Tante, I was able to speak with Bohr only today about the splitting of uranium. The conversation lasted only five minutes as Bohr agreed with us immediately about everything. He just couldn't imagine why he hadn't thought of this before, it is such a direct consequence of the current concept of nuclear structure. He agreed with us completely that this splitting of a heavy nucleus into two big pieces is practically a classical phenomenon, which does not occur at all below a certain energy, but goes readily above it." Frisch then asked American biologist William A. Arnold, working at Bohr's institute, what biologists call it when cells divide. "Binary fission," Arnold said. Three days later, Frisch again met with Bohr and gave him a draft of the *Nature* article.

On January 7, Niels Bohr, his nineteen-year-old son, Erik, and University of Liège physicist Léon Rosenfeld set sail from Göteborg, Sweden, for New York. Bohr was so excited by Frisch and Meitner's theorem that he had a blackboard installed in his cabin for the trip. Rosenfeld: "When we met on the boat, [Bohr] said, 'I have in my pocket a paper that Frisch has given me which contains a tremendous new discovery, but I don't yet understand it. We must look at it.' Bohr accepted the conclusions because it was an argument directly following from the experiments. But he did not understand why the nucleus would split. And then during the trip that took six days or so, he got hold of the solution, and it turned out to be extremely simple. Meanwhile, back in Denmark, Frisch wanted to check by experiment the idea that uranium can split in two. Several methods could be used to study sub-atomic particles. The easiest was to look at electrical effects in an ionization chamber, using an amplifier and oscilloscope. Invisible particles passing through the chamber would show up as pulses on the screen of the oscilloscope. The hallmark of fission would be the size of the pulses: the two halves of a split atom would have far greater energy than any known particle."

Otto Frisch: "I rigged up a pulse amplifier for the special purpose, and I also built a small ionization chamber; but the whole thing only took me about two days, and then I worked most of the night through to do the measurements because the counting rates were very low. But by three in the morning I had the evidence of the big pulses. And I went to bed at three in the morning, and then at seven in the morning I was knocked out of bed by the postman, who brought a telegram to say that my father had been released from the concentration camp." On January 13, Frisch observed uranium pulsing in ionization when hit by neutrons, proving it emitted energy when split. It was the alchemy of transformation, of matter becoming energy, and it was visible.

On the sixteenth, he sent both the article coauthored with his aunt, "Disintegration by Neutrons: A New Type of Nuclear Reaction," and the solo report on his follow-up experiments, "Physical Evidence for the Division of Heavy Nuclei under Neutron Bombardment," dating them both the sixteenth, to *Nature*, the same day that Bohr docked in New York at the Swedish American Line's pier on West Fifty-Seventh Street. Rosenfeld: "We had bad weather through the whole crossing, and Bohr was rather miserable, on the verge of seasickness all the time. Nevertheless, we persevered for nine days, and before the American coast was in sight, Bohr had a full grasp of the new process and its main implications."

So many colleagues were on hand to welcome Bohr to America that he and Fermi did not discuss the Meitner breakthrough at that moment, but when John Wheeler sat next to Rosenfeld on the train to Princeton, Rosenfeld, not knowing that Bohr was keeping the material quiet until publication, told Wheeler all about it. (Rosenfeld would have assumed that the Meitner/Frisch paper would be released immediately; instead, not understanding the extraordinary nature of what they had, it took until February 11 for the editors of *Nature* to publish.) Rosenfeld then repeated the news at the Princeton physics students' weekly Journal Club on Monday. From there, the shocking revelation spread across the American physics community with amazing speed. Isidor Rabi traveled from Princeton back to Columbia, carrying Rosenfeld's news; Fermi remembered hearing it from Willis Lamb, who had also just been at Princeton.

When Otto Hahn published his results on January 28, Frisch-Meitner was barely mentioned as a footnote. On February 5, 1939, Meitner wrote, "Your results present a wonderful close chain of results, and it is marvelous what you accomplished in these few short weeks. Unfortunately I fear from the manner in which you brought up our note that you are personally angered by the lapse in our literature citations. I really am terribly sorry. I had hoped a little that our note would've given you some pleasure also, and it would've been so nice for me if you'd just written that we—independently of your wonderful findings—had come upon the necessity. . . ." The next day she wrote her brother Walter: "Unfortunately I did everything wrong. And I have no self-confidence, and when I once thought I did things well, now I don't trust myself. The Swedes are so superficial; I don't fit in here at all, and although I try not to show it, my inner insecurity is painful and prevents me from thinking calmly. . . . And much as [the Hahn-Strassmann discoveries] make me happy for Hahn, both personally and scientifically, many people here must think I contributed absolutely nothing to it—and I'm so discouraged; although I believe I used to do good work, now I've lost my self-confidence."

On February 7, 1939, Hahn apologized to Lise, but then in the end said, "The uranium work is for me a heaven-sent gift." The explanation for what he had uncovered was actually a Meitner-sent gift, but the erasing of her from history had begun. Because he felt politically weak and his position tenuous as he wasn't an ardent Nazi, Hahn wanted to have sole credit for fission, and he started a campaign to get it. The chemist Hahn so misunderstood the physicist Meitner that he thought his observation of barium trumped Meitner's theory and became enraged when nuclear scientists commonly attributed Meitner-Frisch as scientifically revolutionary . . . and Hahn-Strassmann as mere supporting chemistry. By war's end, Hahn was claiming that fission could only have been discovered with Meitner away from Dahlem, and today, at Munich's Deutsches Museum, the collection of Meitner's instruments and lab equipment is labeled "Worktable of Otto Hahn"—though, in 1983, the museum added a small card of text, mentioning Lise as Otto's assistant. Biographer Ruth Lewin Sime: "Had fission been born into a world at peace, its energy might first have been used to provide light and heat for people's homes. Had fission been discovered in a world free of racial persecution, it might well have been the crowning achievement of Lise Meitner's career."

When *Nature* finally published the Meitner/Frisch paper on February 11, Irène Joliot-Curie raged at her husband, "We've been such dumb assholes!" Bohr immediately went to Columbia to see Fermi. Herb Anderson: "Fermi wasn't in his office at Columbia, so Bohr went down to the basement where the cyclotron was. Fermi wasn't there, either, but I was. Undeterred, he came right over and grabbed me by the shoulder. Bohr doesn't lecture to you, he whispers in your ear. 'Young man,' he said, 'let me explain to you about something new and exciting in physics.' Then he told me about the splitting of the uranium nucleus and how naturally this fit in with the idea of the liquid drop. I was quite enchanted. Here was the great man himself, impressive in his bulk, sharing his excitement with me as if it was of the upmost importance for me to know what he had to say. . . . As soon as he left I rushed off to find Fermi. I found him in his office, but he had anticipated me. He already heard about the fission of uranium from Willis Lamb, who had just heard Bohr talk about it at Princeton. Before I had a chance to say anything he smiled in a friendly fashion and said, 'I think I know what you want to tell me. Let me explain to you about fission.' Then he went to the blackboard in his inimitable graphic way to show how the uranium nucleus will split in two. I have to say that Fermi's explanation was even more dramatic than Bohr's. It made the experimental possibilities even more exciting. . . . So he says, 'Why don't we get the electrode of your ionization chamber, put some uranium on it, let's go down to the cyclotron, and let's see if we can see all this energy release.' And so we got busy just that afternoon. But there was a meeting, a theoretical physics meeting in Washington, the next day. And

Fermi was supposed to go to that. And so he left and I began to wonder what to do and I remembered that Dunning was in and I came to Dunning, and I said, 'Why don't we see if we can see this fission?' "

What Fermi did not mention to Anderson was his deep shame at missing nuclear fission during his epic radioactive study of the periodic table. Emilio Segrè: "One day after the war Fermi and some of his colleagues were studying the architect's sketches for the future institute for nuclear science at the University of Chicago. The drawing showed a vaguely outlined human figure in bas-relief over the entrance door. When the group began speculating as to the significance of the human figure, Fermi immediately interjected that it was probably 'a scientist not discovering fission.' " Physicist Jay Orear: "If Fermi had published that he had seen fission, the half-sized pieces would have an excess of neutrons and these neutrons could give rise to more fissions most likely in a chain reaction. Then both Germany and the United States might have had atom bombs in time for World War II. The world should be grateful for this one mistake by Fermi!"

On the night of January 25, 1939, the first American nuclear fission experiment was conducted in the basement of Columbia's Pupin Hall. Previously that day, John R. Dunning and Enrico Fermi had lunch at the faculty club and discussed the outlines of future experiments with uranium, such as what would be the best state to test—metal, liquid, or gas? Then Fermi left for the train to Washington, and that night a cold front blew in, rattling Pupin's scraggly Ivy League vines. At about seven o'clock, Dunning and Herb Anderson bombarded uranium oxide with neutrons in an ionization chamber, fully expecting nothing would happen. However, instead of the whispery dots that would be expected on the oscilloscope, great wavy lines appeared.

John Dunning: "I went up to the thirteenth floor and brought down one of the old standard stand-by neutron sources, the radon plus beryllium sources that had been used so much before. We put it next to the chamber containing the uranium and in considerable excitement we saw with even this very weak source about one big pulse, a huge pulse, on the oscilloscope every minute. The rate, however, was so slow that I had doubts whether this was really real or whether it was maybe a bad electrical contact. So we had another device, and installing that right next to the chamber, the rate went up according to my notes to something like seven or so with that device, huge pulses. We finally quit about 11 p.m. My notebook contains this phrase: 'Believe we have observed new phenomenon of far reaching consequences.' "

Dunning had never seen anything like it before and took a deep breath. "God!" he said. "This looks like the real thing." They assumed that something was wrong and spent the next two hours making sure the machine was working properly, but finally the truth was evident. They had split atoms and released nuclear power, confirming the fission discovered by Meitner and Frisch.

On January 26, the Fifth Annual Conference on Theoretical Physics opened in Washington, DC. Before anyone could present their findings, Bohr and Fermi took the floor of the conference to announce nuclear fission. The audience erupted, with many immediately abandoning the conference to return to their labs and confirm the Meitner and Frisch findings then and there. On January 29, 1939, an excitable *New York Times* explained this moment to the general reader with the headline "Atomic Explosion Frees 200,000,000 Volts," and on April 29, the head of the physics division of the Reich Research Council, Abraham Esau, assembled a group of German nuclear scientists to form a new organization, the Uranium Club—the Uranverein—and directed them to investigate the potential battlefield uses of fission. At around the same time, the German Army Weapons Bureau started its own rival committee.

When Szilard learned of Meitner and Frisch's theory of fission, he could only think, "All the things which H. G. Wells predicted appeared suddenly real to me." Until Joliot-Curie discovered man-made radiation, no one paid any attention to Szilard's chain-reaction theories, but the French technique, combined with the Meitner/Frisch findings, now indicated a clear candidate for the trigger that would lead to radioactive power and weaponry, and Szilard's H. G. Wells–induced epiphany at a London stoplight would become the wellspring of nuclear science. When uranium fissioned, mass was alchemized into energy . . . and a few stray neutrons were spewed. Enrico Fermi: "It takes one neutron to split one atom of uranium [which then] emits two neutrons. . . . It is conceivable that they might hit two more atoms of uranium, split them, and make them emit two neutrons each. At the end of this second process of fission we would have four neutrons, which would split four atoms. . . . In other words, starting with only a few man-produced neutrons to bombard a certain amount of uranium, we would be able to produce a set of reactions that would continue spontaneously until all uranium atoms were split." If, as Fermi had found with the paraffin, a tamper could slow the neutrons' velocity and maximize their strike efficiency, humans could precipitate a controlled chain and extract its energy—a nuclear reactor producing immense amounts of power from small amounts of ore—as well as isotope by-products for medicine, archaeology, and explosives. If they could trigger an uncontrolled chain reaction releasing massive energy simultaneously, though, it would mean a weapon of unimaginable destructive force.

On February 24, Hungarian Leo Szilard worked with Canadian Walter Zinn at Columbia to determine whether splitting uranium atoms would produce neutrons. At first, their oscilloscope showed nothing whatsoever . . . then they realized they had forgotten to plug it in. Szilard: "We turned the switch and we saw the flashes. We watched them for a little while and then we switched everything off and went home."

It was all true; uranium fission produced neutrons.

Leo: "That night, there was very little doubt in my mind that the world was headed for grief."

PART TWO

THE NEW WORLD

5

The Birth of Radiance

ENRICO FERMI and Leo Szilard now began a series of experiments together at Columbia to create the first sustained chain reaction—the world's first nuclear reactor—with various arrangements of neutron sources and uranium to understand the quantity needed, what medium would work best to slow down the neutrons, what kind of uranium worked effectively, what layout of ore and tamper meant the most likely success, and what could be used to stop the reaction, as well as control its speed.

The two fought constantly. Enrico believed in careful, incremental progress and hard work in the lab; Leo loved debate, being a catalyst, and thinking through original ideas at their earliest, most primitive stages. Eugene Wigner believed that Fermi's most striking trait was "his willingness to accept facts and men as they were" with an approach to research as shoe-plain as the man himself: "One must take experimental data, collect experimental data, organize experimental data, begin to make a working hypothesis, try to correlate so on, until eventually a pattern springs to life and one has only to pick out the results." Szilard challenged conventions, upended every authority in every hierarchy, did not mind charging into battle like a scientific Quixote, and tended to operate independently (to put it politely); while Enrico had spent decades leading teams of scientists; he was so enthusiastic about his employees' efforts and so willing to pitch in at every level that he made it easy to join Team Fermi—and he attracted professional colleagues in droves with his otherworldly mental powers. One said that Enrico had such a feel for neutrons that he could "predict what would happen in any given experiment to within statistical error, and followed his predictions with detailed calculations and these almost always confirmed his intuitions."

Szilard, meanwhile, was impulsive and bold, made intuitive leaps, would spew a torrent of possibilities, and was almost too original. Enrico was particularly stunned that Leo didn't appreciate lab work, seeing one's theories and calculations rendered manifest in the material world. When Enrico described the experiment that won him a Nobel, the bombarding with neutrons of every element of the periodic table, Leo said that, under a similar situation, he would have hired someone else to do such a "boring task." Then at Columbia, Leo took one look at the greasy, dust-spewing graphite that he had determined could be used in immense amounts to slow the neutrons for an industrial-strength reactor, as the paraffin had showed them in Rome, and declined to take part, saying he didn't want to "work and dirty my hands like a painter's assistant." Yet Leo wasn't shy about showing up in others' labs to give them unasked-for advice and interrogating scientists about their work and their findings "with the precision of a prosecuting attorney." When one afternoon the tactless Szilard barged into Isidor Rabi's lab, criticizing his techniques, Rabi told Szilard that he should go do his own work and kicked him out: "You are reinventing the field. You have too many ideas. Please, go away!"

Szilard and Fermi argued over everything—even what the numbers actually meant. Szilard: "We went over to Fermi's office, and Rabi said to Fermi, "Look, Fermi, I told you what Szilard thought and you said 'Nuts!' and Szilard wants to know why you said 'Nuts!' " So Fermi said, "Well . . . there is the remote possibility that neutrons may be emitted in the fission of uranium and then of course perhaps a chain reaction can be made." Rabi said, "What do you mean by 'remote possibility'?" and Fermi said, "Well, ten percent." Rabi said, "Ten percent is not a remote possibility if it means that we may die of it. If I have pneumonia and the doctor tells me that there is a remote possibility that I might die, and it's ten percent, I get excited about it." The relationship was so fraught that a few weeks after this encounter, Szilard wrote a letter to Fermi outlining their difficulties, and concluding that if each had worked separately, a chain reaction would surely have occurred by now. He never sent this letter, but instead on July 4 wrote a more tentative proposal of collaboration, asking Pegram, the department chair, to adjudicate.

After learning that Leo was physically inept, Enrico finally accepted this state of affairs, as the last thing he wanted was an accident-prone physicist working with slippery graphite bricks and uranium. Herbert Anderson: "Szilard was not willing to do his share of experimental work, even the preparation in the conduct of the measurements. He hired an assistant to do what we would've required of him. . . . Fermi's vigor and energy always made it possible for him to contribute somewhat more than his share, so that any dragging of feet on the part of the others stood out the more sharply in contrast. . . . That experiment was important in a number of ways, but it was the first and also the last experiment in which Fermi and Szilard collaborated." *That experiment was important in a number of ways*—as it would be the birth of both nuclear power and atomic bombs.

Eventually, Fermi and Szilard arrived at a satisfactory professional modus operandi. First they would brainstorm theory together, then Fermi would design and execute the experiments with students, share the results with Szilard and the two would debate the next steps. While Fermi was in the lab, Szilard both conceptualized and got the needed graphite and uranium from suppliers. The Italian and the Hungarian essentially worked separately, together, and communicated through a "conduit" by the name of Edward Teller: "During the summer of 1939, I taught summer school at Columbia. I lectured graduate students, but I was invited primarily as a consultant peacemaker on the Fermi-Szilard chain reaction project. Fermi and Szilard both had asked me to work with them. They were barely speaking to each other. Temperamentally, the two men were almost opposites. . . . Fermi seldom said anything that he could not demonstrate. Szilard seldom said anything that was not startling and new. Fermi was humble and self-effacing. Szilard could not talk without giving orders. Only if they had an intermediary could they be in contact with each other for any length of time. Because I admired and enjoyed working with both men and they were comfortable with me, I became a conduit of information, able to solve problems between them unobtrusively, sometimes even before they occurred."

Leo Szilard: "On matters scientific or technical there was rarely any disagreement [but] Fermi and I disagreed from the very start of our collaboration about every issue that involved not science but principles of action in the face of the approaching war. . . . Of all the many occasions which I had to observe Fermi I liked him best on the rather rare occasions when he got mad (except, of course, when he got mad at me). . . . If the nation owes us gratitude—and it may not—it does so for having stuck it out together as long as it was necessary."

At the start, Szilard immediately wanted to make the leap to fission, while Fermi thought it was merely a curiosity and continued his steady, meticulous investigations. Their split in thinking was echoed in the press: On January 29 and 30, 1939, the *Washington Evening Star* reported on page 1, "Power of New Atomic Blast Greatest Achieved on Earth," but that "as a practical power source, the new finding has at present no significance." On February 5 the *New York Times* said, "Hope is revived that we may yet be able to harness the energy of the atom," but called it "remotely possible." The same week, *Newsweek* quoted Einstein, echoing Rutherford on the improbability of nuclear energy: "It is like shooting birds in the dark in a country where there are not many birds."

In Paris, the Joliot-Curies were also investigating fission, and Szilard urged the Americans and the French to keep their studies private to safeguard this information from the Fascists. Beginning at the start of the Enlightenment in the seventeenth century with the *Journal des sçavans* and *Philosophical Transactions of the Royal Society*, publishing scientific findings became a method of announcing one's achievements, expanding the understanding of the natural world, contributing to the public good, and, of course, competing with your professional colleagues—the highway of hive mind. Szilard's insistence that scientists should not publish was such a shocking and contrary notion that a number of physicists in 1939 couldn't comprehend it. I. I. Rabi even took Szilard aside to warn that, by promoting such a bizarre notion, his guest status at Columbia was in danger. Bohr thought there was little future in atomic bombs since it was so difficult to produce the needed U-235 isotope that was known to be an effective source for a chain reaction and so didn't feel this remote possibility meant upending a three-hundred-year scientific tradition of free discourse, even if it involved Hitler. Fermi agreed with Bohr.

The Joliot-Curie team did not honor Szilard's request, publishing in *Nature* on March 18 that the U-235 isotope could produce 3.5 neutrons per fission (later revised to 2.9) and chain-react. Secrecy was not thoroughly maintained in the United States, either, for after Bohr gave a speech at the American Physical Society on fission's potential for explosion, the *Times* said, "The creation of a nuclear explosion which would wreck an area as large as New York City would be comparatively easy," since a remarkably small quantity of uranium could "blow a hole in the earth 100 miles in diameter. It would wipe out the entire City of New York, leaving a deep crater half way to Philadelphia and a third of the way to Albany and out to Long Island as far as Patchogue."

As it turned out, Szilard was right. German scientists brought Joliot-Curie's results to the attentions of both the Berlin War Office and the Reich Ministry of Education. When Fermi's Columbia tests showed that very pure graphite worked but the average industrial product did not, Szilard got Pegram to convince Fermi not to publish. This secrecy may have tricked the Germans, who had failed with impure graphite in their test reactor, to switch to heavy water as a moderator, which crippled their program.

When he heard Fermi had accepted Szilard's argument, Bohr did, too. "Contrary to perhaps what is the most common belief about secrecy, secrecy was not started by generals, was not started by security officers, but was started by physicists," Fermi remembered. "And the man who is most responsible for this certainly extremely novel idea for physicists was Szilard. He is certainly a very peculiar man, extremely intelligent. I see that this is an understatement. He is extremely brilliant and he seemed somewhat to enjoy, at least that is the impression that he gives to me, he seems to enjoy startling people. So he proceeded to startle physicists by proposing to them that given the circumstances of the period—you see it was early 1939 and war was very much in the air—given the circumstances of that period, given the danger that atomic energy and possibly atomic weapons could become the chief tool for the Nazis to enslave the world, it was the duty of the physicists to depart from what had been the tradition of publishing significant results as soon as the physical review or other scientific journals might turn them out, and that instead one had to go easy, keep back some results until it was clear whether these results were potentially dangerous or potentially helpful to our side."

During that summer of 1939, Fermi, Anderson, and Szilard got five hundred pounds of uranium and tried to initiate a chain reaction using water as a moderator. It failed, and secrecy or no secrecy, this would be the last American experiment with fission for nearly a year. While Fermi then spent the rest of the season at an Ann Arbor conference, where he experimented with cosmic rays, Szilard settled on carbon (graphite) or deuterium (heavy water) as promising candidates for the moderator. He and brother Bela met at the Waldorf-Astoria with a group of investors, to whom Szilard described energy generated by what Fermi and Szilard called "piles," and a business plan with a majority of shares controlled by physicists. The financiers regretfully declined.

When a chemist at the Kaiser Wilhelm Institute then published "Can Nuclear Energy Be Utilized for Practical Purposes?" in June 1939, the Hungarians in America became certain that Germany was on the road to atomic weapons and that Washington had to understand the threat. Their fears were valid; three months later, the Nazi War Office began conducting secret conferences on nuclear bombs attended by Bagge, Geiger, Bothe, Hahn, and Heisenberg. In time, the War Office would oversee atomic bomb research and would take over the Kaiser Wilhelm Institute as part of this mission. To drive away curious passersby, they called KWI's nuclear research facility the Virus House.

Eugene Wigner, Ed Teller, and Leo Szilard discussed what would prod the American government into researching and producing atomic bombs before the Nazis could. The Hungarians knew that the Nazis were about to take Belgium, that the Belgian Congo held the world's biggest uranium mine, that the Germans already controlled the Czech mine that had been so useful to Marie Curie, and that suddenly, the Nazis had halted exports of all their Bohemian ore. With nuclear weapons, Szilard worried, Hitler could conquer anyone. A few months before, they tried directly contacting the Pentagon, but got nowhere. George Pegram, the Columbia dean who'd brought Fermi to America, arranged for Enrico to meet with Admiral Stanford Hooper, technical assistant to the chief of naval operations, and Fermi, while waiting for his appointment, got to hear the desk officer receptionist tell the admiral, "There's a wop outside." Wigner, Teller, and Szilard thought they should write to the leaders of Belgium to protect that nation's colonial Congo ore—the biggest source of uranium in the world—from the little housepainter, but wondered if it would be proper to contact a foreign government without the approval of the American State Department. But then Leo remembered that Albert Einstein was friendly with the queen of the Belgians (so friendly that his letters to the queen included such comments as "Princeton is a wonderful little spot, a quaint and ceremonious village of puny demigods on stilts").

Leo and Gene decided they must go see Einstein, vacationing, sailing, and daydreaming the calculus of physics (what he called "thought experiments") in Peconic on the north fork of eastern Long Island. On Wednesday, July 12, 1939, they drove out in Wigner's 1936 Dodge, past the glorious World's Fair, which featured Calvin Coolidge's pet hippo, Billy; England's Magna Carta; pavilions from every major nation (except Nazi Germany); a cat named Hitler with a "mustache" of black under his nose who joined his mistress for the jitterbug contest; and a streamlined moderne architecture that was promised as the World of Tomorrow. In fact, the real world of tomorrow was about to be ignited by two eccentric Hungarians driving by in a rickety Dodge.

Szilard and Wigner got confused by the Native American names of the region and ended up in Patchogue on the south fork instead of Cutchogue on the north. It took two hours to fix that mistake, and then when they asked everyone in Peconic where Dr. Moore, Einstein's friend, lived, no one knew. Finally, Szilard insisted, "Let's give it up and go home. Perhaps fate intended it. We should probably be making a frightful mistake by enlisting Einstein's help in applying to any public authorities on a matter like this. Once a government gets hold of something, it never lets go." But Gene insisted, and they kept driving, this way and that, utterly lost. Then Leo thought, "How would it be if we simply asked where around here Einstein lives? After all, every child knows him." Almost immediately they spotted a little boy, perhaps seven years old, sunburned and playing with a fishing rod by the side of the road. Szilard called out to him from the car, "Do you know where Einstein lives?" "Of course I do," the child said, and gave them directions that took them straight to the doctor's white bungalow.

Einstein had spent the morning sailing and was now relaxing in a screened porch at the back of the house, drinking iced tea. After Szilard described fission, Einstein was quiet for a moment, then explained, "Daran habe ich gar nicht gedacht"—"I hadn't thought of that at all." If it worked, Einstein explained, it would be the first source of energy for human beings that did not derive from the sun. (This is still technically and convolutedly true, as the sun's heat leads to wind, and that wind moves the rain that begins hydro, while the sun's light powers photosynthesis—the origin force of coal, oil, and natural gas—and its heat and light together generate solar, while uranium is derived, like all elements, from supernovas, meaning *a* sun but not *the* sun.) While Szilard drafted, Einstein dictated in German a letter to the Belgian queen, and also one to the US State Department, for its opinion about the protocol for resident aliens' contacting a foreign power.

Returning to the King's Crown, Szilard had second thoughts about the approach they were taking and decided to ask around for opinions on dealing with Washington. A Berlin economist acquaintance put him in touch with Alexander Sachs, a Lehman vice president who was one of FDR's economic consultants. Sachs told Szilard that Einstein shouldn't be writing to the queen of the Belgians about such matters, but to the president of the United States himself, and that Sachs would deliver such a letter to the Oval Office personally. For if any scientist could get FDR's attention, after all, it was Albert Einstein. He was the Franz Liszt of physicists, with young girls mobbing him, people fainting in his presence, and the London Palladium offering a three-week engagement for a one-man show on anything he'd like to say. And, after all, immediately after arriving to exile in America, he'd been invited to sleep over at the presidential mansion by the Roosevelts themselves. Additionally as Szilard noted, "The one thing most scientists are really afraid of is to make fools of themselves. Einstein was free from such a fear and this above all is what made his position unique on this occasion." Laura Fermi: "In the United States of those days there were no links between government and the universities, such as the ministry of education in other countries; and virtually no channels of communications were available. So the scientists took the initiative. In the typical devious Hungarian way, Szilard and Wigner agreed with Einstein, the tallest figure in science, that they would write a letter to President Roosevelt, and he, Einstein, would sign it."

Szilard returned to Peconic to redraft on August 2, this time driven by Ed Teller in his 1935 Plymouth. Back at the King's Crown, he paid Columbia secretary Janet Coatesworth to take dictation for a letter from the world's most famous scientist to Franklin Roosevelt about a bomb that could destroy the world. Szilard: "We did not know just how many words one could put in a letter which a president is supposed to read." Janet thought Leo was a crackpot, and she would continue to think this until 1945, when she learned of Hiroshima and Nagasaki. At war's end, Einstein insisted that, with his letter to FDR, "he really only acted as a mailbox" for Leo Szilard.

Szilard and Einstein then thought they might ask Charles Lindbergh to discuss the matter with FDR, but learned soon enough that the president and the aviator were politically opposed and thus returned to Sachs, who met with the president on October 11 and 12, bringing Einstein's letter for FDR to read directly:

Albert Einstein
Old Grove Rd.
Nassau Point
Peconic, Long Island

August 2nd 1939

F.D. Roosevelt
President of the United States
White House
Washington, D.C.

Sir:

Some recent work by E. Fermi and L. Szilard, which has been communicated to me in manuscript, leads me to expect that the element uranium may be turned into a new and important source of energy in the immediate future. Certain aspects of the situation which has arisen seem to call for watchfulness and, if necessary, quick action on the part of the Administration. I believe therefore that it is my duty to bring to your attention the following facts and recommendations:

In the course of the last four months it has been made probable—through the work of Joliot in France as well as Fermi and Szilard in America—that it may become possible to set up a nuclear chain reaction in a large mass of uranium, by which vast amounts of power and large quantities of new radium-like elements would be generated. Now it appears almost certain that this could be achieved in the immediate future.

This new phenomenon would also lead to the construction of bombs, and it is conceivable—though much less certain—that extremely powerful bombs of a new type may thus be constructed. A single bomb of this type, carried by boat and exploded in a port, might very well destroy the whole port together with some of the surrounding territory. However, such bombs might very well prove to be too heavy for transportation by air.

The United States has only very poor ores of uranium in moderate quantities. There is some good ore in Canada and the former Czechoslovakia, while the most important source of uranium is Belgian Congo.

In view of the situation you may think it desirable to have more permanent contact maintained between the Administration and the group of physicists working on chain reactions in America. One possible way of achieving this might be for you to entrust with this task a person who has your confidence and who could perhaps serve in an inofficial capacity. His task might comprise the following:

a) to approach Government Departments, keep them informed of the further development, and put forward recommendations for Government action, giving particular attention to the problem of securing a supply of uranium ore for the United States;

b) to speed up the experimental work, which is at present being carried on within the limits of the budgets of University laboratories, by providing funds, if such funds be required, through his contacts with private persons who are willing to make contributions for this cause, and perhaps also by obtaining the co-operation of industrial laboratories which have the necessary equipment.

I understand that Germany has actually stopped the sale of uranium from the Czechoslovakian mines which she has taken over. That she should have taken such early action might perhaps be understood on the ground that the son of the German Under-Secretary of State, von Weizsäcker, is attached to the Kaiser-Wilhelm-Institut in Berlin where some of the American work on uranium is now being repeated.

Yours very truly,
(Albert Einstein)

After reading it, the president said, "Alex, what you are after is to see that the Nazis don't blow us up."

"Precisely," Sachs replied.

Roosevelt called in his aide General Edwin "Pa" Watson and said, "Pa! This requires action!"

The White House
Washington
October 19, 1939

MY DEAR PROFESSOR,

I want to thank you for your recent letter and the most interesting and important enclosure.

I found this data of such import that I have convened a board consisting of the head of the Bureau of Standards and a chosen representative of the Army and Navy to thoroughly investigate the possibilities of your suggestion regarding the element of uranium.

I am glad to say that Dr. Sachs will co-operate and work with this committee and I feel this is the most practical and effective method of dealing with the subject.

Please accept my sincere thanks.

Very sincerely yours,
FRANKLIN D. ROOSEVELT

FDR's presidential advisory uranium committee—created six months after the Nazis had begun their own uranium group—would be known as the Briggs Committee, for its chair, Lyman Briggs, head of the Bureau of Standards, the federal physics lab, and a known sloven. On October 21, Wigner, Teller, and Szilard attended the committee's first meeting and tried to convince army and navy ordnance officers to expedite the development of atom-splitting technology before the Nazis beat them to it. Herb Anderson: "Szilard came to the conclusion that a large-scale experiment using carbon for slowing down the neutrons ought to be started without delay. It was a gamble, but the other possibilities looked less practical at the time. The problem was where to get money for the graphite. This was the kind of problem Szilard liked. Fermi, on the other hand, was not very good at the kind of promotion this required. . . . [Szilard] estimated they would need about $10,000 for the graphite. Since this was much more than he could hope to get from any university, he thought of going to the government for support, especially in view of the military implications. . . . A meeting had been arranged with Lyman J. Briggs, Col. Keith R. Adamson of the army and Cmdr. Gilbert C. Hoover of the navy."

Ed Teller: "After the initial presentation, Adamson began by voicing his doubts about novel scientific projects: 'At Aberdeen, we're offering a $10,000 reward to anyone who can use a death ray to kill a goat we have tended to post. That goat is still perfectly healthy.' "

Herb Anderson: "The question of money arose. Szilard thought $6,000 would suffice for the test of graphite in mind. There followed a long declamation from the army representative about the nature of war. In the end he argued it wasn't weapons that won wars, but the morale of the troops. [Eugene Wigner] said, in his high-pitched voice, that it was very interesting to hear this. He had always thought the weapons were very important and that this is what cost money, that this is why the army needed such a large appropriation. But he was interested to hear that he was wrong; it's not weapons but morale which wins the wars. If this was correct, perhaps one should take a second look at the budget of the army, maybe the budget could be cut. Col. Adamson wheeled around to look at Wigner and said, 'Well, as far as those $6,000 are concerned, you can have it.' "

The navy's $6,000 for Szilard's graphite surprisingly did not arrive with due haste from Washington. In early 1940, Peter Debye, who had headed the post-Meitner physics section of the Kaiser Wilhelm Institute and had then been forced to resign by the Nazis for being Dutch, visited his American colleagues and warned everyone he could that the Germans were developing uranium bombs in their Virus House. Fermi guessed that since the Nazis did not bring their atomic scientists into one research facility but left them dispersed, their efforts would go nowhere (and this would turn out to be true). But Debye's report alarmed Szilard, who got Einstein to send a second letter to FDR, and the $6,000 was finally released to Columbia.

By the end of 1939, four Curie Institute associates had patented a uranium oxide chain reaction using deuterium—heavy water—as the moderator. Norway's Norsk Hydro plant in Vemork was the sole industrial European source of heavy water. IG Farben approached Norsk Hydro to buy it, but the Norwegians declined. When, tipped off by the Joliot-Curies, a lieutenant in the French intelligence agency offered to purchase all of the company's supply for FF 36 million, the Norwegians refused. Instead, to "aid France's victory," they shipped it gratis to France in twenty-six cans on March 9, 1940. Three months later Paris fell, the French government fled to Bordeaux, and the British sent in the War Ministry's liaison with the French Ministry of Armaments—Charles Henry George "Mad Jack" Howard—to abscond with as many French scientists, industrial diamonds, machine tools, and other matériel that he could spirit out from under the Nazis, including the deuterium.

Mad Jack and the Curies split up the shipment, with physicist Hans von Halban packing as many cans as possible into a touring car, covering them with cushions and blankets, having his children sit on the blankets, and telling them that no matter what, they had to look happy. In trying to ship the rest out of the country, Mad Jack arrived to find the port in utter chaos, so he corralled the crew of the British coal ship SS *Broompark*, got them too drunk to float, loaded up his cargo, and set off through the Gironde on June 19. With everyone at the Curie Institute save the Joliot-Curies themselves now evacuated, Fred and Irène pretended to be busy with obscure problems in theoretical atomic physics while risking their lives to manufacture explosives and radio equipment for the Resistance. When another French ship hit a mine and sank, Fred Joliot told the Nazis that this was the one with the heavy water. Von Halban's and Mad Jack's supplies, meanwhile, successfully reached Britain.

With the Germans approaching Denmark, Otto Robert Frisch and Rudolf Peierls fled to England, finding jobs at the University of Birmingham. Since the needed isotope for explosives, U-235, was only found in 0.7 percent of natural uranium, many, notably Bohr, believed it was technically unfeasible, even absurd, to attempt fashioning a uranium bomb. But Frisch and Peierls at Birmingham finished calculations proving that a tremendous device would only need a few kilos of uranium-235. Simultaneously, in April 1940, Lise Meitner was visiting the Bohrs when the Nazis invaded Denmark. As they prepared to evacuate Copenhagen, Bohr asked Meitner to send a telegram from Stockholm to British physicist Owen Richardson with the news that the Bohr family was fine. It concluded, "Please inform

Cockroft and Maud Ray Kent Meitner." The British physicists who received this thought "Maud Ray Kent" was a coded message about German atomic weapons research, so the British fission committee called itself the MAUD Committee. At war's end it was revealed that the Bohr's governess, Maud Ray, lived in the town of Kent, and Bohr was hoping to get word to her that all was well. The MAUD report to the British war effort was based on the Frisch and Peierls research: "We have now reached the conclusion that it will be possible to make an effective uranium bomb which, containing some 25 lb of active material, would be equivalent as regards destructive effect to 1,800 tons of T.N.T. and would also release large quantities of radioactive substance, which would make places near to where the bomb exploded dangerous to human life for a long period. . . . The committee considers that the scheme for a uranium bomb is practicable and likely to lead to decisive results in the war."

On May 5, 1940, the *New York Times* reported, "Vast Power Source in Atomic Energy Opened by Science. . . . A chunk of 5 to 10 pounds of the new substance, a close relative of uranium and known as U-234, would drive an ocean liner or an ocean-going submarine for an indefinite period around the oceans of the world without refueling, it was said, for such a chunk would possess the power-output of 25,000,000 to 50,000,000 pounds of coal, or of 15,000,000 to 30,000,000 pounds of gasoline. . . . The main reason why scientists are reluctant to talk about this development, regarded as ushering in the long dreamed of age of atomic power and, therefore, as one of the greatest, if not the greatest, discovery in modern science, is the tremendous implications this discovery bears on the possible outcome of the European war, it was explained. . . . Germany, it was asserted, may regret her act of having sent into exile Dr. Lise Meitner, who, with Professor Hahn, made the first observations that led to the discovery of the fountain-head of atomic energy that German scientists are so feverishly working to harness."

Traveling to the United States from a Britain under siege and a Europe defeated, MAUD's Mark Oliphant was stunned to learn that the apparatchik Lyman Briggs had not passed on the British committee's report to anyone else in the United States, to such effect that, when Oliphant met in secret with the Advisory Committee on Uranium, they were shocked by his openly discussing atomic weaponry as though the technology was imminent. After Vannevar Bush took the MAUD report to FDR, the slow-moving Briggs was quickly replaced by the hard-charging James Bryant Conant, president of Harvard, and on June 12, 1940, the Briggs Advisory Committee on Uranium was disbanded and supplanted by the far more politically connected National

Defense Research Committee (NDRC), soon to be renamed the Office of Scientific Research and Development (OSRD), headed by the protean Vannevar Bush. Before coming to the NDRC, Bush was vice president of MIT and president of the National Advisory Committee for Aeronautics, and one of the few Americans experienced in government-funded research (in years to come, NACA would become NASA). The only purpose of his OSRD was to mobilize science for war, and its first achievement was microwave radar, including a portable radar gun, which was tested by tracking the speed of automobiles. One physicist said, "For the Lord's sake, don't let the cops know about this." With his rimless glasses, military-tight haircut, and lean frame, Bush was the embodiment of can-do American academic management, achieving an apotheosis as head of OSRD. In July 1945, he would even predict the smartphone—"consider a future device for individual use, which is a sort of mechanized private file and library"—and he decided that his outfit should fully investigate the potential military applications of nuclear weapons. The new organizational structure, however, meant that only American citizens were supposed to be involved in classified military research—leaving out Einstein, Fermi, Szilard, Teller, and Wigner, to name just five. One reason why such a large percentage of the physicists involved in America's nuclear project were European émigrés was that the development of radar had first priority among leading scientific administrators of the United States, as radar was deemed immediately useful, unlike the pie in the sky of splitting atoms to make munitions.

In July 1940, the army asked the FBI for its opinion on whether Einstein should receive a security clearance. J. Edgar Hoover replied with a collection of incorrect information, scurrilous letters, and half-truths that concluded the physicist was an "extreme radical" who had written for Communist journals. The navy cleared Einstein anyway, but the army would not. He was not asked to join the Manhattan Project, not officially told of it, and as a devout pacifist had no interest in being involved in any case. However, when Harold Urey's gaseous diffusion producing uranium-235 at Columbia ran into trouble, Bush asked for Einstein's help, and he worked up a chemical process he would recommend. Any more involvement, though, would require a security clearance, and Bush knew enough not even to ask. Security would continue to be a menace throughout the history of the Manhattan Project, and the great irony in all this was summed up by biographer William Lanouette: "The only secrets worth protecting were in the minds of the very scientists the authorities wanted to exclude."

As the federal government was now supporting Fermi and Szilard, the army had them investigated to see if they were loyal enough to work on defense projects. Intelligence officers reported that Fermi was "undoubtedly a Fascist" and Szilard "very pro-German, and to have remarked on many occasions that he thinks the Germans will win the war," and concluded for each, "Employment of this person on secret work is not recommended." On August 22, the War Department forwarded a copy to J. Edgar Hoover, asking that the FBI "verify their loyalty to the United States." Neither department could uncover any real anti-American evidence, though, so the two physicists were allowed to keep working, for the time being.

Enrico, Laura, the children, and their maid then left the King's Crown for an apartment on Riverside and 116th Street, with Laura's fears about a cultural chasm seemingly confirmed daily. Nella barely got into Horace Mann School after failing a question on the entry intelligence test about skunks since, in Europe, there are no skunks. Laura had terrible difficulties with her English—every time she tried to buy something over the phone, something else arrived instead—and the other Italians she met in New York came from Naples or Sicily, with dialects so different from Rome's that they were incomprehensible.

Enrico did not believe in renting, but no houses or apartments were for sale near Columbia. During a visit with Harold Urey in Leonia, New Jersey, Urey—the chemistry laureate who would engineer one of America's "atomic secrets," a breakthrough in separating fissionable uranium-235 to fuel the Manhattan Project's nuclear weapons—couldn't stop praising the wonders of the Palisades. Four months later, Fermi bought a house nearby, with a big yard and a pond for the kids, and a wet basement. He had repeatedly told Laura how he couldn't wait for the chance to return to tending the soil, the work of his forefathers, but apparently this yearning did not include the horticulture of New Jersey—when Laura asked Harold to come help her get rid of her crabgrass, he told her there was a problem, as her entire lawn was crabgrass. Enrico did fall madly in love with American gadgets, so much so that, for her first New Jersey Christmas, Laura got from her husband a garbage can whose lid raised with a foot lever. She never forgot the thoughtfulness of that gift.

After decades of being a physicist's wife, Laura decided that Enrico was unusual among his peers as he "oscillated between theoretical and experimental physics, conveniently adapting to changing needs. Whenever there seems to be no chance for an interesting experiment, Enrico withdraws to his office and fills sheet upon sheet with calculations. . . . But as soon as he gets an idea for a piece of experimental research or whenever a new apparatus is being devised and completed, he lets his paper become covered with dust and spends all his time in the laboratory." Graduate student Leona Woods remembered a favorite Fermi bit of humor: "He frequently said, he was amazed when he thought how modest he was."

In July 1941, Fermi and Walter Zinn begin experimenting with materials to slow down the neutrons and produce enough fission to start a chain reaction, the factor known as k. If $k > 1$, uranium's neutrons will keep splitting nuclei, producing more free-ranging neutrons, and a chain reaction will sustain; if $k < 1$, it will not. For civilians, that atomic power and thermonuclear bombs are based on $k > 1$ is Greek . . . but not knowing the mathematics of physics means we are deaf to the real music of the spheres.

With the navy-supplied thirty tons of graphite and uranium finally at hand, Fermi had to figure out how to engineer chain-reaction test runs at Schermerhorn Hall. These first experiments included the sprinting technique refined in Rome. Herb Anderson: "Cartons of carefully wrapped graphite bricks began to arrive at the Pupin laboratory until $1\frac{1}{2}$ tons had come, enough for the experiment. . . . We stacked the graphite bricks into a neat pile. We cut narrow slots in some of the bricks for the rhodium foil detection we wanted to insert, and soon we were ready to make measurements. The radioactivity induced in rhodium by slow neutrons has a quite short half-life, 44 seconds. The Geiger counter had to be separated from the neutron source and was installed in Fermi's office some distance down the hall from the room with the graphite pile. . . . To get the rhodium foil under the Geiger counter in the allotted 20 seconds took coordination and some fast legwork. The division of labor was typical. I removed the source on signal; Fermi, stopwatch in hand, grabbed the rhodium and raced down the hall at top speed. He had just enough time to place the foil carefully into position, close the lead shield and, at the prescribed moment, start the count."

Enrico Fermi: "Physicists on the seventh floor of Pupin laboratories started looking like coal miners and the wives to whom these physicists came back tired at night were wondering what was happening. . . . What was happening was that in those days we were trying to learn something about the absorption properties of graphite, because perhaps graphite was no good. So, we built columns of graphite, maybe 4 feet on the side or something like that, maybe 10 feet high. It was the first time an apparatus in physics, and these graphite columns were apparatus, was so big you could climb on top of it—and you had to climb on top of it. Well, cyclotrons were the same way too, but anyway that was the first time when I started climbing on top of my equipment because it was just too tall—I'm not a tall man. . . . Graphite is a black substance, as you probably know. So is uranium oxide. And to handle many tons of both make people very black. In fact it requires even strong people. And so, well, we were reasonably strong, but I mean we were, after all, thinkers. So Dean Pegram again looked around and said that seems to be a job a little beyond your feeble strength, but there is a football squad at Columbia that contains a dozen or so of very husky boys who take jobs by the hour just to carry them through college. Why don't you hire them? And it was a marvelous idea; it was really a pleasure for once to direct the work of these husky boys, canning uranium—just shoving it in—handling packs of 50 or 100 pounds with the same ease as another person would've handled 3 or 4 pounds. In passing these cans fumes of all sorts of colors, mostly black, would go in the air." Physicist Jay Orear: "The workers by the end of the day turned from white to black. There is a scene of Fermi wearing goggles and stripped to the waist machining a block of graphite and creating a black cloud that rises up and hits him in the face. It was typical of Fermi to participate in all phases of an experiment—even the dirty parts. It is easy to understand why his machinists especially praised him."

After a lunch in September of 1941, Fermi offhandedly told Edward Teller that he wasn't sure why the Americans had chosen to pursue fission bombs, since the temperatures of an atomic bomb might be over 400 million degrees C, meaning it could fuse hydrogen with helium—the same process that creates starlight. If that proved to be correct, a bomb made from fusion would be three times stronger than a fission device, but much, much cheaper to produce. Teller was so inspired by this offhand, innocuous comment that he spent the ensuing decades trying to design what he called "the Super"—a thermonuclear fusion weapon, the hydrogen bomb.

After sending a memo to the president in November 1941 that "within a few years the use of bombs such as described here, or something similar using uranium fission, may determine military superiority. Adequate care for our national defense seems to demand urgent development of this program," Bush's team decided that "urgent development" meant a two-to-three-year program at a cost of $133 million. But at the first meeting to plan a nuclear future in Schenectady, the science was so new that the attending physicists couldn't give a sincere estimate of either time or cost (when German physicists did something similar in a meeting with Nazi military chiefs, the generals judged them too incompetent to be trusted). As the atomic project's finances then mushroomed into the heavens, Bush discovered that he could no longer sneak the money through the War Department's "discretionary" account. The only Pentagon department with a flexible enough financial structure and a grand enough budget to support what would in time be called the Manhattan Project was the Army Corps of Engineers.

By December 1941, Szilard and Anderson had found seven possible sites for the first nuclear reactor—which they code-named "the egg boiling experiment"—including a Yonkers golf course, and a New Jersey hangar for blimps. That same month, physicist Arthur Holly Compton, who'd won his Nobel for studying the scattering of X-rays' radiant energy by free electrons, was named director of American nuclear research—under Conant, who was in turn under Bush—and he consolidated the various efforts at Princeton, Berkeley, and Columbia into one department at his own University of Chicago. In line with American thinking that dreary names are good for security—think Manhattan Engineer District versus the Virus House—Compton named his uranium program the Metallurgical Laboratory, or Met Lab. The one secret Laura Fermi knew about her husband's new employer was that no metallurgists were working at the University of Chicago's Metallurgical Laboratory.

Physicist Isabella Karle remembered the dangers of the Met Lab's soda dispenser, "a style of machine that dropped a paper cup, which was then filled with carbonated water and Coca-Cola syrup. The man who came to service the machine at our lunch time forgot to bring his hose for filling the syrup reservoir. He walked into the neighboring laboratory where wet chemistry was being performed and borrowed a rubber hose from an aspirator, filled the reservoir with the syrup, returned the hose and left. Some time after lunch a technician was carrying an alpha counter and noticed the meter went off the scale as he passed by the Coke machine. By the next day, the Coke machine was replaced with one that dispensed bottles rather than liquids. We never did know how many, if any, employees drank the radioactive Coca-Cola."

By now, the dangers of Röntgen rays had been known for nearly forty years, but radioactive science remained as perilous as it was for the Curies. George Cowan: "My supervisor at the Met Lab was Herbert Anderson, Fermi's right-hand man. He required a neutron source to measure reactivity of the Fermi pile. Neutron sources were usually made with a mixture of radium and beryllium. The alpha particles emitted by radium hit the beryllium and made neutrons. The sources were prepared by drying a solution of radium on beryllium metal powder and sealing the mixture in a leak-proof brass capsule. I was coached on what to do and sent to New York carrying beryllium powder, a brass capsule, and a gamma survey meter to make sure that once the capsule was soldered shut, it didn't leak radioactive radon. We didn't know it at the time, but the danger posed by the beryllium was greater than the potential damage from radiation. Herb Anderson eventually died of berylliosis, a lung disease caused by breathing beryllium or beryllium oxide. I traveled by train to New York and took a cab to a big building on Sixth Avenue that housed the offices of Radium Chemical Company. I brought a portable survey meter with me that measured gamma radiation. It started to register when I entered the building. It went berserk when I checked out the primitive chemical hood I was directed to use to make the neutron source. Out of a mixture of curiosity and alarm I took the elevator to the rooftop. Air from the hood was discharged there. Even by the low standards we used at that time, the roof was unacceptably radioactive. I spent two days at the Radium Chemical Company making and checking the neutron source. I was anxious to leave as soon as possible. I later found that the owners of the company operated under a different name in New Jersey, where they employed young women to paint luminous watch dials with radium loaded brushes which they tipped with their mouths." In 1989, the Environmental Protection Agency inspected the abandoned Radium Chemical Company's site in New York City and discovered radiation levels so high that in the most contaminated parts of the building a person could exceed the yearly occupational exposure limit after only one hour.

Another Chicago Met Lab employee was Leo Szilard, who had the title of chief physicist and an annual salary of $6,600. Before, Leo had consorted with Albert Einstein to get this whole megillah going; now, he was just another worker-bee scientist given a series of assignments. Szilard was hardly the kind of employee bosses adore—in years to come, he would develop a serious enemy in Brigadier General Leslie Groves—but beyond his distinct lack of social graces, he did himself no favors with higher-ups when he started politically organizing his colleagues as early as September 21, 1942, with a startling series of brilliant premonitions. This was the birth of Szilard's antinuclear activism, which he would pursue for the rest of his life, even though, if anyone birthed the Atomic Age, it was Szilard:

> These lines are primarily addressed to those with whom I have shared for years the knowledge that it is within our power to construct atomic bombs. What the existence of these bombs will mean we all know. It will bring disaster upon the world if the Germans are ready before we

are. It may bring disaster upon the world even if we anticipate them and win the war, but lose the peace that will follow.

We may take the stand that the responsibility for the success of this work has been delegated by the President to Dr. Bush. It has been delegated by Dr. Bush to Dr. Conant. Dr. Conant delegates this responsibility (accompanied by only part of the necessary authority) to Compton. Compton delegates to each of us some particular task, and we can lead a very pleasant life while we do our duty. We live in a pleasant part of a pleasant city [Chicago] in the pleasant company of each other, and have in Dr. Compton the most pleasant "boss" [at the Metallurgical Laboratory] we could wish to have. There is every reason why we should be happy, and since there is a war on, we are even willing to work overtime.

Alternatively, we may take the stand that those who have originated the work on this terrible weapon and those who have materially contributed to its development have, before God and the World, the duty to see to it that it should be ready to be used at the proper time and in the proper way.

I believe that each of us has now to decide where he feels that his responsibility lies.

When the Americans entered the war in December of 1941, Enrico, Laura, and a wide swath of their friends and colleagues were classified as enemy aliens and had to relinquish their binoculars, cameras, and shortwave radios. The Fermis burned their daughter's Italian second-grade reader (since it was filled with pictures of Mussolini) and had to do something about Giulio, who at the age of five decided to start telling all the neighbors that he hoped the Axis Powers would win. Worried that they might be facing another need to flee with their assets frozen or seized, Enrico withdrew a goodly amount of his Nobel Prize money from the bank, stuffed it in a lead pipe, and buried it in the concrete floor of the basement coal bin in Leonia.

Enrico was especially hurt that, as part of his official wartime status, the post office was reading all his mail. He complained to higher-ups at the postal service and was officially informed that the surveillance had ended. He then opened his mailbox to discover a note to the Leonia carrier ordering that all Fermi mail be opened and reports of its contents submitted. When Enrico complained again, officials said that there was absolutely no order from the Postal Service to read his mail, so the note must be part of a scheme by Axis spies. He burst out laughing.

Since enemy aliens were not permitted to travel by airplane, Enrico now had to regularly commute to the Met Lab in Chicago by rail, always carrying with him an "endorsement of the United States Attorney" letter from the state capital, which enemy aliens needed for any travel. But his work was part of the war effort and top secret—he traveled under the name Eugene Farmer—so he wasn't able to tell state officials in Trenton exactly why he needed to constantly go to Chicago (not until October 12, 1942, Columbus Day, did the US government declare that Italians were no longer enemy aliens, and on July 11, 1944, after completing the required five years of residence, Enrico and Laura Fermi were sworn in as American citizens). In April 1941, Arthur Compton decided that he needed Fermi in Chicago full-time, so the family left New Jersey and found a nice rental close to the U of C campus. Unfortunately, the apartment came with a lavish radio that included a shortwave band, and the building's other renters were two Japanese American girls. The radio manufacturer sent a serviceman to disable the feature, but it was against the law for so many enemy aliens to be living together, and the girls were forced to leave.

By July 1942, Fermi and Szilard had enough data from their experimental piles to try to design one that could go critical—chain-react. Wally Zinn's designs for the uranium oxide cores were pressed into dies, with 250 tons of graphite and 6 tons of uranium cut by hand. They still needed more information on how to control the chain, and what the size of a critical pile would be, and would go through thirty piles at this intermediate stage before they were ready to birth the forge both of nuclear power and nuclear weapons.

History's first atomic power-generating nuclear reactor was to have been built in a facility in the Argonne Forest, which Arthur and Betty Compton had come across while out horseback riding. But when the contractor, Stone & Webster, suffered a strike, delaying construction, Fermi talked Compton into building the pile right on the University of Chicago campus. He insisted that he knew how to stop the reaction before what in time would be called a meltdown could destroy the city in an atomic conflagration.

The first pile's location was odd in another way. University of Chicago president Robert Hutchins had decided in 1939 that football was a distraction from higher education and canceled it. Twenty-two years later, after football returned to the University of Chicago, its students created a unique cheer:

Cosine, secant, tangent, sine,
3.14159.
Square root, cube root, BTU,
Sequence, series, limits too.
Rah.

But in that autumn of 1942 when Fermi was looking for his great pile's location, Chicago's campus football stadium, Stagg Field, in the style of a faux–English Gothic castle, with ivy-bedecked stucco turrets, battlements, a keeplike bell tower, and gargoyles, had been abandoned, its field now weedy and returning to the wild. Underneath the building's eaves was a little-used squash court, sixty feet long, thirty feet wide, and twenty-six feet high, with slate walls. It was perfect.

Construction began November 16, 1942, on CP-1—Chicago Pile-1—the most dangerous scientific experiment in American history. A team of high school dropout toughs waiting on their draft notices were recruited to set 771,000 pounds of pure graphite bricks in a lattice array with 2.25-inch uranium cores (80,590 pounds of oxide and 12,400 pounds of metal). The reactor would grow into an oval shape, twenty-five feet wide and twenty feet high, controlled by rods coated in cadmium. The graphite slowed the neutrons enough so that they would more efficiently hit their nuclei, freeing more neutrons, while the cadmium absorbed enough of the free neutrons to slow their barrage and quiet the pile.

Nuclear scientists had never seen an atom, much less a nucleus, the object of their profound fascination; all of their experiments, and their resulting theories, arrived secondhand, through the response of instruments. Fermi was reading *Winnie-the-Pooh* to improve his English, so his CP-1 instruments were given names of characters in the *Pooh* stories—Tigger, Piglet, Kanga, and Roo. In this case, measurements would be taken with a boron trifluoride counter that clicked just like a Geiger. When the first pile chain reacted and went critical, however, the counters would make a noise like nothing these physicists had ever heard before.

The graphite bricks were machined in the West Stands, overseen by millwright August Knuth. Physicist Albert Wattenberg: "We found out how coal miners feel. After eight hours of machining graphite, we looked as if we were made up for a minstrel. One shower would remove only the surface graphite dust. About a half-hour after the first shower the dust in the pores of your skin would start oozing. Walking around the room where we cut the graphite was like walking on a dance floor. Graphite is a dry lubricant, you know, and the cement floor covered with graphite dust was slippery." Fermi, stripped to the waist, was black and glistening. One colleague remarked that he could have played Othello. . . . In the late afternoon as the sun faded, the squash court was so filled with a smog of black dust that the scientists and the workers could only be seen by the whites of their teeth and their eyes.

Herb Anderson: "For the construction of the pile Fermi assigned the responsibility jointly to [Walter] Zinn and me. Our two groups combined for a concerted effort. Two special crews were organized: one machined the graphite, the other pressed the uranium oxide powder using specially made dies in a large hydraulic press. Both crews managed to keep their output up to the rate of the deliveries. Thus in our report for the month ending October 15, Zinn and I could state that 210 tons of graphite had been machined. . . . We organized into two shifts: Wally Zinn took the day shift, mine was the night shift. A simple design for a control rod was developed which could be made on the spot: cadmium sheet nailed to a flat wood strip was inserted in a slot machined in the graphite for this purpose. One special, particularly simple, control rod was built by Zinn; it operated by gravity through weights and a pulley and was called 'Zip.' It was to be pulled out before the pile went into operation and held by hand [Zinn's] with the rope. In case of an emergency or if Zinn collapsed, the rope would be released and Zip would be drawn into the pile by gravity."

Day after day, the pile grew. Fermi protégé Leona Woods—twenty-three years old, and the only woman on the project—took careful measurements. As the finish line drew nigh, the tension in the room grew along with the tonnage of graphite and uranium. Everyone who worked in the squash court knew this would be it. But, what if it wasn't . . . or what if something dreadful happened? In Washington, their bosses had already gone forward as though Fermi and Szilard would succeed perfectly in this revolutionary endeavor, having contracted with DuPont to build a $350 million nuclear reactor based on Fermi's design on the Columbia River outside Pasco, Washington.

There was more than meltdown to fear. Al Wattenberg: "One aspect of Fermi that wasn't so fortunate for me was that he was also my chauffeur. I would drive the car over to his house twice a week for a year and a half or so. Sometimes we picked up Leona and we would drive out to work together. And we kept talking about doing estimates of things—he was constantly estimating things. I tend to be an unbeliever in his intuition. It was that he had calculated things, and he remembered what he'd calculated. These little tiny estimates of things—he was always doing it, and he remembered them. Anyway, one of the days we were at a railroad crossing, and he was the driver.

I forget what we were calculating in our heads at the time, but anyway, the train went by and, of course, we went ahead because we have the calculation on our minds. It was two tracks, not one, and we almost got hit by the train coming in the other direction—missed by three seconds. And I don't know whether he was jocular about his fatalism or not, but he said, 'You see? It is exceedingly important that you always be with me when I drive.' "

Fermi had to create an environment that would focus the neutrons onto their atomic targets until they struck regularly enough to initiate a continuously explosive cascade of splintering nuclei. He tried submerging the pile in water, enclosing it in metal and vacuuming out the air, but none of these were helpful. Emilio Segrè: "The pile, according to plan, was shaped as a rotational ellipsoid with a polar radius of 309 centimeters and an equatorial radius of 388 centimeters. Most of the uranium lumps had to be placed in the central region for better utilization. The weight of the uranium was approximately six tons. To use the material efficiently the purest fuel had to be located more centrally and one had to watch carefully the details of the geometry because they could affect the reproduction factor. The whole structure was supported by a wooden frame. Fermi feared that there might not be enough material to reach criticality, and to reduce parasitic neutron absorption by the nitrogen of the air, he ordered a huge rubber balloon to enclose the pile." Anderson's request for an enormous square balloon from Goodyear Tire & Rubber led to a great deal of joking since the aerodynamics of a square balloon are poor, and Goodyear wasn't told of its purpose. Experiments in the final stages indicated that the last layer of uranium and graphite didn't need to be laid, and that the balloon didn't need to be sealed.

Physicist Harold Agnew: "Graphite was an awful material; it's heavy, and dense, and very slippery. Those things are heavy, and you could really get your fingers pinched and also hurt your knees because you had to crawl on this pile of graphite. . . . All during this time, it was very precise, we always stopped for lunch, and we had a sort of a team of us who always went to lunch together at the Commons, and we talked about things. Not about work things, but I remember one thing that really impressed me was our fear of where the Germans were. This was a real thing that maybe every third lunch would come up. Where do we think we are, where do we think they are; it was a concern during those days."

On the early afternoon of December 1, tests indicated that critical size was approaching. At 4:00 p.m., Zinn's group was relieved, and Anderson's stepped in. Then, the fifty-seventh layer of graphite and uranium bricks was set in place. So exact were Fermi's calculations, based on the ever-growing pile, that Enrico had already predicted to the exact brick the point at which the reactor would become self-sustaining, and he made the graduate student promise this would not be a repeat of the success of the fission experiment back at Columbia, which only Anderson and Dunning had witnessed. Zinn joined Anderson to measure the activity; they were both convinced that, when the cadmium rods were fully removed, the pile would chain-react and split its own nuclei until its atomic energies were exhausted. Anderson and Zinn had the control rods locked and the workers had the rest of the day off. Anderson: "It was a great temptation for me to pull the final cadmium strip and be the first to make a pile chain react. But Fermi had anticipated this possibility. He had made me promise that I would make the measurement, record the result, lock them all in place, go to bed, and nothing more."

Down the Great Lakes, winter rolled in; university squash courts from the 1930s were unheated. The men casting the uranium, sawing the graphite, and assembling the pile were kept warm by their physical labors, but the military boys outside guarding the world's biggest secret were freezing to death. A wonderful surprise was discovered in the locker room: the long-gone football team had left behind their outerwear. The key experiment that would be the birth of nuclear power and pave the way for the Atomic Age was thus guarded from enemy attack and Axis spies by men in raccoon coats.

The night before the test, Szilard ate two dinners back-to-back, explaining to his companion that the second was "just in case an important experiment doesn't succeed."

On Wednesday, December 2, 1942, every train, elevated, bus, and trolley was packed and suffocating—wartime gas rationing had begun. The temperature in Chicago rose all the way to 10°F. That day, the US State Department announced that the Nazis had already murdered 2 million Jews and that 5 million more were endangered. German infantryman Willy Peter Reese: "Marched into Russia. Murdered the Jews. Strangled the women. Killed the children. Everyone knows what we bring." That night would also be the first night of Hanukkah.

Beginning at 8:30 a.m., Fermi and Szilard's entire team assembled at Stagg Field to see if history would (or would not) be made. At the squash court's northern end, a viewing-stand balcony had been converted into a control center, where Fermi, Zinn, Anderson, and Compton monitored the instruments, and everyone else who wanted to watch crowded together.

Before them lay 380 tons of graphite, 40 tons of uranium oxide, 6 tons of uranium metal, 22,000 uranium slugs surrounded by 57 layers of graphite bricks, all at an inflation-adjusted cost of $2.7 million.

Crawford Greenewalt: "The whole atmosphere there was one of calmly observing an experiment being made. To be sure there was a suicide squad that you could see on the other end of the platform with their cadmium nitrate ready to pour in if it didn't work. But it became obvious very quickly that it was going to be controlled."

Within the environs of a reactor, uranium eats itself by throwing pieces of its nuclei against each other, like an organic 3-D pinball machine crowded with a seemingly infinite mass of pinballs. As this happens, the instruments indicating the subatomic activity rise steadily. Then, at fission, the free neutrons are so numerous and their attacks so great that the bombardment turns exponential, into a cascade . . . and a reactor can run away with itself and melt down.

To all who worked with him, Fermi appeared supernaturally confident, but he knew full well how dangerous CP-1 was and had gone step-by-step with his thirty piles preliminary to acquire data and feedback until he knew exactly what he was doing. It was a deliberate, careful, and meticulous process.

For the first reactor, he created a series of three safeguards to make absolutely certain the worst could not happen. Besides the main control rod operated manually on the floor by George Weil, a second, known as ZIP, was attached to a solenoid and an ionization chamber that was set to trigger automatically in the event of high, sustained neutron counts. Untouched by human hands, the solenoid would release ZIP, and gravity would drop it into the pile, stopping the reactor.

Another safeguard ZIP hung over the pile, this one tied by a rope to the balcony, where graduate student Norman Hilberry was standing by with an ax, ready at Fermi's command to cut the rope and let the rod fall.

Waiting in the wings, meanwhile, was Fermi's third safety measure, a "suicide squad" of Harold Lichtenberger, W. Nyer, and A. C. Graves. These three stood gravely on a platform over the pile accompanied by buckets of cadmium-salt solution. If the various rod safety devices failed, they would flood the pile.

Today, reactor control panels around the world have the same button for emergency shutdowns, the SCRAM button, and a member of Fermi's team, Volney Wilson, is credited with coining the term. For decades, insiders believed this referred to Safety Control Rod Ax Man, an homage to Norman

Hilberry. Decades after the fact, Wilson revealed the true story. When an electrician had finished wiring CP-1's emergency button, he asked Wilson and another physicist, Wilcox Overbeck, what its label should say.

Overbeck: "Well, what do you do when you push the button?"

Wilson: "You scram out of here as fast as you can."

And after twenty years of reactors being called the Fermi-Szilard scientific term *piles*, their operators are known as pile drivers.

At 9:45, Fermi announced (for visitors unfamiliar with the mechanism), "The pile is not performing now because inside it there are rods of cadmium which absorb neutrons. One single rod is sufficient to prevent a chain reaction. So our first step will be to pull out of the pile all control rods, but the one that George Weil will man."

A recorder's twitching pen inscribed the permanent data record onto a roll of graph paper, like a lie detector. "This pen will trace a line indicating the intensity of the radiation. When the pile chain-reacts, the pen will trace a line . . . that will not tend to level off. In other words, it'll be an exponential line. Presently we shall begin our experiment. George will pull out his rod a little at a time. We shall take measurements and verify that the pile will keep on acting as we've calculated."

A little after 10:00, Fermi said, "ZIP out," and Zinn pulled the rope controlling Hilberry's manual emergency rod by hand and tied it to the balcony.

Then at 10:37, with all eyes on his instruments, Fermi said, "Pull it to thirteen feet, George." Weil threw his switch, a motor buzzed, and the main rod began to withdraw. Marked with a vernier scale to show how much of it remained within the pile, it was now halfway extracted. Everyone in the balcony watched the panel where lights showed the amount of the rod's penetration, while listening to the staccato tempi of the boron trifluoride counters, barely noticing their clocklike faces. Their click rate increased rapidly, until it stuck a steady beat.

Fermi and his team in the balcony wrote down their findings and computed the results with slide rules. "This is not it," Fermi said, pointing to the area on the graph paper where the pen would reach when the pile went critical. "The trace will go to this point and level off."

At 10:42, Fermi ordered the rod pulled to fourteen feet. The counters' chatter and the pen's twitch rose once again and settled. The chain had not yet been reached.

At 11:00, he had it extracted to 14.5 feet. The clicks rose and the pen twitched higher, but still the pile had not turned critical.

Norman Hilberry: "Fermi had, the night before, sat down and computed what the trace on the recording galvanometer would be for every single position of the control rod. Clearly, if there were any new law of physics, it would begin to show up in an actual deviation of the observed graphs from those he had computed, and each time it hit absolutely right on the nose. I am sure that long before Fermi finally said, 'George pull it out another ten inches,' the question had long since been settled in his mind, and it had long since settled in mine, too."

At 11:15 and 11:25, the rod was inched out again. Each time, Fermi showed his viewers where the instruments would fall, and each time he was correct. The controlled and deliberate experiments he had been known for all his life could not be more apparent than now. Everything was being double-checked. At any moment, the pile would self-sustain.

At 11:35, the automatic, solenoid-controlled ZIP rod was removed. The ratcheting of the counters sounded like a motor come to life. The team watched the graph paper's spot where Fermi had said critical would be recorded.

Suddenly there was a crash, and the entire team fell into shock. Then, all at once, they realized that the automatic rod had fallen into place. The solenoid's activation threshold had been set too low. After that problem was fixed, Fermi said, "I'm hungry. Let's go to lunch."

While eating, the team and their visitors discussed everything but the experiment.

At 2:00, they reassembled, and at 2:20, Fermi had Weil move the control rod back to its previous spot. The instruments were rechecked.

Thirty minutes later, the control rod was pulled another foot, and the counters ratcheted up into a chatter. The pen was thrown off its chart. But this turned out to be another false alarm; like the automatic rod, the pen's ratios needed to be reset to accurately indicate fission.

Herb Anderson: "At first you could hear the sounds of the neutron counter, clickety-clack, clickety-clack. Then the clicks came more and more rapidly, and after a while they began to merge into a roar; the counter couldn't follow anymore. That was the moment to switch to the chart recorder. But when the switch was made, everyone watched in the sudden silence the mounting deflection of the recorder's pen. It was an awesome silence. Everyone realized the significance of that switch; we were in the high-intensity region and the counters were unable to cope with the situation anymore. Again and again, the scale of the recorder had to be changed to accommodate the neutron intensity, which was increasing more and more rapidly."

At 3:20, George Weil at Fermi's instruction moved the rod another six inches, and five minutes later, Fermi asked for another foot, and it happened: Weil withdrew the rod.

Turning to Compton, his boss, Enrico explained, "This is going to do it. Now it will become self-sustaining. The trace will climb and continue to climb. It will not level off."

But as he worked his six-inch ivory slide rule, Fermi's expression seemed to turn grim. He waited a minute, then reran his rule, looking at some numbers he'd jotted earlier on its back side. A few minutes later he looked over the instruments and ran the calculations again. By now, the individual clicks of the counters could not be heard; there was just an insistent buzz, of neutrons attacking nuclei.

Weil: "I couldn't see the instruments [so] I had to watch Fermi every second, waiting for orders. His face was motionless. His eyes darted from one dial to another. His expression was so calm it was hard to read. But suddenly, his whole face broke into a broad smile."

Fermi closed his slide rule and announced with a pleased thrill in his tone, "The reaction is self-sustaining. The curve is exponential."

For four and a half minutes, the group watched the first nuclear chain reactor producing half a watt of power, their eyes focused on the graph pen, which swept upward and never leveled off. Grad student Leona Woods asked Fermi, in the tone of confirming an instrument's readout, "When do we become scared?"

Fermi then turned to Zinn: "Okay, ZIP in." The counters slowed to a fizzle, and the pen stopped its frantic wavering. At 3:53 p.m., it was over. Fermi and Szilard had succeeded in splitting atomic nuclei to produce an immense force. Chicago Pile-1 was the beginning of nuclear medicine, of atomic power and propulsion, and of course the critical start of the Manhattan Project. The Atomic Age was born.

Physicist James Mahaffey: "Fermi's demonstration of controlled, sustained chain-reacting fission in uranium is the most strangely flawless experiment on record. It was run with 42 witnesses in attendance, dressed in business suits, who watched the world's 1st operating nuclear reactor do exactly as it was supposed to do. There was no ambiguous evidence, no competing team in another country, no contradictory data, no fudge numbers, and there was no reason to run it a second time to confirm anything. It was simply perfect."

Eugene Wigner: "Nothing very spectacular had happened. Nothing had moved and the pile itself had given no sound. Nevertheless, when the rods were pushed back in and the clicking died down, we suddenly experienced a letdown feeling, for all of us understood the language of the counter. Even though we had anticipated the success of the experiment, its accomplishment had a deep impact on us. For some time we had known that we were about to unlock a giant; still, we could not escape an eerie feeling when we knew we had actually done it. We felt as, I presume, everyone feels who has done something that he knows will have very far-reaching consequences which he cannot foresee."

Emilio Segrè: "Probably for Fermi, however, the real victory in the making of a natural uranium reactor had come a few months earlier when he succeeded in building a lattice with $k > 1$, which was tantamount to reaching criticality. In October 1942 while I was in Chicago on a laboratory errand, he locked me up in a room alone to read a few reports on his work. After an hour or two he returned and found me rather speechless and with bulging eyes. Of course I knew of the attempts to obtain a chain reaction with natural uranium, but I had no precise idea of how far the work had proceeded, although my own work was dependent on the production of plutonium and hence on a functioning nuclear reactor. The progress reports I read impressed me as though I had seen a critical pile with my own eyes."

Eugene Wigner brought out a bottle of Chianti, which he'd been hiding behind his back, to honor the Italian's success. The bottle was quite a sign of confidence; Wigner had bought it before America entered the war and Italian imports were banned. Enrico popped it open and passed out paper cups for everyone to have a sip. There were no toasts. Later, however, everyone would autograph the bottle's straw casket.

Arthur Compton called James Conant at Harvard on the phone. "The Italian navigator has landed in the New World," he said, in prearranged code, barely able to conceal his excitement.

"How were the natives?" Conant asked.

"Very friendly."

Arthur Compton: "One of the things that I shall not forget is the expressions on the faces of some of the men. There was Fermi's face—one saw in him no sign of elation. The experiment had worked just as he had expected and that was that. But I remember best of all the face of Crawford Greenewalt. His eyes were shining. He had seen a miracle, and a miracle it was indeed. The dawn of a new age. As we walked back across the campus, he talked of his vision: endless supplies of power to turn the wheels of industry, new research techniques that would enrich the life of man, vast new possibilities yet hidden."

Leo Szilard: "There was a crowd there, and when it dispersed, Enrico Fermi and I remained. I shook hands with Fermi and I said that I thought this day would go down as a black day in the history of mankind. I was quite aware of the dangers. . . . But I was also aware of the fact that something had to be done if the Germans get the bomb before we have it. They had knowledge. They had the people to do it and would have forced us to surrender if we didn't have bombs also.

"We had no choice, or we thought we had no choice."

6

The Secret of All Secrets

DURING that autumn of 1938 when Enrico Fermi won his Nobel and exiled his family to America, fellow *ragazzo Corbino* Emilio Segrè was a visiting professor at Berkeley, who learned through the newspapers about Italy's new anti-Semitic laws, which meant he was now both a man without a job, and a man without a country. In Palermo the year before, Emilio had done what Enrico had failed to do by discovering technetium, the first of what would be an avalanche of human-engineered additions to the periodic table, with Segrè an essential figure in many of those breakthroughs. The head of Berkeley's Radiation Laboratory, Ernest Orlando Lawrence, gave Emilio a job. But what Segrè called "the Cyclotron Republic" paid him so little that he soon left for the Republic's competitor, the University of California at Berkeley physics department, even though Lawrence's work was very, very interesting.

Cyclotrons were a racetrack of vacuum tubes edged with magnets and coils—similar to the cat toy that traps a Ping-Pong ball in a spinner track—which pushed and pulled subatomic particles faster and faster, until they were a bright blue beam. Ernest Lawrence described it to would-be investors as a "proton merry-go-round." Every lab tool in the Cyclotron Republic, from wrench to ruler, had to be made of rust-resistant silicon bronze, with all other steel items banned, from watches to belt buckles, key chains, tie clips, boot tips, and even buttons, since the machine's magnets were so powerful they could snatch anything steel with immense greed, potentially damaging the beam window, or the accelerator's delicate mechanisms, or various graduate students in the way.

Known as Maestro or Boss by his subjects, Lawrence was a South Dakota–bred Norwegian Lutheran who'd financed his college education by selling pans door-to-door. In a commencement address, he borrowed from Pasteur and called laboratories "temples of the future—temples of well-being and happiness." When his beloved cyclotrons failed, however, as they often did, veins would pop out in his temples and he would bellow, "Oh, sugar!" Besides inventing the proton merry-go-round, he was even better than Marie Curie at getting financed. With money from California banker William Crocker, he built the sixty-inch "Crocker Cracker," but then Ernest found the golden ticket when he convinced Wall Street tycoon Alfred Lee Loomis to turn away from MIT's original Rad Lab (which pioneered radar and developed the first worldwide radio navigation system—LRN, Loomis Radio Navigation—the most widely used navigation system until GPS) to finance and create the biggest cyclotron on the planet. It would be named for its state of birth—the calutron—and one of its cyclotroneers would be Frank Oppenheimer.

At this time, though, Lawrence and his Republic were less than successful in getting results out of their big science machines. The Berkeley cyclotron had been creating artificial radiation for at least a year when the Joliot-Curies announced their breakthrough, but no one had looked at the instruments after the machine was fired down, so no one noticed that the cyclotrons were irradiating everything, including, they now discovered, the cyclotroneers' spare change and tooth fillings. Ernest wasted no time in bandwagoning himself onto the Curies' discovery, curving his machine's focus to generate medical radioisotopes, with his brother John developing cyclotron isotope science into a profit center in an annex known for its resident test subjects—not the Rad Lab, but the Rat Lab. Ernest had become by this time so financially well-endowed and so politically powerful that Berkeley's physics department chairman said that they were less a university with a cyclotron than a cyclotron with a university.

Ernest Orlando Lawrence's best friend, Julius Robert Oppenheimer, was his polar opposite—wealthy, Jewish, of Riverside Drive and Harvard. The two shared bright blue eyes and frequently double-dated, camping in Yosemite and horseback riding through the Berkeley hills. When Oppenheimer—called Bob by his close friends as a child, and Robert or Oppie as an adult—chalked up a notice for a Spanish Loyalists benefit on the Rad Lab blackboard in 1940, Lawrence erased it, yelling that the laboratory was no place for politics. The incident was noted by Luis Alvarez—who was not

Hispanic, but of Irish descent, his name pronounced Louie, and who would in time be best known as one of those theorizing the extinction of dinosaurs from extraterrestrial collision—as the first time he had ever seen the two men fight. They were so close that Lawrence even named his second son Robert, much against the wishes of his wife, Molly—she judged her husband's best friend as lightweight, callow, and fundamentally lacking in character.

By calculating the collapse of dying stars, Robert Oppenheimer predicted what would become the pulsar, and his interest in particle physics would influence the next generation of American physicists. Like the *ragazzi Corbino* in Rome, Berkeley physics students and visiting acolytes imitated Robert's style of speech—he murmured while thinking of what to say—as well as his shambling walk, and his baroque chain-smoking. Friend Haakon Chevalier: "He was tall, nervous and intent, and he moved with an odd gait, a kind of jog, with a great deal of swinging of his limbs, his head always a little to one side, one shoulder higher than the other. But it was the head that was most striking: the halo of wispy black curly hair, the fine, sharp nose, and especially the eyes, surprisingly blue, having a strange depth and intensity, and yet expressive of a candor that was altogether disarming." "He wanted everything and everyone to be special, and his enthusiasms communicated themselves and made these people feel special," Frank, Oppenheimer's younger brother, remembered. "He couldn't be humdrum. He would even work up those enthusiasms for a brand of cigarettes, even elevating them to something special. His sunsets were always the best."

Another of Oppie's closest friends was Isidor Rabi, who enjoyed introducing himself to Germans as an Austrian Jew since he knew Austrian Jews were the most hated (he was in fact from an Orthodox Hungarian-émigré family raised on New York's Lower East Side). Rabi: "Oppenheimer was Jewish, but wished he wasn't and pretended he wasn't. . . . [He] never got to be an integrated personality. . . . I remember once saying to him how I found the Christian religion so puzzling, such combination of blood and gentleness. He said that is what attracted him to it. . . . God knows I'm not the simplest person, but compared to Oppenheimer, I'm very, very simple. . . . In Oppenheimer, the element of earthiness was feeble." There was also what Hans Bethe noted, that "Robert could make people feel that they were fools," a point echoed by Emilio Segrè: "Oppenheimer's prestige and ascendancy were great among his close entourage, but he sometimes appeared amateurish and snobbish to people more remote from him, who were not under the spell of his personality. For all his brilliance and solid merits, he had some great defects, which in part account for the mortal enmity by which he was later unjustly victimized. Very conscious of his intellectual distinction, he was occasionally arrogant and thereby stung scientific colleagues when they were most sensitive; furthermore, he was sometimes devious in his actions. All this bore ugly fruit years later." Then there was this item in the local paper:

> J. Robert Oppenheimer, 30, associate professor of physics at the University of California, took Miss Melba Phillips, research assistant in physics . . . for an automobile ride in the Berkeley Hills at 3 o'clock this morning.
>
> He stopped his machine on Spruce Street at Alta Street and tucked a large robe about his passenger.
>
> "Are you comfortable?" Prof. Oppenheimer asked.
>
> Miss Phillips replied that she was.
>
> "Mind if I get out and walk for a few minutes?" he queried.
>
> Miss Phillips didn't mind, so the professor climbed from the auto and started to walk.
>
> One hour and 45 minutes later Patrolman C. T. Nevins found the professor's car and Miss Phillips, still comfortable, dozing in the front seat. He woke her up and asked for an explanation of her early morning nap.
>
> Miss Phillips told her story. Police headquarters was notified that Prof. Oppenheimer was missing and a search was launched.
>
> A short time later the professor was awakened from a sound sleep in his room at the Faculty Club, two miles distant from his auto, and asked to explain.
>
> "I am eccentric," he said.

In 1921, a young woman by the name of Katherine Chaves was told that she was not long for this earth, that soon she would die. Katherine decided to spend the rest of her days as a wife to Winthrop Page, a Chicago millionaire as old as her father, and live out West on the Page family ranch, which lay in a desert of lavender, mariposa, bluebirds, and deer, between the Pecos River and the Sangre de Cristos Mountains, named for their sunsets, when the peaks' snowcaps burned a red both corporeal and incandescent, like the sacred yet potable blood of the Lord. The following year, a pale, neurasthenic Jewish boy with such a serious cough his doctors suspected

TB (but not chain-smoking) showed up to stay at Katherine's Los Piños ranch, and she taught him how to ride a horse through the canyons and across the mesas in every kind of weather. Bob returned to New Mexico with brother Frank, and this time, Katherine Page—whose death would not come for decades, and whose husband would never come West—took them ninety-five hundred feet into the peaks, to a cabin with a fireplace made from clay, surrounded by 154 acres of alpine meadow, fields of clover, and heart-stopping views of the Pecos River and the Sangre de Cristos. "Hot dog!" Robert said. "No, Perro Caliente!" Katherine explained. The two boys convinced their father to rent it, a lease Robert would continue as an adult, until he could buy Perro Caliente for $10,000 in 1947. He and Frank went there every chance they could, living the great guy dream of the American West, riding for thousands of miles all the way to Colorado, living on Vienna sausages, chocolate-covered raisins, cheese, and whiskey. During one stay, Oppie wrote to a friend, "My two great loves are physics and New Mexico. It's a pity they can't be combined," and one trek he took with Katherine was through a volcano crater, the Jemez Caldera, and then through a canyon with a stream, along which cottonwood flourished. The canyon was named for the trees: Los Alamos.

In the spring of 1940, Robert invited Dr. Richard Stewart Harrison and his wife, Kitty, for a vacation at Perro Caliente. At the last minute, the doctor had to regretfully decline, but Kitty went anyway and stayed for two months. At summer's end, Robert called Dr. Harrison to tell him that his wife was pregnant, and the two men agreed that the right thing to do was for Harrison to divorce Kitty and Robert to marry her. When Robert's best friend, Bob Serber, heard the news, he was so shocked that he wasn't sure if Oppenheimer had said he would be marrying Jean Tatlock, the great love of his life for decades, or Kitty. It could've been either.

In 1939, when Niels Bohr publicly revealed Meitner and Frisch's discovery of fission, California's Rad Lab boys knew they'd blown another chance. Luis Alvarez: "I remember exactly how I heard about it. I was sitting in the barber chair in Stevens Union having my hair cut, reading the *Chronicle*, and in the second section, buried away someplace, was an announcement that some German chemists had found that the uranium atom split into pieces when it was bombarded with neutrons—that's all there was to it. So I remember telling the barber to stop cutting my hair and I got right out of the barber chair and ran as fast as I could up to the Radiation Laboratory. And my student Phil Abelson had been working very hard to try and find out what transuranium elements were produced when neutrons hit uranium. And he was so close to discovering fission that it was almost pitiful. I mean, he would have been there, guaranteed, in another few weeks—when I arrived panting from the Student Union with my news about fission, and I played it kind of dramatically. I saw Phil there and I said, 'Phil, I've got something to tell you but I want you to lie down first.' So, he lay on the table (right alongside the control room of the cyclotron). 'Phil, what you are looking for are not transuranium elements, but they are elements in the middle of the periodic table.' I showed him what was in the *Chronicle*, and, of course, he was terribly depressed."

Philip Abelson: "When Alvarez told me the news, I almost went numb as I realized that I had come close but had missed a great discovery. During that day, other members of the laboratory, including Alvarez, prepared experiments to check on the validity of the fission process, of resample, by measuring the energy liberated in a linear amplifier when uranium was exposed to neutrons. For nearly twenty-four hours I remained numb, not functioning very well." Chemist Glenn Seaborg spent all night wandering in a daze through the streets, shocked that they had missed what was so, in retrospect, obvious.

When Alvarez told Oppie about Meitner's fission, Oppie immediately said, "That's impossible," and ran to the blackboard to prove it. But the next day, Alvarez "invited Robert over to see the very small natural alpha-particle pulses on our oscilloscope and the tall, spiking fission pulses, twenty-five times larger. In less than fifteen minutes he not only agreed that the reaction was authentic but also speculated that in the process extra neutrons would boil off that could be used to split more uranium atoms and thereby generate power or make bombs. It was amazing to see how rapidly his mind worked." Ernest Lawrence's response? He needed an even bigger cyclotron than ever as now "prospects for useful nuclear energy become very real!"

On September 3, 1939, news arrived that a Nazi submarine had torpedoed the ocean liner *Athenia*, which was sinking off the shores of Scotland. John Lawrence, Ernest's brother, had been in Britain lecturing on the Rad Lab's medical achievements. He was on that ship and it took two days to learn that he was safe. Previously, Ernest was a die-hard midwestern noninterventionist firmly in the camp of Charles Lindbergh, who believed Europe's troubles were none of America's business. Nearly losing his brother made Ernest fervently change course. Then, on November 9, he won his Nobel for inventing the cyclotron.

In December 1940, Segrè and Lawrence met with Fermi and Columbia's Pegram to mull over the cyclotron's recent breakthrough of forcing U-238 to capture a neutron and produce a new element—#93, neptunium. Besides this new human-made element, neptunium appeared to be regularly half-living up to another new element, #94, as yet undiscovered, but the results were inconclusive. Segrè recommended trying to produce enough #94 to determine its nuclear properties, Lawrence agreed, and on February 23, 1941, Segrè and the Seaborg chemistry group bombarded a series of three hundred pounds of irradiated uranyl nitrate hexahydrate, yellowish crystals similar to rock salt, which had been shipped to California from St. Louis. Along the way, the compound's plywood boxes cracked open in their trucks, spilling out a trail of hot rocks.

With their three hundred pounds, Segrè and Seaborg produced enough #94 to determine a mass of 239. They wondered what it should be called, toying with the names ultimium, extremium, and pandemonium. Paid for by the military, the discovery could not be published in the scientific journals, as it was classified. Simultaneously, Egon Bretscher and Norman Feather at Cavendish Labs in Cambridge also discovered element 94 and also kept it secret, under Britain's Tube Alloys classified fission-research program. Remarkably and in secret for a decade, both teams gave the new element the exact same name (as it follows uranium and neptunium): plutonium. After the war, America beat the British in declassifying the discovery, meaning Seaborg got the 1951 Nobel.

The great majority of the world's uranium has an atomic weight of 238. Much rarer is the variant, or isotope, that can be used to make bombs: 235. It is so rare that Bohr for five years would categorically insist there was no potential in nuclear arms—until he came to America and saw it all with his own eyes. Ernest Lawrence thought his Republic could use magnetic fields to split a beam of ionized uranium in half in his cyclotron track, directing the mildly lighter isotope, U-235, onto a metal collector where its vapor would condense into a green slurry of U-235 and carbon tetrachloride. The slurry was reduced into a soft, silvery U-235 metal, which would then be re-ionized and recollected, "enriching" the uranium, atom by atom. After four months, the Rad Lab had produced two hundred micrograms of 35 percent U-235, and on March 9, 1942, Vannevar Bush reported to FDR that the country should build a $20 million centrifuge, so that by the end of 1943 they could be making a bomb a month that was the equivalent of two thousand tons of TNT. Roosevelt's only reply: "Time is very much of the essence."

Two months later, Oppenheimer was anointed the Coordinator of Rapid Rupture under Compton's Chicago Met Lab and assembled a team to research nuclear explosives, including Hans Bethe, Ed Teller, and Bob Serber. Teller sidetracked the committee's work by constantly promoting the idea Fermi had given him the year before, that a fission trigger of deuterium could initiate fusion, promising a thermonuclear weapon of 1 million tons TNT, a result both much cheaper and much bigger than the fission bomb everyone else wanted to pursue. As part of his argument, Teller then warned that fission weapons might ignite the nitrogen in the earth's atmosphere or the hydrogen in its ocean waters, causing a global holocaust. Hans Bethe was left to do the calculations proving this scenario unlikely.

Ed Teller, meanwhile, now began keeping a chart of thermonuclear-bomb ideas in his office. At the bottom of the list was his most ambitious weapon—the Backyard. The Backyard was so massive that it would probably kill every single living thing on earth, so there was no reason to take it anywhere and drop it on anyone. You could just set it off in your backyard. This was the beginning of the thinking and behavior that would inspire Isidor Rabi to muse, "The world would be better without an Edward Teller."

On September 13, 1942, at 20601 Bohemian Avenue, Monte Rio, California, under a portal that claimed WEAVING SPIDERS COME NOT HERE, a committee of Ernest Lawrence, Lyman Briggs, James Conant, Arthur Compton, and Harold Urey met to discuss the MAUD Committee report—which Briggs had finally passed on to others in the Bush empire—as well as who should lead the American effort to create the atomic bomb. They settled on the 250-pound, eternally enraged bulldog with a distinctively dead-fish handshake, Colonel Leslie Groves. Groves was exhausted by having just built the Pentagon and had accepted orders for overseas combat when Army Services of Supply chief General Brehon Somervell asked him to take on this new and overwhelming assignment.

Somervell: "The Secretary of War has selected you for a very important assignment, and the President has approved the selection. . . . If you do the job right, it will win the war."

Groves, upset over being reassigned away from combat, replied, "Oh. That thing."

Newly promoted to brigadier general, Leslie Groves arrived at the Rad Lab expecting to hire Ernest Lawrence to be his head of scientific research and development. But Lawrence didn't want the job, and someone else did. Robert Oppenheimer understood that, at thirty-eight, his best years as a physicist were behind him, and that as a scientist he was first-rate but not first rank among the giants of his era. If he could not be a genius, then at least he could be a handmaid to genius.

Even with the Met Lab's consolidation under Arthur Compton, the Coordinator of Rapid Rupture told Groves that having scientists all across the United States working separately on fission, with none of them communicating due to security, was uncoordinated, inefficient, diffused, and certain to slow progress. Oppenheimer insisted that R&D needed to be consolidated, in a compound isolated from the world. He then sang to the general what was, to Groves, the most beautiful of music: in such a location, security would be drastically improved.

Secretary of War Henry Stimson called Groves the most security-conscious man he'd ever met, and Groves's use of compartmentalization, secret budgets, and lack of financial or other control by Congress would be implemented by his many employees who went on to careers in the CIA at the dawn of the Cold War. Emilio Segrè: "When Groves saw that the usual security rules would preclude recruiting those he wanted, he invented new rules. Each of us was to guarantee some colleague he knew well. 'Guarantee' sounded good, but how? Somebody proposed an oath on the Bible, but Groves objected: 'Most of them are unbelievers.' An Intelligence officer then proposed an oath of personal honor, but Groves replied, 'They do not have any sense of honor. Rather,' he concluded, 'let them swear on their scientific reputation. It seems to me that is the only thing they care for.' "

Oppenheimer did everything he could to get the job and make the partnership work—significantly, when Groves asked stupid questions, Oppie never made him feel like an idiot, as he would have with anyone else. Famous at Berkeley for an office floor covered in enormous piles of paper, Robert knew Leslie thought a tidy desk meant a tidy mind and maintained a spick-and-span charade for his boss. They could even josh. When Groves complained that Robert's porkpie hat was becoming so well known it was a security risk, for their next meeting Oppenheimer wore an eagle-feathered Indian headdress and asked, "Is this better, sir?" (Oppie's hat became so famous that, after the war, it was featured on a magazine cover, by itself.) Groves said of Oppenheimer, "He's a genius. A real genius. While Lawrence is very bright, he's not a genius, just a hard worker. Why, Oppenheimer knows about everything. He can talk to you about anything you can bring up. Well, not exactly. I guess there are a few things he doesn't know about. He doesn't know anything about sports." But perhaps the greatest reason for their success was that Groves could not pull rank on titans such as Lawrence, Fermi, Bohr, von Neumann, or any of the

cs whose work was crucial to Groves's success. But knowing
other sci f Oppenheimer's Communist past—Kitty Oppenheimer was a "[illegible]ying" Communist during the 1930s; Frank and Jackie Oppen-[illegible] were members from 1937 to 1941; Robert contributed without [illegible]g on, calling himself a "fellow traveler," until 1942—Groves could [illegible]xactly that with his director.

The Army Corps of Engineers had an office for building airfields and shipping ports on 270 Broadway at Chambers Street in New York City, and when Vannevar Bush put the atom bomb under the Corps for financial reasons, the headquarters were set up in the same building on the eighteenth floor. Originally named the Department for the Development of Substitute Materials, Groves feared such a provocative title might draw the attention of foreign agents and instead followed Corps procedure by generically calling it the Manhattan Engineer District, eventually known as the Manhattan Project. The building of the Pentagon had required 1,300 men; Manhattan would employ 130,000. As the war progressed, Oppenheimer chain-smoked and grew more wizened; Groves ate chocolate pecan turtles and enlarged.

After all their difficulties in Chicago, Enrico Fermi never worked with Leo Szilard again on a joint experiment, though both of their names would be on the 1955 patent for the nuclear reactor. To prove how prophets are without honor in modern times, quickly in his new assignment Groves became absurdly convinced Szilard was a Nazi spy and kept him under twenty-four-hour surveillance. When agents complained that following Leo around was pointless, Groves insisted, "The investigation of Szilard should continue despite the barrenness of the results." After hundreds of man-hours, the FBI uncovered such damning facts as that Szilard ate breakfast in drugstores and other meals in restaurants, got shaved in a barbershop, walked often when taxis weren't available, and read *Newsweek*.

But this was not the end of Szilard's troubles with authority. When a raging conflict over the engineering of piles among nuclear physicists, the DuPont, Stone + Webster contractors, and army engineers escalated—with Szilard as reagent—Compton was forced on October 26, 1943, to ask Leo to work for Met Lab as a consultant from Columbia, leaving Chicago in forty-eight hours. In reply, Szilard asked for patents and royalties on all his inventions being used by the government. Two days later, fearing that Szilard's exit would create a scientific brain drain of his colleagues, Compton accepted a truce with his gadfly and his engineers—Leo becoming chief physicist at Met Lab with a monthly salary of $950 and a year's back pay, the army getting his patents in exchange for his costs of $15,416—and cabled Groves

that all had been resolved. Groves, meanwhile, on October 28, d a letter for Secretary of War Stimson to sign, having Szilard interned fd
of the war as an enemy alien. Stimson refused, which only further int st
the always-infuriated Groves.

The Coordinator of Rapid Rupture found for his secluded compound a boys school out West the Army Corps of Engineers could buy, which would coincidentally realize Oppenheimer's lifelong dream of uniting physics with New Mexico. But what did Groves think? It was in the middle of nowhere, connected to the world with one dirt road and one decrepit phone line, and in terms of security was absolutely perfect. When the town opened for business on March 15, 1943, it was home to a hundred employees. Six months later, a thousand; one year later, thirty-five hundred; and by the summer of 1945, six thousand. Physicists and their families arrived to find an entire burg painted army green, and homes furnished with refrigerators, woodstoves, water heaters, and fireplaces, but no telephones. It would be known as the Project, the Mesa, the Hill, Site Y, and Shangri-La—Los Alamos.

Robert Oppenheimer: "The prospect of coming to Los Alamos aroused great misgivings. It was to be a military post; men were asked to sign up more or less for the duration; restrictions on travel and on the freedom of families to move about would be severe. . . . But there was another side. Almost everyone knew that if it was completed successfully and rapidly enough, it might determine the outcome of the war. Almost everyone knew that it was an unparalleled opportunity to bring to bear the basic knowledge and art of science for the benefit of his country. Almost everyone knew that this job, if it were achieved, would be part of history. This sense of excitement, of devotion and of patriotism, in the end prevailed."

Mathematician Stanislaw Ulam: "People I knew well began to vanish one after the other, without saying where. . . . Finally I learned that we were going to New Mexico, [so] I went to the library and borrowed the *Federal Writers' Project Guide to New Mexico*. At the back of the book, in the slip of paper on which borrowers signed their names, I read the names of Joan Hinton, David Frisch, Joseph McKibben, and all the other people who had been mysteriously disappearing to hush-hush war jobs without saying where." Hans Bethe: "I went by train to a place called Lamy, New Mexico, which was the railroad station for Santa Fe. Later on, there was a story about some people who went to the railroad station in Princeton to buy tickets to Lamy, and the ticket seller told them, 'Don't go there. Twenty people have already gone there, and not one of them has ever come back.' "

In Lamy, the Super Chief streamliner stopped at a one-room adobe station sitting hard by an empty road of tumbleweeds and an eternal hot, dry wind. Two scientists arrived, couldn't find the deliberately inconspicuous Army Corps of Engineers offices at 109 East Palace Avenue they'd been told to report to, and had to call back to Met Lab: "Where the hell are we?" Civil engineer Joe Lehman: "I drove right through the square and kept on going looking for the main part of town and I found myself out in the desert again. So I turned around and came back and finally pulled into a service station right off the square and asked where downtown Santa Fe was. The guy said, 'This is it. You're in it.' "

The scientists and engineers arrived to find a situation so primitive that the telephone network was a single line to the outside world shared by twelve parties, gnawed at by chipmunks, so static-filled a caller needed to shout to be heard, and unusable when it rained for fear of being electrocuted. The quality was so poor that at one point, the secretaries were once flummoxed by a call requesting eight extra-large trucks. It was actually for eight extra lunches.

It was, however, all that Oppie had said it would be. "Here at Los Alamos," one tough British physicist said, "I found a spirit of Athens, of Plato, of an ideal Republic." Ed Teller: "In spite of the difficulties, I (and many others) consider the wartime years at Los Alamos the most wonderful time in our lives." Hans Bethe: "It was an unforgettable experience for all the members of the laboratory. There were other wartime laboratories of high achievement. . . . But I have never observed in any one of these other groups quite the spirit of belonging together, quite the urge to reminisce about the days of the laboratory, quite the feeling that this was really the great time of their lives. That this was true of Los Alamos was mainly due to Oppenheimer." Dick Feynman: "Oppenheimer was very patient. He paid attention to everybody's problems. He worried about my wife, who had TB, and whether there would be a hospital out there, and everything. . . . He was a wonderful man." Well, not everyone thought this was the great time, as Leslie Groves told his officers: "At great expense, we have gathered on this mesa the largest collection of crackpots ever seen."

They were all so young, and all so full of vim—not representing the grand old men of science, the average age of those at Los Alamos was twenty-seven. Many of the accompanying wives were upset to see all the bedding stamped USED, until it was explained that the organization running the compound was the United States Engineer Detachment. The water that came out of the tap could be accompanied by rust, algae, dirt, or worms, and sometimes was so chlorinated that it dissolved the wives' stockings. Regardless, in eighteen months the water supply was so thinning that cars could not be washed, lawns could not be watered, showers were supposed to last only a few minutes, and all were warned to "REPORT LEAKY FAUCETS IMMEDIATELY!" Men gave up shaving, and women made do with less laundry and shampooing by switching to overalls, and pulling their hair back into bandannas. Everyone joked that they all started looking like the pictures of the first down-on-their-luck Santa Fe prospectors, and if the base were open to the public, they would think it was all an operation run by and for cowboys. Bethe: "The water supply had been built for the Los Alamos school, which had maybe fifty people, and there were several thousand of us. There just wasn't any water. We got water delivered in trucks that had previously been used to transport gasoline, so the water generally tasted of gasoline."

Tewa-speaking Indian women in deerskin boots, Hopi shawls, and glossy hair piled up or pigtailed were bused in to the mesa to work as maids, cooks, and nannies. One had a mother-in-law with a reputation as a good potter, and the secret spread; her talent and her connections made her a fortune in blackware.

Emilio Segrè sent a letter to his wife back in California and included a strand of his hair. She opened the letter, the hair was missing, and that was how the physicists discovered their mail was being read. A number of the émigrés were additionally unhappy to be surrounded by a razor-wire fence, which reminded them of Nazi prison camps. Junior physicist Richard Feynman discovered, however, that workers had put a hole in the Los Alamos fence as a shortcut, and he enjoyed exiting the property through that hole and then coming in through the security entrance, so the guards would always see him arriving, but never departing. The kids, too, discovered every hole the fence had to offer, as well as a treasure pile of toys to play with—lab equipment left out in the trash.

In November 1943, James Chadwick invited Otto Robert Frisch to join the thirty British scientists who would be going to America. Otto replied, "I would like that very much," and when it was explained that he would need to first become a British citizen, he said, "I would like that even more." Chadwick then invited Lise Meitner from Sweden, and the discoverer of fission said, "I will have nothing to do with a bomb." Accompanied by Rudolf Peierls and Klaus Fuchs on December 3, 1943, the group of thirty found no available taxis in London and had to pay a hearse driver to take them to the wharves. When they then crossed the United States by rail and stopped in Richmond, Virginia, for a meal, Frisch became ecstatic at the sight of fruit stands blooming with immense pyramids of oranges. He hadn't seen an orange in three years. Arriving in New Mexico, they received the standard hello from Oppenheimer: 'Welcome to Los Alamos. Who the devil are you?' "

Oppenheimer did everything he could to get his friend Isidor Rabi to join, but Rabi, like Meitner, refused, telling Oppenheimer that he did not want to make "the culmination of three centuries of physics" a horrible weapon. He did not believe in bombs, feeling that they killed both the innocent and the guilty, and continued his work on radar at Columbia. He did, however, travel to the Los Alamos sporadically to offer advice as a visiting consultant, as did Einstein.

North-central New Mexico is today as green as any artificially irrigated desert in the world, but during World War II it was home to columbines, gentians, ponderosa, aspens, indifferent porcupines, tender marmot, determined badger, and fearless skunk. Ruth Marshak: "Behind us lay the Sangre de Cristo Mountains, at sunset bathed in changing waves of colors—scarlets and lavenders. Below was the desert with its flatness broken by majestic palisades that seemed like the ruined cathedrals and palaces of some old, great, vanished race. Ahead was Los Alamos, and beyond the flat plateau on which it sat was its backdrop, the Jemez Mountain Range. Whenever things went wrong at Los Alamos, and there was never a day when they didn't, we had this one consolation—we had a view." Emilio Segrè: "The Mesa was indented by deep canyons, which in time came to be occupied by special laboratories. The region was extraordinarily beautiful. This beauty was to have profound influence on many scientists who by inclination and long habit were sensitive to nature and could appreciate the noble countryside surrounding Los Alamos. The view of the mountains; the ever-bearing clouds in the sky; the colorful flowers blooming profusely from early spring to late autumn; the possibility of walking on interesting trails leading to fishing streams, skiing slopes, mineral beds, or Indian ruins. . . . Often at the end of a strenuous period of work one was completely exhausted, but the out-of-doors was always a source of renewed strength. Any number of spirited discussions were conducted during hikes in the beautiful and savage countryside."

Back in May 1940, what would become the Manhattan Project had been handed a phenomenal twitch of luck. The Congo uranium mine that Szilard and Einstein had warned Roosevelt about had been discovered and managed by a Belgian corporation, the Union Minière du Haut Katanga. When Hitler came to power, Union Minière chief Edgar Sengier assumed the Nazis would have to take Belgium on their way to France, and he didn't want his cobalt—a key ingredient in airplane engines—to fall into Fascist hands. Just before his nation fell in May 1940, Sengier moved himself and his wife to a suite in New York's Ambassador Hotel and renamed Union Minière the African Metals Corporation. Under the nose of the German occupation, he then shipped 1,250 tons of uranium—the whole of the mine's on-hand inventory—out of the Congo and into a vegetable-oil plant storage facility operated by Archer Daniels Midland on Staten Island, New York, as suggested by the Joliot-Curies in Paris and Henry Tizard in Britain. For two years, Sengier worked to sell the whole of it to the American government for the war effort, but no one was interested . . . until September 18, 1942, when Leslie Groves signed an eighteen-year, four-hundred-tons-a-month contract with AMC, which ore would supply two-thirds of the enriched uranium for Hiroshima's Little Boy and a substantial amount of the plutonium for Fat Man at Nagasaki. When the Belgians declared Congo independent in the 1960s, Union filled the mine with cement, but as of this writing illegal miners are digging away and exporting their ore through Zambia. The mine is known as Shinkolobwe, named after a fruit that has to be boiled before it can be eaten. If you try to eat some without letting it cool, you will get burned, and *shinkolobwe* is also the term for a man who is, like uranium, cool on the outside but boiling from anger within.

During the war, over a million people would know that the Manhattan Project existed; it was the job of the Counter-Intelligence Corps (known by their colleagues as the Creeps) to keep those people from knowing anything else. The cops created posters—DON'T BE A BLABOTEUR—and invented a code:

topic boat	atomic bomb
urchin fashion	nuclear fission
igloo of urchin	isotope of uranium
tenure	U-235 (2+3+5 = 10)
Henry Farmer	Enrico Fermi
Eugene Samson	Emilio Segrè
Oscar Wilde (who wrote *The Importance of Being Earnest*)	Ernest Lawrence

The Los Alamos physicists came to believe that their Creep bodyguards had remarkable powers. "They had this notion that we could know whatever they were doing even if they didn't tell us," one agent was surprised to understand. Many of the scientists behaved erratically, even life-threateningly—notably Fermi, with his perilous driving skills—and the agents became alarmed, since besides spying on their scientists, a key part of their assignment was to keep their charges alive until the Bomb was born.

Like all the rest of the Creeps, Fermi's bodyguard and chauffeur, the Italian-speaking American, halfback-size lawyer John Baudino, filed weekly reports on his subject with army intelligence. To keep Fermi from talking about his work to others, Baudino got him to talk about it to him. So Fermi started saying that, "Soon Johnny will know so much about the project he will need a bodyguard, too."

When agent Charles Campbell, who hated physics but pretended to like it as part of his job, mentioned to John von Neumann that he was too busy to study, von Neumann got upset: "It is my fault! You will come with me and together we will study theoretical physics in New Mexico!" FBI surveillance teams used walkie-talkies disguised as hearing aids, and any assembled together looked conspicuously like an outing of deaf people.

The same month that Los Alamos opened, the FBI suspended its surveillance of Oppenheimer at the insistence of the army. No more help was needed from the Feebs as the Creeps had placed agent Andrew Walker as Robert's driver and bodyguard, had wiretapped his home and office, and read every piece of his mail before he did. By the summer of 1943, JRO was so sick of this relentless snooping that he thought he should quit.

The military police stationed outside the Oppenheimer house on Bathtub Row insisted on seeing everyone's pass before letting him or her enter, including, repeatedly, Kitty Oppenheimer, who often forgot to take her pass with her when she left the house and made a huge fuss when the men interfered with her coming back. She got the detail canceled, however, when the sergeant in charge learned she was using the MPs to babysit the Oppenheimers' son, Peter.

Phil Morrison's wife, Emily: "Kitty was a very strange woman. . . . She could be a very bewitching person, but she was someone to be wary of. . . . She would pick a pet, one of the wives, and be extraordinarily friendly with her, and then drop her for no reason. She had temporary favorites. That's the way she was. She did it to one person after another." Oppenheimer did nothing in the face of his wife's caprices, even when she "threw a hate on Charlotte" Serber, whose husband was one of his closest friends.

When Elinor Pulitzer, newspaper heiress, married Los Alamos medical director Louis Hempelmann, she decided to throw a big dinner party on the mesa, but didn't know that the high altitude meant longer roasting times. While all her guests waited for their food for hours, they drank and drank until Elinor developed a reputation as a spectacular hostess.

Single men and women lived in dormitories, where parties were marked by lab-created 200-proof rotgut ladled from reagent jars, Mexican vodka brought in from Santa Fe, and, if they were lucky, visiting consultant Isidor Rabi playing the comb. The marrieds dressed up and went square dancing; the singles gathered at the two PXs for hamburgers and the jukebox; there were group drives, horseback riding, fishing, mountain climbing, poker, amateur theatricals and concerts, golf, baseball, softball, basketball, and a women's dorm that for a time became a brothel. So many young people trapped behind barbed wire led to a baby boom. When Groves realized that one-fifth of the married women on the base were pregnant, he insisted that Oppenheimer do something about it. But, as Kitty herself was expecting, there was little he could do, other than inspire mesa-wide sing-alongs:

The General's in a stew
He trusted you and you
He thought you'd be scientific
Instead you're just prolific
And what is he to do?

When a study was published linking a decrease in male fertility with long, hot baths, copies were immediately forwarded to Groves in the hopes of getting more tubs. But all the new children led in turn to an outpouring of social effects: a newspaper, library, barbershop, radio broadcast, church services, golf course, baseball field, and, on a military site, a democratically elected town council. Family life brought civilization to the desert, and a real sense of community to Los Alamos.

Naval officer Deke Parsons: "You know, usually on a military post the commander is the social arbiter and top dog. It's really sort of hard for the military here because everybody looks down on them." Good news from the fronts would cheer the civilians and dispirit the junior military, who felt gypped at being stationed in the desert and missing out on the action. Later in the war, when men arrived with ribbons from Anzio, though, they were clearly grateful to be stationed at the peaceable mesa.

In March 1943, Hans Bethe was at MIT's Rad Lab working on radar, after escaping from Germany through London: "Though I had an excellent time with my colleagues and my friends in England, it was clear there that I was a foreigner and would remain a foreigner. In America, people made me feel at once that I was going to be an American—that maybe I was one already. I felt that Germany was much stranger than America—that it was a weird country. It was clear that according to the laws, I could not hold a university position, because two of my grandparents were Jewish. The first I heard about this directly was when one of my two Ph.D. students in Tübingen wrote me a letter saying, 'I read in the papers that you have been dismissed. Tell me, what shall I do?' What should he do? I had not heard of my dismissal, but it had been published in the papers."

To help with the war effort, in 1941 Bethe had produced a theory of armor penetration, which was published by Philadelphia's Frankford Arsenal as confidential, meaning Bethe himself wasn't allowed to read it. Two years later, he got the call from Oppenheimer and agreed to come to Los Alamos. On the way, Bethe passed through Chicago, picked up Ed Teller, and visited Enrico Fermi. By April, Teller's wife, Mici, and two-month-old son, Paul, arrived, with a Steinway concert grand acquired from a hotel sale and a Bendix automatic washer. The Hungarian physicist played his remarkable instrument at all hours of the day and night, which alternately enthralled and irritated his neighbors. Teller described himself as "choleric"—tending to anger—and when he played Mozart, it was always fortissimo.

Ed Teller: "The Army routinely leveled every bush and tree within two hundred feet of a building site before beginning construction, thereby destroying whatever attractiveness the immediate surroundings had. Yet, throughout the war, our apartment building had a pretty stand of pines growing between us and the Canyon. On the morning the bulldozers appeared to level the area near our apartment as part of the next construction project, my wife had spread a blanket under the trees and settled herself, Paul, his diapers and bottles, and a picnic lunch on it. The young soldier responsible for the bulldozer asked her politely to move; she, just as politely, refused. He leveled the rest of the surroundings and returned to ask again—to no avail. Finally the soldier went to Gen. Groves for advice. 'Leave the trees,' Groves grumbled on hearing about the situation."

With a face like an abdominal muscle foreshadowed by a prow of beetle brows, Hungarian Quartet member Edward Teller would, over the decades, become the Dr. Strangelove of this history and the Saruman of this real-life *Lord of the Rings*, with a fairly distinct personality: "Fermi once told me with hardly a trace of a smile that I was the only monomaniac he knew with several manias. My grandson, my son, and my editor-collaborator all claim that the film character *ET* and I have more in common than our initials." In 1928, Teller fell on the street, and a trolley ran over his foot, leaving him with a lifelong limp and a lifelong friend and nemesis, Hans Bethe, who visited him in the hospital. In the 1930s, Ed got a grant to study in Rome after Fermi "wrote an official letter to the Hungarian government. He called me a great physicist (which I certainly was not), asked for the privilege of collaborating with me (which could hardly have been a privilege for him considering my ignorance of his area of study), and expressed the hope that some means could be found by the Hungarian government to make my stay possible. . . . I do not believe that as a total stranger I have ever been more warmly welcomed."

Edward Teller had helped Oppie organize Los Alamos and recruit, but then became notably upset when the sparkling Bethe was named head of the Theoretical Division, as Bethe himself admitted: "Teller had worked on the bomb project almost from the day of its inception and considered himself, quite rightly, as having seniority over everyone then at Los Alamos, including Oppenheimer." Teller didn't like working under Bethe, didn't like military secrecy, didn't like collaborating, and didn't like the lack of support for his thermonuclear fusion design, the Super. Still, at that time, he was a true Oppenheimer acolyte. Ed Teller: "Throughout the ten years, Oppie knew in detail what was going on in every part of the Laboratory. . . . He knew how to organize, cajole, humor, soothe feelings—how to lead powerfully without seeming to do so. He was an exemplar of dedication, a hero who never lost his humanness. Disappointing him somehow carried with it a sense of wrongdoing. Los Alamos' amazing success grew out of the brilliance, enthusiasm and charisma with which Oppenheimer led it." Yet, after one outrageous incident with the prima donna Teller and his obsession with fusion over fission, Robert said to Charles Critchfield, "God protect us from the enemy without and the Hungarians within."

In August of 1944, the Fermis were told that they, too, would be joining the exodus to Site Y in New Mexico's Pecos Valley, with J. Robert Oppenheimer himself visiting them in Chicago to explain what it was like, and what they could expect. Enrico would be heading up the F division, the *F* standing for "Fermi," which would be a team of freelance troubleshooters, including Herb Anderson and Ed Teller—getting Teller out from under Bethe—working on any advanced problems as required. Laura remembered Oppenheimer's saying that all his friends called him Oppie and they should, too. "He made it all sound just wonderful." Oppenheimer: "Fermi was simply unable to let things be foggy. Since they always are, this kept him pretty active." In turn, Enrico said that Robert had a remarkable talent for appearing far more knowledgeable about a given topic than he actually was.

At the last minute in getting the family ready to head west, Enrico was called away to oversee the final building of his next-generation piles in Hanford, Washington, so Laura and the children had to go on without him. But when they arrived at Lamy in August 1944, Laura hadn't been told of her alias, "Mrs. Farmer," and so kept telling the WAC driver sent to meet them that, though she wasn't as important as this Mrs. Farmer, maybe she could get a ride in the car since the VIP hadn't shown up.

When Enrico then arrived in September, he and Laura turned down the school faculty cottage that had been set aside for them to take an ordinary apartment, to make a statement about the castes forming in the mesa. Physicist Bob Wilson's wife, Jane: "Below her, she had Rudolf and Gennia Peierls. They were very good friends, but Mrs. Peierls had a very loud and piercing voice, and her voice came wafting up through the floorboards—her laughter, too. Actually, the first time I met Laura, she was with Gennia Peierls, they were good friends, and I noted in my diary that every time Gennia spoke (and Gennia was a person of many words) Laura visibly shuddered. . . . I was somewhat surprised, having known Enrico off and on for quite a while, to find, after this ebullient, extroverted man, this very serious, shy, reticent woman. I think Enrico prided himself on being matter-of-fact and pragmatic, and Laura was unabashedly idealistic. . . . [She would say,] 'We wanted to become genuine Americans.' I don't think her American dream was the Mercedes in the garage or the mansion in the suburbs; it really was the Jeffersonian ideal. . . . In Italy, this lady had been blessed with two maids; she had been raised in a wealthy household where the maids did the housework, her mother picked out her clothes, and she didn't know anything about money. Even when she married Enrico, on a salary of $90 a month, she had a maid. Then here she was at Los Alamos, with amenities few and far between, which Chadwick had characterized as 'pigging it,' and we had famous water shortages always, incredible dust, and limited supplies. As for maids, she was lucky to have an Indian maid a couple hours, several times a week. Even under those circumstances, she had one dinner party after another, with good food, and really, it was remarkable. I think she liked it."

Mici Teller took Enrico Fermi for his first ride on a horse, and "he told the horse in a firm voice, 'I am the boss.' It worked."

Emilio Segrè: "I required a special small laboratory for measuring spontaneous fission, the like of which I have never seen before or since. It was in a log cabin that had been occupied by a ranger and was located in a secluded valley a few miles from Los Alamos. It could be reached only by a Jeep trail that passed through fields of purple and yellow asters and the canyon whose walls were marked with Indian carvings. On this trail we once found a large rattlesnake. The cabin laboratory, in a grove protected by huge broadleaf trees, occupied one of the most picturesque settings one could dream of. Fermi was very fond of the site and visited us there several times."

The great secret that made America the leader in nuclear warfare and gave it an atomic monopoly for many years took place far, far away from the glories of New Mexico. Though the scientists and engineers of that desert paradise engineered the method of uranium and plutonium bombs and ensured their place in history, America's real nuclear breakthrough was taking place in Godzilla-size buildings in the hollows of Tennessee and the inland gulches of Washington State. Not every nation in the world has an atomic stockpile because, though it is simple enough to make a bomb, it is not so easy to fashion weapons-grade fissile fuel in the form of plutonium or highly enriched uranium (HEU)—the isotope U-235. That was America's great "atomic secret"—the secret that kept others from developing their own nuclear arsenals—the making of plutonium from reactors and enriched uranium by separating U-235 from U-238, using the 2 percent difference of their weights.

Jim Conant was told he needed to pick between three methods of isolating U-235 (gas, electromagnetic, or centrifuge), or of building a nuclear reactor to generate plutonium. He chose all four, the U-235 done at Oak Ridge, Tennessee, and the plutonium at Oak Ridge, in the Chicago suburb of Argonne, and at Hanford, Washington.

Fermi's squash-court reactor, now called Chicago Pile-1, was dismantled and reassembled at Red Gate Woods in February 1943—the spot in the Argonne Forest twenty-five miles southwest of Chicago where it was supposed to have been built in the first place. Now known as Chicago Pile-2, it was still overseen by Fermi and produced plutonium for Los Alamos. After the war, her descendants at Argonne would be the research vessels for modern-day reactors and the power plants of nuclear submarines.

In one corner at Columbia, Harold Urey (Enrico and Laura's Leonia neighbor) and a team of chemists combined natural uranium with fluorine to produce a uranium gas—uranium hexafluoride—which they centrifuged to separate out the U-235. Months of trials determined that making this centrifuge process work at an industrial level would mean fifty thousand one-meter-rotor machines operating continuously at state-of-the-art speeds to produce a kilo a day. That plan was abandoned; the technology has since so evolved that Urey-style centrifuges are now the most common method of producing HEU; but at the same time that one Urey-Columbia team was working on centrifuges, another was developing diffusion, which used membrane-fine filters to separate the lighter U-235 from 238. It worked, but barely, as the uranium hexafluoride needed to be diffused again and again to produce even a tiny amount of isotope—through thousands of membranes, known as cascades, requiring plants of hundreds of acres in size, with bicycles and cars driven inside to get from pipe to pipe.

Starting around the year 1900, an Elza, Tennessee, prophet named John Hendrix proselytized to his neighbors that the Bear Creek Valley—their four little hamlets of tobacco, corn, and coal in the Cumberland Mountains by the Clinch River forty miles west of Knoxville—"someday will be filled with great buildings and factories, and they will help toward winning the greatest war that ever will be." When in 1942 FDR held a secret meeting with congressional leaders to discuss the Manhattan Project's finances, he asked them to put aside petty local concerns to ensure victory. Senate Appropriations chair and senior senator from Tennessee Kenneth McKellar replied, "Mr. President, I agree that the future of our civilization may depend on the success of this project. Where in Tennessee are we going to build it?" Soon after, the Army Corps of Engineers requisitioned all fifty-nine thousand acres of Bear Creek Valley, with a court order for three thousand to vacate their homes. Groves built at Oak Ridge, Tennessee, the world's biggest building, the U-shaped K-25, finished in the last months of 1943, which cost $500 million and employed twelve thousand to diffuse fissile uranium with the Urey cascade method. Physicist James Mahaffey rhapsodized, "The sight of the inside of the K-25 building, with its clean and orderly maze of gleaming nickel plumbing [the hexafluoride would immediately corrode any other metal] seeming to extend forever and disappearing into the haze at semi-infinite distance, was beautiful." After the war, the sealant developed for the cascades' pumps appeared in American home kitchens, where it was known as Teflon.

At Berkeley, chemist Glenn Seaborg had tried various techniques to extract plutonium from uranium, getting a yield of 250 parts per million, or two tons producing an amount the size of a penny. Testing this required microchemistry, where under 30x power microscopes, capillary straw pipettes became test tubes and a single quartz fiber with a platinum-foil tray was a weighing scale. Next, about twelve miles from K-25, rose the $427 million Y-12 Plant, where thirteen thousand workers ran Lawrence's Alpha calutrons: 3,000-to-10,000-ton, 250-foot-long magnets across ninety-six vacuum tanks in five merry-go-rounds to separate uranium-235. His next-generation Beta calutrons had twenty-five hundred tanks, eighty-five thousand vacuum tubes, were the lengths of four football fields, and employed twenty-five thousand.

Building his magnets required five thousand tons of copper, which in wartime couldn't even be gotten by Leslie Groves, but a substitute was discovered—silver. When Undersecretary of the Treasury Daniel W. Bell was asked for six thousand tons of silver bullion, he erupted, "You may think of silver in tons, but the Treasury will always think of silver in troy ounces!" Eventually, 14,700 tons were needed, costing Groves $300 million. A pickling plant had to be built on-site for cleaning the pipes.

By bending the path of speeding uranium ions, the lighter 235s would pull slightly ahead of the 238s in the proton merry-go-rounds and splat against a target in a slightly different location. This low-production method, with less than 5 percent of the distilled 235 hitting the sweet spot, required constant monitoring of the magnetic pulse to correct for the wavering voltages of the factory's power supply. At first the calutrons were run by Berkeley scientists to work out the bugs; then they were turned over to local women sitting on stools, most of whom were high school dropouts. The two teams had a race . . . and the women won. The management believed it was because they were "trained like soldiers not to reason why," while "the scientists could not refrain from time-consuming investigation of the cause of even minor fluctuations of the dials."

Calutron operator Theodore Rockwell: "If you walked along the wooden catwalk over the magnet, you could feel the tug of the magnetic field on the nails in your shoes. It was like walking through glue. People who worked on the calutrons would take their watch into the watchmaker and discover that it was all smashed inside. The magnetic field had grabbed the steel parts and yanked them out by the roots. . . . One time they were bringing a big steel plate in and got too close to the magnetic field. The plate pinned some poor guy like a butterfly against the magnetic field. So the guys ran over to the boss and said, 'Shut down the magnet! Shut down the magnet! We got to get that guy off.' And the boss replied, 'I've been told the war is killing three hundred people an hour. If we shut down the magnet, it will take days to get restabilized and get production back up again, and that's hundreds of lives. I'm not going to do that. You're going to have to pry him off with two-by-fours.' Which is what they did. Luckily he wasn't badly hurt, but that showed what our priorities were."

By the start of 1944, Urey's membrane cascades were still corroding viciously. The Naval Research Lab's Phil Abelson told Groves about their work with thermal diffusion, which had been pioneered by the omnipresent yet underappreciated fission codiscoverer and U-235 cocalculator, Otto Robert Frisch, and Groves decided to try that method as well. Construction of S-50, a thermal-diffusion plant, started on June 24, 1944, with twenty-one racks of 2,142 forty-eight-foot-tall columns housing three concentric tubes ferrying 545°F steam on the outside, cool water on the inside, and uranium hexafluoride in the middle—the lighter 235 floated up with the rising heat, while the heavier 238 settled down at the colder bottom.

By March 1945, Oppenheimer understood that feeding the various outputs into each other dramatically increased the yield, so the thermal-diffusion output was sent to be gaseously diffused, and those results were in turn racetracked in the calutron merry-go-rounds. The final product was 89 percent enriched—good enough to power the Bomb. Yet, by the summer of 1944, even with this massive effort, the calutrons were producing only enough U-235 to make a single explosive. Because they had so little HEU and the uranium gun bomb was so simple in design, Oppenheimer decided it could be dropped on the enemy without even being tested.

The third Oak Ridge plant, the $12 million pilot nuclear reactor X-10, employed 1,513 people and was built ten miles from Y-12. It was known as the Black Barn and joined Chicago Pile-2 in creating plutonium-239 for the implosion bomb. The Oak Ridge reactor was modified from Fermi's Chicago pile by using channels for replacing exhausted uranium rods with fresh fuel—the depleted uranium used to make armor-penetrating bullets—and pressurized helium as a coolant. Coolant was a topic sure to cause an argument, with Fermi insisting on air, Szilard wanting liquid bismuth (as he and Einstein had used in their refrigerator patent), and Wigner believing in ordinary water from the river. In time, Szilard's technique would be used with breeder reactors, and Wigner's with standard burners.

With Fermi in attendance, X-10 went critical on November 4, 1943. One of its engineers, Arthur Rupp, had not believed in Wigner's calculations of what would happen, yet the results almost exactly matched the Hungarian's predictions. "I knew then," Rupp said, "the atomic bomb was going to work!" X-10 would, for the implosion bomb, produce 326.4 grams of plutonium, and Eugene Wigner would become Oak Ridge's "patron saint."

Though the army was fully against it for security reasons, Emilio Segrè insisted on touring Oak Ridge to make sure it was operating properly. He found that the Tennessee reactor employees thought they could use a small amount of water to contain the extract, having no idea the water would then become lethally radiant. They also didn't realize that storing fuel in adjacent rooms against a shared wall could start a chain reaction. Segrè was horrified, and Oppenheimer sent in Richard Feynman to do a follow-up inspection. Feynman was told by his boss that, if the army refused to listen to the physicist, he should say, "Los Alamos cannot accept the responsibility for the Oak Ridge plant." That suitably alarmed the army bureaucrats, and the physicists' recommendations were followed.

The Oak Ridge reactor was a proving ground for a much bigger operation. The small towns of Hanford, White Bluffs, and Richland in the dry sheep and vineyard inlands of Washington State were evacuated so that fifty-one thousand people—forty-seven thousand of them men—living in a city of barracks and the largest trailer park in history, could build seven nuclear power plants requiring twenty-five thousand gallons a minute of cooling Columbia River water on a half million acres to be known as Hanford Engineer Works. Their bakery made twenty thousand pies a day, and their auditorium seated five thousand for movie nights; the other entertainment was sitting on blankets on the dusty streets, playing cards, gambling, and fighting, the last stopped by security with fire hoses. The work was finished in eighteen months, and their camp was destroyed.

Those going to Hanford imagined it would be like the Washington State of postcards—snow-peaked mountains, crystal mountain streams, great camping, hunting, and fishing. Instead, they arrived at the Columbia Basin desert of sand dunes, saw grass, tumbleweeds, and pygmy rabbits, next to a barren lava plateau—the Scablands. One ordnance émigré from Denver said, "It was so darned bleak. If I'd had the price of a ticket, I wouldn't have stayed." The darned bleak was accompanied by dust storms so ferocious they were known as "termination winds" because so many employees gave up and resigned after suffering through one.

On September 13, 1944, Leona Marshall, Crawford Greenewalt, and Enrico Fermi climbed a twelve-story ladder to survey the two-thirds-finished Hanford site, where Fermi would insert the first uranium slug into the first of the three plutonium-generating reactors. It ran perfectly for twelve hours, then the chain faltered and died. The next morning it was back chain-reacting, but twelve hours later it died all over again. Princeton's John Wheeler theorized that a by-product in the chain was absorbing neutrons, and this turned out to be xenon gas. Wheeler advised DuPont to add more uranium channels, and they tried amending Wigner's 1,500 with an additional 504. This kept the xenon from overwhelming the uranium, and the plutonium was now generated on schedule.

Hanford operations manager Walter Simon: "Fermi was very discreet about disagreements. He was a very pleasant person. His mind raced all the time. For instance, if there was a little time to kill while they were loading the reactor, he would do equations in his head, with someone next to him with a calculator. You know, multiply 999 by 62 and divide this by that, and he did that for amusement. His mind raced so much the only way he could relax was to walk on the desert. They would try to take him to a movie, and he would sit there, and in five minutes he would have the whole plot figured out."

One evening, Sam Allison, Arthur Compton, and Enrico Fermi were taking the train to Hanford, a ride that seemed to go on for a purgatorial eternity. To pass the dead hours, Compton said, "Enrico, when I was in the Andes Mountains on my cosmic-ray trips, I noticed that at very high altitudes my watch did not keep good time. I thought about this considerably and finally came to an explanation which satisfied me. Let's hear you discourse on the subject." Enrico took out his slide rule, a piece of paper, and a pencil and, within a few minutes, had totted up the formula for the interactions of air pressure on a watch balance wheel's inertia, quickly producing a figure that matched the dissatisfactions of high-elevation timekeeping. Allison said he would never forget Compton's amazed look.

Shortly after that first Hanford reactor began to be tested early in 1944, a balloon appeared in the sky. It was one of thousands carrying incendiary bombs sent by Japan to incinerate the American West. Though some of these fire balloons did cause forest fires in Northern California, Oregon, and Washington, this one struck the electric line carrying power to the reactor building and shut it down.

By February 4, 1945, Hanford, Oak Ridge, and Chicago reached their target monthly yield of twenty-one kilos of the flourlike yellow-green plutonium oxide, and a few months later, Oak Ridge began sending its HEU to Los Alamos. When Oak Ridge's U-235 was first unwrapped after reaching the mesa, it was a silvery metal. Contact with the air turned it dawn-sky blue, deepening to cobalt, and then purple. Like the warm, silvery puppy that was plutonium, and the blue-green thrill that was radium, the U-235 also seemed to be, in some way, alive.

Groves now knew that everything was working as he'd hoped and prayed, and that by the end of the year, they would have enough fissile core for eighteen bombs. That is, if the implosion bomb worked at all.

Robert Serber gave the introductory lecture to new arrivals at Los Alamos and, worried about omnipresent USED construction crews able to eavesdrop on every meeting, he called the Bomb "the gadget." Serber estimated critical mass could be arrived at with fifteen kilos (thirty-three pounds) of uranium-235 or five kilos (eleven pounds) of plutonium-239, surrounded by a shell of ordinary uranium. In the two years it would take Hanford, Oak Ridge, and Chicago to amass that quantity of fissile ordnance, Los Alamos would have to design an atomic trigger. Dick Feynman: "All science stopped during the war except the little bit that was done at Los Alamos. And that was not much science; it was mostly engineering."

The labors of the mesa were far more dangerous than any civilian knew. The MAUD report included an atom-bomb gun design, which fired a plug of uranium at a bowl of uranium. This would become Little Boy, used against Hiroshima, and was a very basic idea—U-235 molded into six-inch stackable washers with a four-inch hole in the middle that would fit a four-inch-round plug of the same, also made of disks. One design question was, how much of the rare and precious uranium would the Little Boy gun-style bomb need to work? In 1944, Otto Robert Frisch was assigned this investigation, and his method was, by any standards, hair-raising. In Omega Canyon, so remote as to result in as few casualties as possible if the worst happened, Otto built a small tower, the "guillotine," which dropped plugs of uranium metal through blocks of uranium metal surrounded by tamper—the basic design for the Hiroshima bomb—which went supercritical for less than a second and immediately showed the quantity needed. Dick Feynman said it "was like tickling the tail of a sleeping dragon . . . as near as we could possibly go towards starting an atomic explosion without actually being blown up." After testing with Frisch's guillotine, the amount of each segment could be shimmed by adding additional washers of isotope. In the bomb, a gunlike explosion of cordite would unite the two and could be triggered by pressure or proximity fuse or just about anything else in the Pentagon trigger basket. Little Boy was easy to set off accidentally, such as by dropping it on its nose, so small studs were put in place that would be sheared away by the cordite blow. The design was so foolproof that Oppenheimer didn't bother testing it before it was dropped over Japanese skies, and it was considered so dangerously easy for a foe to build that, after Japan's surrender, the design files were destroyed in a Los Alamos bonfire.

In another experiment, Frisch was stacking cores of enriched uranium when he leaned over the pile, which he called Lady Godiva, to yell to an assisting grad student. The neutron counters' red lights started up and did not fade, and Frisch realized that the water from his body and the reflection of his white lab coat were exciting Godiva's neutrons. He swept his arm, throwing several blocks of Oak Ridge's HEU onto the floor. Two seconds more, and he would have been dead.

One mesa family had a cat that developed a strange wound in its jaw that wouldn't heal. Army vets realized that somehow it'd gotten contaminated and was losing bone to radiation; they kept it alive to see what would happen next, but after its hair fell out and its tongue swelled, the family asked that it be put down. Walter Zinn opened a can of thorium and it exploded, severely burning his hands and face. Herbert Anderson was drying radium on a hot plate when it began to burn; he rushed into the room to stop the destruction, and six years later he was diagnosed with berylliosis—the sweet-to-the-tongue but poisonous beryllium had deposited in his lungs. Los Alamos chemists developed a policy that, if cut skin came in contact with plutonium, "immediate high amputation" was necessary.

Research Division (R-Division) chief Robert Wilson: "The Critical Assemblies Group decided not to have the elaborate safety devices that were used, for example, with cyclotrons. Instead they decided to depend on their wits alone. . . . I was the crew member whose turn it was to help the single physicist who showed up. His equipment consisted of a small wooden table, a single neutron counter, and boxes containing the small cubes of enriched uranium hydride. I was impressed by the simplicity of the equipment, as advertised, 'So simple nothing could go wrong.' Not quite. The physicist began stacking the uranium cubes as I stood next to him and watched with considerable interest. It was my first experience with a prompt neutron reactor approaching criticality, and I was thrilled in expectation. After a while, as the stack got quite large, I asked why the neutron counter was not counting. I was assured that this was regular and that it would not start counting until we were closer to the critical point. Uncomfortably, I gave the neutron counter a hard going over and asked if the signal light on the high-voltage supply was operative or if it was burned out—as is often the case. The voltage was indeed turned off, so the neutron counter was not working. When the voltage was turned on, the counter to my horror started blazing away. A few more cubes and the stack would have exceeded criticality and could well have become lethal. I was outraged. This incident was my closest brush with death. The reason given was that a wooden table instead of a metal table was being used for the first time, so thermal neutrons were reducing the critical point."

The plutonium bomb—the one tested at Trinity and dropped over

Nagasaki—turned out to be a very different creature from the simple Little Boy. In April 1944, Emilio Segrè determined that the plutonium coming from the Hanford reactors was less pure than the material envisioned in the design calculations, that Hanford's product would spontaneously fission. This meant that in a bomb, the chain reaction would start before critical mass was reached—what scientists refer to technically as "fizzle."

An Oppie protégé from Caltech, Seth Neddermeyer, was struck when an army ordnance expert told the scientists they were wrong to use the word *explosion* to mean detonation; the correct concept was *implosion*. Seth imagined a ball of plutonium the size of an orange, not critical in that state, compressed to a ball the size of a shooter marble. The pressure would combine with the merger of two plutonium halves to trigger a chain reaction. Oppie, Fermi, and Bethe all argued against Neddermeyer's proposal, but Johnny von Neumann insisted that, with the right explosive lenses, it would work. Edward Teller: "Just as a beam of light changes its direction when it passes through water and can be focused by passing it through a glass lens, the direction of a shock wave produced by the explosive can be focused and redirected by passing it through different explosive materials. Explosive lenses consist of various pieces of explosives that are fitted together so that the detonation waves move at different speeds as it passes through different portions of explosive."

Born in Budapest in 1903, Janos von Neumann was so eminently brilliant that fellow Hungarian Quartet member Eugene Wigner remarked, "Only he was fully awake." In 1928, he published the *Theory of Parlor Games*, a breakthrough in game theory, and with 1932's *Mathematical Foundations of Quantum Mechanics*, he reconciled the competing mathematics of Erwin Schrödinger and Werner Heisenberg. In 1933, he was selected with Albert Einstein and Kurt Gödel for Princeton's Institute for Advanced Study, where he remained until his death. There, his colleagues complained about his penchant for enjoying extremely loud German marching music on the gramophone, but far more problematic was that he liked to read books while driving cars. His many collisions inspired a Princeton intersection—the "von Neumann corner." The mathematician was also famously natty. At his 1926 doctoral interview, one of the judges asked, "Pray, what is the candidate's tailor?" If anyone remembers anything about him at Los Alamos, it was the suit and tie he wore to hike the desert, or the time he rode a Grand Canyon mule in a three-piece pinstripe. Von Neumann threw lavish, outrageous parties; slept a bare four hours a night; and loved to eat and loved to drink. His devoted wife, Klari, said that he could count everything except calories and had "an almost primitive lack of ability to handle his emotions."

For Neddermeyer's plutonium-orange-into-shooter-marble notion, von Neumann did shock-wave calculations that proved implosion would work if it did not deviate by more than 5 percent of perfect spherical symmetry. He and explosives expert George Kistiakowsky (known as Kisty) invented a lens implosion system of blocks, in the shape of a pie cut into pieces, the size of car batteries. The fast-burning, wider edge was focused by the slower-burning narrow end of the pie piece so that the detonation waves would arrive together, achieving that perfect sphere.

Just as the mesa was to achieve all of its astonishing goals, personal resentment ushered in sabotage. Hans Bethe:

> At the start I had regarded Teller as one of my best friends and as the most valuable member of my division. Our relation cooled when Teller did not contribute much to the work of this division [the theoretical division, which had the main responsibility for the conceptual design of weapons]. More important perhaps for a disturbance of relations was his wish to spend long hours discussing alternative schemes which he had invented for assembling an atomic bomb or to argue about some remote possibilities why our chief design might fail. . . . [Then] he declined to take charge of the group which would perform the detailed calculations of the implosion. Only after two failures to accomplish the expected and necessary work, and only on Teller's own request, was he, together with his group, relieved of further responsibility for work on the wartime development of the atomic bomb. . . . Since the theoretical division was very shorthanded, it was necessary to bring in new scientists to do the work that Teller declined to do. Partly for this reason, some members of the British Atomic Energy team, already working in the U.S. on other aspects of the Manhattan District project, were brought to Los Alamos and asked to help with this problem. The leader of the British theoretical group was Rudolf Peierls, and another very hardworking member was Klaus Fuchs.

A smidgen of a man in comically oversize eyewear, Klaus Fuchs was perhaps the least physically imposing of any human being on the mesa. He had come to New Mexico as a member of the British Tube Alloys team, all of whom had forgone oversight by American intelligence after Groves was assured that British intelligence had thoroughly security-checked its members. Fuchs would become such a devoted worker that he would rise in the organization until he was a key player on the plutonium-design team, the fundamental group behind the gadget that was Trinity.

A security officer casually mentioned to Klaus Fuchs that army intelligence knew the Soviets had a number of spies working in London and the United States, but that the Anglo-Americans only had one agent in all of Russia. Fuchs thought this was especially funny as he happened to be a spy working for Russia. Laura Fermi: "The first Sunday I was there, a group of friends organized a picnic in a canyon. Our car was needed, but I wouldn't drive in that unknown, wild territory. So the Peierls asked Fuchs, their friend and protégé, to drive my car. He was an attractive young man, German-born, with a quiet look through round eyeglasses, who answered sparingly to my questions. Even as he spoke to me, he was leading a double life, that of a competent physicist appreciated by his colleagues, and that of spy. As he was to confess in 1950, he was giving secret information to the Russians on the progress of the atomic bomb. Fermi was to say Fuchs had made it possible for the Russians to make an atomic bomb five to ten years earlier than they would have otherwise."

The great Cold War slander that would tear apart the global physics community and destroy the Los Alamos alumni's Edenic spirit—that Robert Oppenheimer might not be loyal to the United States, as he was not unswervingly devoted to developing thermonuclear weapons—would be in part advanced by Edward Teller, who could easily be considered treasonous for his own wartime dereliction of duty. Teller's refusal to do his assigned implosion calculations inserted a Russian agent at the center of Los Alamos, while army intelligence's and the FBI's obsession with Oppenheimer's jejune Communist flirtations left them blind to one of the greatest feats of espionage in modern times.

Klaus Fuchs was originally sent to America as one of fifteen British scientists helping to develop Harold Urey's gaseous-diffusion technique for producing weapons-grade uranium. Fuchs was already a refugee from the Nazis, as his Soviet handler, Harry Gold, reported: "While Klaus was a mere boy of eighteen, he was head of the student chapter of the Communist Party at the University of Kiel . . . and Klaus, a frail, thin boy, led these boys in deadly street combat against the Nazi storm troopers . . . and later, when the Nazis had put a price on his head, he barely managed to escape with his life to England." His sister Elizabeth had also been a member of the Communist Party fighting Hitler; as she was about to be arrested, she jumped in front of a train; while Fuchs's mother committed suicide, grotesquely, by drinking hydrochloric acid.

Klaus Fuchs was, like many other agents operating in post-Depression Britain and America, a true believer in communism as the best system of economics and government, as committed to the cause as Jean Tatlock, the great love of Oppenheimer's life, who told a friend she couldn't bear to go on living if she believed Soviet life was inferior. Ethel Rosenberg's elder brother, Samuel Greenglass, got so tired of hearing about her and Julius's love of the Soviet system that he "offered to pay the transportation to Russia . . . if they would agree to stay there." After Ethel's little brother, David, was drafted into the army as a machinist and assigned to Los Alamos on August 4, 1943, he worked with implosion under demolition expert George Kistiakowsky, machining high-explosive lenses. From basic training, David wrote to his new bride, Ruth, that world socialism was inevitable, even in America, and wasn't that glorious? When the lonely Ruth then had dinner with her in-laws, Julius Rosenberg explained that David should help Moscow by passing along information, since it was unfair that the United States and United Kingdom weren't sharing their technology with their Russian allies. When Ruth went to New Mexico for a second honeymoon and relayed the information to her husband, David said that Julius was his hero, and that absolutely he would do it. Another Los Alamos employee, the nineteen-year-old Theodore Hall, agreed with Julius that it was repugnant for America and Britain to keep atomic secrets from the Soviets. While on vacation in New York, Hall figured out how to contact Soviet authorities and became the third enemy agent at the epicenter of nuclear research. His NKVD handler, "Helen," Lona Cohen, became a legend in NKVD history for her epic sangfroid. One story: Helen was aboard a train and waiting to depart with Hall's Manhattan Project materials hidden in a Kleenex box. Some undercover FBI agents approached. She began fumbling around with her bags, pretending to look for her ticket. A conductor came over; she handed him the Kleenex box, in the middle of which the Feebs interrogated her. They left, she got her Kleenex box back, and went on to New York.

The Russian search for Anglo-American nuclear secrets began when, after independently discovering spontaneous fission in 1940, Georgi Flerov and Konstantin Petrzhak were nominated for a Stalin Prize, but their work was not validated or cited by scientists in the rest of Europe, so the prize was never awarded. During the mass evacuations before the Wehrmacht invasion, a still-upset Flerov decided to investigate the scientific journals at one abandoned university's library. Not only was his work not cited, the whole of nuclear physics had vanished from the world's publications. Flerov suddenly realized what this meant: Britain and the United States were developing nuclear weapons. The lieutenant, all of twenty-eight years old, wrote directly to warn Joseph Stalin.

The NKVD maintained a global network of spies operating out of Soviet embassies, and one field of great expertise was industrial espionage, stealing patents, processes, and formulas so the Russians wouldn't have to pay licensing fees. American chemist Harry Gold, the Soviet handler for both Fuchs and Greenglass, became an agent for the Soviets in 1935 by providing the formulas for anesthetics and lacquer solvents. During the 1940s, Moscow's American network was so extensive it seemed to be omnipresent. Kitty Oppenheimer's first husband, Joe Dallet, was killed while volunteering for the Spanish Civil War; one of his good friends in Spain, Steve Nelson, became an agent in Berkeley, gathering information on Lawrence's cyclotrons, isotope separation, and other technologies that would become significant in producing nuclear weapons. Canadian Alan Nunn May had been recruited by Donald Maclean in Cambridge; he worked with Tube Alloys; Chicago's Met Lab; and the heavy-water reactor in Canada's Chalk River. The Russian network was so good that, when the KGB's files were opened after the fall of the USSR, it was revealed that the Russians had known quite a bit about Fermi's first nuclear reactor in Chicago, but they had translated the reactor's location—squash court—as "pumpkin patch."

Since 1938, the NKVD—which oversaw the civil police, the secret police, and domestic and foreign espionage, as well as managing the Great Terror—had been run by the brutal, sadistic, and perverted Lavrenti Pavlovich Beria, who was described by a diplomat as "somewhat plump, greenish pale, and with soft damp hands . . . a square-cut mouth and bulging eyes behind his pince-nez . . . a certain self-satisfaction and irony mingled with a clerk's obsequiousness and solicitude," and by a Russian as "placed in control for the precise purpose of inspiring deadly fear. I often asked myself—as others assuredly did in their secret hearts—why Stalin had decided to take this step. I could find only one plausible answer. It was that he lacked faith in the patriotism and national honor of the Russian people and was therefore compelled to rely primarily on the whip. Beria was his whip."

On February 16, 1945, Harry Gold told his superiors that Klaus Fuchs had determined what was needed for the Soviet atomic program to succeed, and by June the diminutive agent submitted "a description of the plutonium bomb, which had been designed and was soon planned to be tested at Alamogordo; a sketch of the bomb and its components with important dimensions indicated; the type of core; a description of the initiator; details as to the tamper"; and "the names of the types of explosives to be used in the bomb [information important to the design of high-explosive lenses]; the fact that the Trinity test explosion was to be made, with the approximate site indicated, soon, in July, 1945, and that this test was expected to establish that the atom bomb would produce an explosion vastly greater than TNT and the comparative estimated force of this explosion was indicated in detail with relation to TNT." On Sunday morning, June 3, 1945, Gold went to the Greenglasses to pick up sketches of Kisty's various implosive lenses, including schematic views of their layers and detonators. It was the first meeting between agent and handler, one that Gold remembered well: "Greenglass was not only young, but at once impressed me as being frighteningly naive, particularly in his eager volunteering of the idea of approaching other people at Los Alamos as potential sources of data. I was horrified at his total inexperience in espionage, especially considering what we were after."

On January 25, 1943, Niels Bohr was given a key in Copenhagen. Inside was a microdot, which revealed a letter from James Chadwick, asking Bohr to join the Anglo-American fission project. Bohr said that he still thought nuclear weapons were implausible because of the difficulty of producing U-235, but that he might change his mind in the future. This reply was also converted into a microdot and carried back to England, inside the filling of a courier's tooth.

At that time, Bohr's Denmark was under German control, but in a very different manner from the rest of occupied Europe. The Nazis so depended on Danish butter, meat, and other foodstuffs that they allowed the country to govern itself and left their eight thousand Jews alone. But after the shocking German defeat at Stalingrad revealed the Fascists as less than omnipotent, Danes began organizing regular labor strikes and acts of sabotage. The Germans retaliated by taking control of the royal palace and arresting Jews. A friend of the Bohr family's was working as a clerk for the Gestapo in Copenhagen and came across the warrants for Niels and his brother, Harald. In the middle of the night of September 29, 1943, under cloud cover that rendered the darkness visible, around a dozen people, including the two Bohr brothers, Niels's wife, Margrethe, and Harald's son Ole, rode a fishing boat to cross the sound and find safe harbor in Malmö, Sweden. Simultaneously, two German freighters arrived in Denmark to begin ferrying Danish Jews to concentration camps.

Bohr went to Stockholm, where he soon learned that Sweden was overrun with Gestapo agents whose careers would be made if they caught or killed him. Though he needed to leave for England as soon as possible, the great Bohr instead went to the palace to beg King Gustav V to give refuge to Denmark's Jews, saying he had learned that the Nazis planned to arrest all of them the following day. The Swedish government questioned the Germans, who insisted nothing of the sort was happening. This was a lie, but the Danes had been so well informed ahead of time that they hid nearly eight thousand Jews, and the countrywide Nazi anti-Semitic campaign resulted in 284 men and women being taken from a nursing home. On October 2, Sweden announced that it would grant refuge, accepting 7,220 Danish Jews over the next two months. At the same time, rumors began circulating that Niels Bohr was going to be assassinated.

Diplomatic pouches were flown between Stockholm and Westminster in a two-engine plywood Mosquito that cruised above the twenty-thousand-foot ceiling of German antiaircraft cannon on the shores of Norway. On October 6, the fifty-eight-year-old Bohr was suited up, given a stick of flares and a parachute, and strapped into the plane's bomb bay. If the plane went down, he could use the flares to help rescuers find him in the North Sea. Bohr's son Aage (pronounced *Awa*): "The Mosquito flew at a great height and it was necessary to use an oxygen mask; the pilot gave word on the intercom that the supply of oxygen should be turned on, but as the helmet with the earphones did not fit my father's head, he did not hear the order and soon fainted because of lack of oxygen. The pilot realized that something was wrong when he received no answer to his inquiries, and as soon as they passed over Norway, he came down and flew low over the North Sea. When the plane landed in Scotland, my father was conscious again."

At the Savoy in London, Chadwick told Bohr about the progress made with Tube Alloys and the Manhattan Project. The physicist was speechless. Then in New York, an entire contingent of FBI agents escorted Niels and his son and future Nobel laureate Aage from the docks to their hotel and were so pleased in getting Niels across town incognito. Then they noticed his suitcase, which had huge black letters on its side: NIELS BOHR.

Bohr was such a terrible jaywalker—a habit shared by Ernest Lawrence—that when Groves had a security team follow Niels and Aage in Washington, one reported, "Both the father and the son appear to be extremely absentminded individuals, engrossed in themselves, and go about paying little attention to any external influences. As they did a great deal of walking, this Agent had occasion to spend considerable time behind them and observe that it was rare when either of them paid much attention to stop lights or signs, but proceeded on their way much the same as if they were walking in the woods. On one occasion, subjects proceeded across a busy intersection against the red light in a diagonal fashion, taking the longest route possible and one of greatest danger. The resourceful work of Agent Maiers in blocking out one-half of the stream of automobile traffic with his car prevented their possibly incurring serious injury." The physicist met with British ambassador Lord Halifax and Supreme Court justice Felix Frankfurter and insisted that, unless the Russians were told and involved with the atomic weapons program, a terrifying postwar arms race would ignite. Groves, at first leery of having the voluble and famous Bohr at Los Alamos, now wanted him there as soon as possible to keep him isolated, and quiet. Groves was so alarmed that he personally escorted Bohr on the train to New Mexico.

Using the name Nicholas Baker and called by everyone Uncle Nick, Bohr arrived at the mesa on December 30, 1943. He fell in love with the American Southwest, especially remembering his first encounter with a skunk, and listening, at dusk, to the hissing tails of rattlesnakes as they scissored through the brush. After learning of the immensities of the Oak Ridge and Hanford operations, Uncle Nick insisted to Edward Teller that he'd been right all along in doubting the prospect of nuclear weapons: "You see, I told you it couldn't be done without turning the whole country into a factory. You have done just that."

Bohr was so used to having wife Margrethe taking care of his day-to-day needs that his life at Los Alamos alternated between the inspired and the comic. One day he came to work wearing a rope to hold up his pants since he'd forgotten where he'd put his belts. He returned from one party with a hugely oversize coat, saying he must have taken it by mistake since there were keys in the pockets, and he didn't have any keys.

Emilio Segrè: "When Niels and Aage Bohr arrived, one night in Oppenheimer's house he told a few European scientists of the conditions prevailing in Denmark and his escape. For many of us this was the first eyewitness account of what was really happening in a Nazi-occupied country. Although conditions in Denmark at that time were relatively tolerable and the worst horrors of Nazism were unknown to Bohr, the account left us depressed and worried, and more determined than ever that the bomb should be ready at the earliest date possible." Bohr's meeting with Heisenberg had frightened him—Bohr showed everyone Heisenberg's sketches of a heavy-water reactor that the German had brought with him to a meeting in Copenhagen—

and Bohr told Oppenheimer that Germany planned to end the war with an atomic bomb.

While the Los Alamos refugees fretted almost continuously over Hitler's taking control of nuclear weapons, Vannevar Bush and Franklin Roosevelt, however, left no trace to history that they thought about it whatsoever, beyond the initial meeting over Einstein's letter between Alexander Sachs and the president. Neither made much effort to uncover information on the Nazi program. One officer involved at the time, Major Francis Smith, believed that "the Nazis' work was held in such poor esteem by our military authorities that certain German laboratories whose locations we knew were left unbombed to enable Hitler's experts to continue their failures."

The German effort to produce atomic weapons centered on Werner Heisenberg, whose role was equivocal, peculiar, and damning. Nazi physicist Johannes Stark had published an article in the July 15, 1937, SS journal *Das Schwarze Korps*, "'Weisse Juden' in der Wissenschaft" (White Jews in Science). Its "main theme was that it was not sufficient to exclude all Jews from sharing in the political, cultural and economic life of the nation, but to exterminate the Jewish spirit, which is stated to be most clearly recognizable in the field of physics, and its most significant representative Professor Einstein. . . . Several men of science of international reputation were named in the article as followers of Judaism in German intellectual life, and it was remarked that 'They must be gotten rid of as much as the Jews themselves.'" Stark called Heisenberg a "Jew lover" and a "Jewish pawn." Heisenberg wrote a letter to Himmler—which his mother gave to Himmler's mother, as they were good friends—asking that Himmler publicly approve or disapprove of these attacks. Himmler launched an SS investigation that lasted eight months and included interrogations of Heisenberg, recording devices in his home, and spies in his classroom. Himmler then cleared him.

When Heisenberg toured America in the summer of 1939, he repeatedly defended remaining in Germany, even though Columbia had offered him a post, telling Ed Teller, "Even if my brother steals a silver spoon, he is still my brother," and Laura Fermi, "People must learn to prevent catastrophes, not to run away from them." When on September 26 he was then called to join the Uranverein in Berlin, he thought that he could use the Reich's interest in physics for the purposes of science and eagerly agreed. In December 1939 he gave the Nazi War Office a report, "The Possibility of Technical Energy Production from Uranium Fission," which called reactors "uranium burners" and calculated a successful burner of a ton of uranium and a ton of heavy water in a spherical container producing 800°C (1,472°F).

In January 1940, the Nazis replaced Peter Debye as head of KWI Physics with Werner Heisenberg; by October 1940, the Virus House construction was finished, and the Uranverein experimented with paraffin as a moderator, which did not work. Neither did graphite since, due to Szilard's silence campaign, they didn't know that the reason was its impurities. They then tried heavy water, which did work, but they could only now get eight liters of it from Norway's Vemork. They needed fifteen tons.

By May of 1941, the British came to understand why the Nazis were so keenly interested in Norwegian heavy water. The construction superintendent of the dam that powered Vemork, Einar Skinnarland, took a two-week vacation in England, during which he was trained by the Special Operations Executive to be a key member of a commando team, the Norwegian Independent Company. On October 13, 1942, Combined Operations put into effect Operation Freshman. Advance ground party Grouse, nine natives of the region experienced in the local wilderness, were trained by the SOE in pistols, knives, poisons, explosives, hand-to-hand combat, breaking locks, and cracking safes. Led by the twenty-four-year-old Jens Pousson, Grouse would find a landing spot to welcome commandos floating in on gliders, and the combined team would disable the heavy-water plant and then hike 250 miles to the Swedish border. On October 18, Grouse was parachuted in, but the drop turned out to be thirty miles from where they were supposed to be. Terrible weather meant it took them fifteen days to reach their base near the dam. At the same time, Norsk Hydro's chief of hydrogen research, Jomar Brun, was running a campaign of sabotage within his own plant, putting castor oil in the electrolyte, producing a foam that would stop the process for hours and sometimes days. He didn't know that others were themselves adding cod-liver oil to achieve the same purpose. Of the five tons of heavy water that Heisenberg wanted by June 1942, Norway delivered one.

On November 11, Grouse informed SOE that they'd found a landing site, and on November 19, two British Halifax bombers towing Horsa Mk.I gliders each carrying seventeen men flew into Norway. These fragile plywood gliders landed in a "controlled crash." The first Halifax released too soon, and her glider crashed into a mountain, killing seven. The second glider's towline broke, and it crashed, killing eight. All of the survivors were captured by the Nazis, interrogated, and executed by firing squad. On one officer's body, the Germans found a map of the Vemork plant. They stationed nine hundred troops in the vicinity, and now the Grouse team was cut off from rescue.

A six-man reinforcement, Operation Gunnerside, parachuted in on February 16, 1943. The Nazis had heavily reinforced the plant, but not its rail-line depot. From that direction, leader Joachim Ronnenberg snuck into the building, set up demolition charges, and blew up the tanks. Evading the Germans, they raced to an icy plateau and skied to the Swedish border. By August, the Germans had repaired the damages and had the plant running again.

On November 16, 1943, under orders from Leslie Groves, three hundred B-17 Flying Fortresses and B-24 Liberators from the Eighth Air Force took off from East Anglia to drop 700 thousand-pound and 295 five-hundred-pound bombs on Vemork, to end its production of heavy water once and for all. But only two bombs hit the plant directly, and neither took out the electrolysis cells that were the real target, while twenty-two civilians died. SOE agent Knut Haukelid learned that the Nazis were planning to move the whole of the plant and its heavy water to Germany on February 20, 1944, the cargo being railed and ferried through Denmark to Berlin. The SOE arranged an operation to sink the cargo's Norwegian ferry in the middle of a fjord, ensuring the Nazis could never resurrect their cargo. On February 18, Haukelid snuck onto the ferry and improvised a bomb with two alarm clocks. It worked, and Hitler's entire stock of heavy water was finally destroyed.

But before this difficult victory, there was the meeting that had so terrified Niels Bohr. On September 15, 1941, accompanied by fellow Uranverein member Carl Friedrich Freiherr von Weizsäcker, Werner Heisenberg took the train from Berlin to Copenhagen to visit his great mentor. After the war, a controversy arose over who said what in Denmark that autumn, a controversy that turned out to be one of many German attempts to rewrite history.

The rewrite began with letters Heisenberg sent to Robert Jungk after reading Jungk's 1956 history of nuclear science, *Brighter than a Thousand Suns*. In these letters, Heisenberg claimed that he had gone to Copenhagen in 1941 to discuss with Bohr his moral objections about scientists working on nuclear weapons, but that he had failed to say this clearly before the conversation came to a sudden stop. Jungk published portions of the letters in the Danish edition of the book, conveying that Heisenberg suggested he had sabotaged the German bomb project on moral grounds:

> With the beginning of the war there arose of course for every German physicist the dreadful dilemma that each of his actions meant either a victory for Hitler or a defeat of Germany, and of course both alternatives presented themselves to us as appalling. . . .
>
> My visit to Copenhagen took place in the fall of 1941; I seem to

remember that it was about the end of October. At that time, as a result of our experiments with uranium and heavy water, we in our "Uranium Club" had come to the following conclusion: It will definitely be possible to build a reactor from uranium and heavy water which produces energy [but] the production of nuclear explosives from reactors obviously could only be achieved by running huge reactors for years on end. . . . This situation seemed to us to be an especially favorable precondition as it enabled the physicists to influence further developments. For, had the production of atomic bombs been impossible, the problem would not have arisen at all; but had it been easy, then the physicists definitely could not have prevented their production. The actual givens of the situation, however, gave the physicists at that moment in time a decisive amount of influence over the subsequent events, since they had good arguments for their administrations—atomic bombs probably would not come into play in the course of the war, or else that using every conceivable effort it might yet be possible to bring them into play. That both kinds of arguments were factually fully justified was shown by the subsequent development; for, in fact, the Americans could not employ the atomic bomb against Germany. . . .

Because I knew that Bohr was under surveillance by German political operatives and that statements Bohr made about me would most likely be reported back to Germany, I tried to keep the conversation at a level of allusions that would not immediately endanger my life. The conversation probably started by me asking somewhat casually whether it were justifiable that physicists were devoting themselves to the Uranium problem right now during times of war, when one had to at least consider the possibility that progress in this field might lead to very grave consequences for war technology. Bohr immediately grasped the meaning of this question as I gathered from his somewhat startled reaction. He answered, as far as I can remember, with a counter-question: "Do you really believe one can utilize Uranium fission for the construction of weapons?" I may have replied, "I know that this is possible in principle, but a terrific technical effort might be necessary, which one can hope, will not be realized anymore in this war." Bohr was apparently so shocked by this answer that he assumed I was trying to tell him Germany had made great progress towards manufacturing atomic weapons.

> In my subsequent attempt to correct this false impression I must not have wholly succeeded in winning Bohr's trust, especially because I only dared to speak in very cautious allusions (which definitely was a mistake on my part) out of fear that later on a particular choice of words could be held against me. I then asked Bohr once more whether, in view of the obvious moral concerns, it might be possible to get all physicists to agree not to attempt work on atomic bombs, since they could only be produced with a huge technical effort anyhow. But Bohr thought it would be hopeless to exert influence on the actions in the individual countries, and that it was, so to speak, the natural course in this world that the physicists were working in their countries on the production of weapons. . . . Since 1933 Germany had lost a number of excellent German physicists through emigration, the laboratories at universities were ancient and poor due to neglect by the government, the gifted young people often were pushed into other professions. In the United States, however, many university institutes since 1932 had been given completely new and modern equipment, and been switched over to nuclear physics. Larger and smaller cyclotrons had been started up in various places, many capable physicists had immigrated, and the interest in nuclear physics even on the part of the public was very great. Our proposition that the physicists on both sides should not advance the production of atomic bombs was thus indirectly, if one wants to exaggerate the point, a proposition in favor of Hitler. The instinctive human position "As a decent human being one cannot make atomic weapons" thus coincided with an advantage for Germany.

Niels Bohr was outraged after reading this in Jungk's book, for he remembered that in 1941 Heisenberg was elated to be making nuclear weapons for Hitler. Bohr's wife, Margrethe, remembered the meeting in detail: "Heisenberg stated that he was working on the release of atomic energy and expressed his conviction that the war, if it did not end with a German victory, would be decided by such means. Heisenberg said explicitly that he did not wish to enter into technical details but that Bohr should understand that he knew what he was talking about as he had spent two years working exclusively on this question. Bohr restrained himself from any comment but understood that this was important information which he was obliged to try to bring to the attention of the English. . . . Heisenberg and Weizsäcker sought to explain that the attitude of the Danish people towards Germany, and that of the Danish physicists in particular, was unreasonable and indefensible since a German victory was already guaranteed and that any resistance against cooperation could only bring disaster to Denmark. . . . Weizsäcker further stated how fortunate it was that Heisenberg's work would mean so much for the war since it would mean that, after the expected great victory, the Nazis would adopt a more understanding attitude towards German scientific efforts." Other Danish physicists at the meeting said that Heisenberg "stressed how important it was that Germany should win the war. . . . The occupation of Denmark, Norway, Belgium and Holland was a sad thing but as regards the countries in Eastern Europe it was a good development because these countries were not able to govern themselves."

After reading Jungk, Bohr wrote several letters to Heisenberg about this meeting but had never mailed them, and after his death Bohr's private papers, including these letters, were to remain sealed until the year 2012. But when Michael Frayn's 1998 play *Copenhagen* turned the incident into a global conversation about the morality of science in a time of war, the Niels Bohr Archive decided to release the correspondence to protect Bohr's reputation . . . and seal Heisenberg's:

> Dear Heisenberg,
>
> I have seen a book by Robert Jungk, recently published in Danish, and I think that I owe it to you to tell you that I am greatly amazed to see how much your memory has deceived you in your letter to the author of the book, excerpts of which are printed in the Danish edition.
>
> Personally, I remember every word of our conversations, which took place on a background of extreme sorrow and tension for us here in Denmark. In particular, it made a strong impression both on Margrethe and me, and on everyone at the Institute that the two of you spoke to, that you and Weizsäcker expressed your definite conviction that Germany would win and that it was therefore quite foolish for us to maintain the hope of a different outcome of the war and to be reticent as regards all German offers of cooperation. I also remember quite clearly our conversation in my room at the Institute, where in vague terms you spoke in a manner that could only give me the firm impression that, under your leadership, everything was being done in Germany to develop atomic weapons and that you said that there was no need to talk about details since you were completely familiar with them and had spent the past two years working more or

> less exclusively on such preparations. I listened to this without speaking since [a] great matter for mankind was at issue in which, despite our personal friendship, we had to be regarded as representatives of two sides engaged in mortal combat. That my silence and gravity, as you write in the letter, could be taken as an expression of shock at your reports that it was possible to make an atomic bomb is a quite peculiar misunderstanding, which must be due to the great tension in your own mind. From the day three years earlier when I realized that slow neutrons could only cause fission in Uranium 235 and not 238, it was of course obvious to me that a bomb with certain effect could be produced by separating the uraniums. In June 1939 I had even given a public lecture in Birmingham about uranium fission, where I talked about the effects of such a bomb but of course added that the technical preparations would be so large that one did not know how soon they could be overcome. If anything in my behavior could be interpreted as shock, it did not derive from such reports but rather from the news, as I had to understand it, that Germany was participating vigorously in a race to be the first with atomic weapons.

In another undated letter never sent to Heisenberg, Bohr wrote: "I remember quite clearly the impression it made on me when, at the beginning of the conversation, you told me without preparation that you were certain that the war, if it lasted long enough, would be decided with atomic weapons. I did not respond to this at all, but as you perhaps regarded this as an expression of doubt, you related how in the preceding years you had devoted yourself almost exclusively to the question and were quite certain that it could be done, but you gave no hint about efforts on the part of German scientists to prevent such a development."

When Bethe returned to Germany after the war in 1948, Heisenberg told him he'd been developing an atomic bomb for the Nazis and came up with another series of reasons for his quisling behavior. Hans Bethe: "He said that his main aim had been to save the lives of German physicists. . . . The second reason he gave was that he believed making an atomic bomb was far beyond the means of any country during the war. . . . His third reason was that in 1942 he had come to the conclusion that the Germans should win the war. This struck me as a very naive statement. He said he knew that the Germans had committed terrible atrocities against the populations on the Eastern Front—in Poland and Russia—and to some extent in the West as well. He concluded that the Allies would never forgive this and would destroy

Germany as a nation—that they would treat Germany about the way that Romans had treated Carthage. This, he said to himself, should not happen; therefore, Germany should win the war, and then the good Germans would take care of the Nazis."

Things were in fact much less fruitful in Berlin than Heisenberg had implied in Copenhagen. On December 16, 1941, Bothe, Hahn, Harteck, and Heisenberg reported a series of uranium research disappointments, and the army ordnance department responded by withdrawing entirely from the nuclear program, as well as from KWI itself, turning everything back over to the (far less moneyed and influential) Reich Research Council. On February 26, 1942, the Uranverein tried to reverse its fall from grace by inviting every significant military and government official to a crucial technical progress announcement on nuclear weapons in Berlin. But along with the invitation came the news that everyone would be treated to an experimental flash-frozen lunch deep-fried in synthetic lard. Not a single influential Nazi attended, and the Uranverein fell further behind in funding and influence.

Then, at the end of April 1942, Albert Speer, Hitler's second-in-command, was told by General Friedrich Fromm, chief of the army's training and reservists, that the only way the Germans could win would be through a new weapon some scientists were working on that could "annihilate whole cities." Speer asked Hitler to appoint Göring as head of the Reich Research Council to elevate the program (which Hitler did) and called for a meeting with the Uranverein on June 4, joined by the heads of ordnance, navy armaments, and the air force. Speer asked the physicists what kind of funding they needed; Weizsäcker responded with 40,000 reichsmarks, a ridiculously low number. Speer asked again for a figure, and they responded with 350,000 marks ($80,000). Speer thought this, too, was absurd (almost this exact same meeting had played out in 1941 in Schenectady, when American physicists had no idea how to estimate the cost of developing nuclear arms and responded to the US War Department's questions with comically flailing sums). On June 23, Speer told Hitler that nuclear science would reap benefits in the distant future, but no superbomb could be produced in time to affect the war. The German military once again abandoned nuclear research.

The same day that Speer and Hitler gave up on physics, Heisenberg almost died. His Leipzig experimental reactor developed a leak, and water reached the uranium metal, creating hydrogen bubbles. Robert Döpel had the pile pulled out of its water tank, and a technician opened one of the sphere's inlet valves. Air immediately flowed in, setting the uranium powder on fire, which sprayed out. Döpel and the technician were able to put out the fire and return the sphere to its tank, but a little later, he and Heisenberg were in the lab when they noticed the sphere was beginning to swell. They ran outside as fast as they could, and the reactor exploded, destroying the uranium, the lab, and the heavy water. When the two scientists had to explain to the city fire department what they were doing, the firemen sarcastically congratulated them on their remarkable achievement in science. A rumor, however, spread across Germany that scientists had been killed in a uranium bomb explosion, and it reached Leo Szilard. To those inside the Manhattan Project, this could only mean that the Germans had already initiated a sustained chain reaction, meaning they were at least a year ahead of the Allies in developing nuclear arms.

Believing the Nazis were equal or better when it came to research and development, in the heat of the war and the New Mexican desert of Los Alamos, the scientists and engineers rushed pell-mell forward to beat Hitler at the mastery of nuclear arms. Isidor Rabi: "The big problem was: Where was the enemy in the field of work? We finally arrived at the conclusion that they could be exactly up to us, or perhaps further. We felt very solemn. One didn't know what the enemy had. One didn't want to lose a single day, a single week. And certainly, a month would be a calamity." Germany's visible military losses during this period did not encourage the expatriates in New Mexico; they assumed the Reich's leaders would become desperate and turn to desperate solutions.

A number became convinced that the Nazis had a two-year lead on the Allies. Bethe and Teller wrote Oppenheimer on August 21, 1943, of their worries that the American effort wasn't moving fast enough: "Recent reports both through the newspapers and through secret service, have given indications that the Germans may be in possession of a powerful new weapon which is expected to be ready between November and January. There seems to be a considerable probability that this new weapon is tube alloy [i.e., uranium]. It is not necessary to describe the probable consequences which would result if this proves to be the case. It is possible that the Germans will have, by the end of this year, enough material accumulated to make a large number of gadgets which they will release at the same time on England, Russia and this country. In this case there would be little hope for any counter-action. However, it is also possible that they will have a production, let us say, of two gadgets a month. This would place particularly Britain in an extremely serious position but there would be hope for counter-action from our side before the war is lost, provided our own tube alloy program is drastically accelerated in the next few weeks."

Others at the mesa thought Allied efforts were lagging because of the slow-moving corporations managing production and suggested that Fermi and Urey lead a crash course in R&D on heavy-water piles that could be overseen directly by the physicists in New Mexico. Then in the autumn of 1944, the émigrés were terrified to learn that the Nazis had taken over the French company Terres-Rares and had shipped a notable amount of thorium back to Germany. Could they have developed a highly advanced thorium reactor? As it turned out, the German chemical company Auer, sensing that the war was coming to a close, had forcibly cornered the thorium market, hoping to cash in on the craze for radioactive cosmetics, especially its thorium toothpaste, which promised to give you "sparkling, brilliant teeth—radioactive brilliance!"

But there were few reasons to laugh at the time. Ed Teller: "The thought of how far the Germans might have come in the years since the discovery of fission was enough to give us all nightmares."

Philip Morrison remembered believing that "the only way we could lose the war was if we failed in our jobs."

As the Third Reich sputtered to its gruesome final months, a combination of gossip, competing evidence, and the urgent need to get security officer Boris Pash to stop bothering him about Robert Oppenheimer's youth of Communist flirtation inspired Leslie Groves to create Project Alsos. Greek for "grove" and a play on the general's name, Alsos was a scientific intelligence mission that followed victorious Allied troops across Europe to assess the status of the German nuclear program, as well as prevent the Russians from amassing nuclear materials and scientific manpower. Simultaneously with Alsos, Groves set up AZUSA with the OSS, sending in onetime Boston Red Sox catcher, attorney, and linguist Moe Berg to interview Amaldi, Wick, and other scientists whose names were provided by Fermi. Groves ordered Colonel Carl Eifler to kidnap Heisenberg and bring him to America, depriving the Nazis of their chief nuclear scientist. If the kidnapping was not successful, Eifler was told to "deny the enemy his brain." But Eifler's plan was so convoluted, with so many things that could go wrong, that it was eventually canceled. Instead, Moe Berg was sent to listen to a December 1944 lecture given by Heisenberg in Zurich. If Berg thought the Germans were close to having nuclear weapons, he was to kill Heisenberg in what would have become a suicide mission. Berg wasn't schooled enough in physics to make a decision on German weaponry based on the lecture, but he was then invited to join a group for dinner, a group that included Heisenberg. When the physicist admitted that the Germans were lost, Berg understood that there could be no Hitler-triggered nuclear holocaust, and Heisenberg's life was saved.

Groves now asked for known scientific institutions to be included in Allied bombing runs, and on February 15, 1944, the Kaiser Wilhelm Institute for Chemistry, the Dahlem home to Meitner, Hahn, and Strassmann, was fully destroyed. In December of 1943, Heisenberg's Leipzig house had been decimated in an aerial assault; another air raid completely destroyed Max Planck's house. Then Planck's first son, Erwin, was arrested for taking part in von Stauffenberg's failed assassination of Hitler. He was tortured by the Gestapo and hanged to death.

The Uranverein moved, with KWI, away from heavily targeted Berlin to a small town in the Black Forest. When Strasbourg was liberated in November 1944, Alsos leader Samuel Goudsmit uncovered a letter dated three months before, from one of the Uranverein to Heisenberg, discussing piles, but not bombs, and the intelligence team knew how far behind the Nazi scientific effort had fallen. Even so, it was believed in America's interest to ensure that none of the German scientists or their documents and equipment fell into Russian hands. When Alsos discovered eleven hundred tons of Belgian Congo uranium ore in a factory that would fall into the postwar Soviet zone, it was packed up and shipped to the United States. When Groves's overseas team then determined that a metal-refining plant fifteen miles north of Berlin, also to be included in the Soviet sector, was the source of uranium processing for a reactor, "since there was not even the remotest possibility that Alsos could seize the works, I recommended to General Marshall that the plant be destroyed by air attack," Groves recalled. It was destroyed, keeping at least that fuel away from the Soviets as well.

Just as the American army with Paperclip swept in to mop up Wernher von Braun and his team of scientists and engineers who developed the V-2 rocket—an effort strategic for both NASA and the Cold War's missile race—so Alsos, headed by Pash and American physicist Sam Goudsmit, dismantled Heisenberg's toylike reactor and captured Bagge, von Laue (who had nothing to do with the program), Hahn, Heisenberg, and six other scientists from the Nazi bomb group. Just as with von Braun and his team, the Uranverein had fled west in the war's last months to avoid being taken prisoner by the Russians and believed their work would save them. A few tried in the last moments to get a reactor burning, but it was too late. When Werner Heisenberg was caught, he thought the British and the Americans would be fascinated by his research, having no idea how long ago he had been eclipsed by Los Alamos. After finishing his mission, meanwhile, Sam Goudsmit would come to believe that America had spent more on Alsos than the Nazis had spent on the whole of Germany's wartime nuclear research.

Under Allied guard, the Uranverein were taken to Farm Hall in England, the SOE's country house. Unlike America with its habeas corpus, interrogation under British law allows detaining anyone for six months "at His Majesty's pleasure," and Farm Hall was bugged inch by inch. Not, however, until February 24, 1992, did the British release the transcripts, and before that time the Germans were able to rewrite much of this history.

As captured by SOE's microphones, the Nazi scientists were initially fearful that they would be turned over to the Russians. They frequently discussed how their work, at the forefront of science before Hitler, was probably being deeply studied by the Allies at that moment, with scientists across the United States re-creating their signal experiments. Then they were told that America had spent 500 million pounds and employed 125,000 people to develop an atomic bomb with the power of twenty thousand tons of TNT, which they dropped on Japan. Twice. The Germans fell into a profound shock. Werner Heisenberg thought that it was so impossible, it had to be a lie. Otto Hahn thought his role in discovering fission left him responsible and considered the only honorable action would be to commit suicide.

Then they started rationalizing their failures as both Germans and as scientists. Heisenberg: "We wouldn't have had the moral courage to recommend to the government in the spring of 1942 that they should employ 120,000 men just for building the thing up." Weizsäcker: "I think it was dreadful of the Americans to have done it. . . . I believe the reason we didn't do it was because all the physicists didn't want to do it, on principle. If we had all wanted Germany to win the war, we would have succeeded. . . . History will record that the Americans and the English made a bomb, and that at the same time the Germans, under the Hitler regime, produced a workable [reactor]."

This would become the party line of postwar German scientists for decades to come. After seeing the results of Hiroshima and Nagasaki, many American and British scientists were anguished, and even those who had nothing to do with Los Alamos were disturbed that science had been used to create such a horrific weapon. There was no similar introspection, though, among the German scientists, who congratulated themselves on not having built a nuclear arsenal, and who never admitted to the outside world how much effort they had put into giving Hitler the ultimate threat. In his reworking of what had happened at his meeting with Bohr in Copenhagen, Heisenberg would expand on all these themes. Von Laue alone seemed to understand that Germany was in fact responsible, writing his son Theo that the "émigrés passionate hatred of Hitler was . . . the thing that set it all in motion."

All this time, Leo Szilard remained in Chicago designing traditional and breeder reactors, while nearly everyone he knew was in Los Alamos, Hanford, or Oak Ridge. Since he wasn't part of the immense no-thought-but-Trinity rush to successfully engineer the first nuclear weapons, he had time to think, and in March 1944 he tried to convince Vannevar Bush that the nation's leaders needed to prepare for a future of "armed peace," with several countries having atomic arsenals able to stalemate each other. Bush, and the others Leo approached, weren't interested in what seemed so distant, but of course international stalemate would become the Cold War tenet Mutual Assured Destruction—MAD—the seemingly crazy notion that would keep the Cold War cold.

The only French resident of the Hill, Françoise Ulam, was nursing her baby when she heard Rabi playing "La Marseillaise" outside her window on his comb. It was August 24, 1944, and Paris had been liberated.

While pretending to be a scientist divorced from the politics of the real world in Vichy Paris, Frédéric Joliot made explosives and radio equipment for the Resistance and became president of the National Front. At the same time, he proved how radioactive iodine could be used as a tracer for the thyroid gland. After liberation, he was given the Legion of Honor, the Croix de Guerre, elected to the French Academy of Sciences, and named head of the French Atomic Energy Commission. But in the spring of 1942, Joliot had become a central-committee member of France's Communist Party, and during the Cold War, both he and Irène were (like Oppenheimer in America) purged from the French AEC. Beginning in 1953, Frédéric began suffering from severe hepatitis, perhaps incurred by radiation poisoning. Irène was working with polonium in 1946 when it exploded, showering her with a staggering dose of radiation, which led to her death of leukemia—the same disease that had killed her mother—at the Curie Hospital on March 17, 1956. Two years later on August 14, 1958, Frédéric would succumb to polonium-induced liver cirrhosis. The Curies' great ally, German chemist Friedrich Giesel, lived to seventy-five before dying of lung cancer. His lungs were so suffused with radon that when he exhaled, it would set his gold-leaf electroscopes aquiver.

On November 16, 1945, Otto Hahn, still interned and monitored at Farm Hall, was told that he had won the 1944 Nobel in Chemistry for discovering fission, and the rewriting of Lise Meitner out of scientific history began in earnest. During the Nazi era, Hahn could never have admitted that his momentous discovery was made through correspondence with a Jewess. By the end of the war, the chemist would deny the whole of Meitner's career, and of their collaboration. He wrote on August 8, 1945, "As long as Prof Meitner was in Germany the fission of uranium was out of the question. It was considered impossible," and the press surrounding Hahn's Nobel referred repeatedly to Lise as Otto's assistant. But it wasn't solely Hahn's doing. In Sweden, many believed that her boss Siegbahn's jealousy of Meitner's international stature kept the prize from her, so much so that her friends became convinced that, if she had emigrated anywhere but Sweden, she would have been a laureate.

On August 9, 1945, Eleanor Roosevelt, working to begin the United Nations, arranged to interview Lise Meitner on NBC radio. After an introduction that compared her historic importance to Marie Cure, Eleanor and Lise both expressed the need for world cooperation, with Meitner calling on women to help create a lasting peace and a responsible use of nuclear power: "They are obliged to try, so far as they can, to prevent another war." Later that day, Nagasaki was bombed. Meitner's refusal to then be interviewed by an invasive press led to a fanciful story, beginning with William Laurence in the *Saturday Evening Post*, of the "fleeing Jewess" who stole the secret of atomic bombs from Hitler and gave it to America. Magazine and newspaper articles now regularly referred to her as "the Jewish mother of the atomic bomb," even though she had nothing to do with any weapons and was only Jewish in the eyes of German racists.

On January 27, 1946, Meitner followed in Curie's footsteps with a six-month lecture tour of the United States, the National Press Club's Woman of the Year award, and a meal with Harry Truman, who joked, "Ah, so you're the little lady who got us into this mess!" Her lectures included such advice as that a professional woman needed to be well groomed and "never to let her petticoat show." She saw, for the first time in many years, her sisters Lola in Washington and Frida in New York, was reunited with her nephew and historic partner, Otto Robert, in Los Alamos, and appeared in a *San Francisco News* crossword puzzle and an *American Scholar* sonnet. Metro-Goldwyn-Mayer had a biopic developed, *The Beginning of the End*, and showed her the script. She told Frisch that it was "nonsense from the first word to the last . . . based on the stupid newspaper story that I left Germany with the bomb in my purse." When she refused, MGM raised its offer; she threatened to sue, saying, "I would rather walk naked down Broadway."

Otto Hahn was released from internment in February of 1946 to become the new president of KWI, which during the Allied occupation was renamed the Max Planck Institute, as it is called to this day. During his Nobel ceremony in 1946, he repeatedly insisted that Germans were victims, first of the Nazis, and now of the Allies. The Allies policing of Germany, according to Hahn among others, was the same as Hitler's takeover of Poland. In all his public statements as a laureate, he never once mentioned Meitner and gave less than 10 percent of the prize moneys to Strassmann. He did give Lise some money, and she immediately donated it to Einstein's Emergency Committee of Concerned Scientists. As one biographer described it, Hahn "was Germany's anointed postwar scientific icon, the decent German, great scientist, Nobel laureate, discoverer of fission but against the bomb, nationalist but never a Nazi, decorated veteran of WWI, personable, affable, witty, photogenic, a symbol of all that was good about the pre- and post-war Germany, made an honorary member of nearly every scientific society on earth, his face on a stamp, his name on buildings, institutes, schools, libraries, trains, a nuclear-powered ship, a moon crater, coins, an element, and an Antarctic island." Meitner meanwhile, who had begun and headed KWI's physics section for twenty-one years, was now depicted as his assistant. It was so egregious that on June 22, 1953, she wrote Hahn directly: "After the last 15 years which I wouldn't wish on any good friend, shall my scientific past also be taken from me? Is that fair? And why is it happening?" He never replied. She wrote to sister Lola about *Hahnchen* that "perhaps one cannot be such a charming person and also very deep." Niels's wife, Margrethe, said to Lise on February 8, 1948, "It is a difficult problem with the Germans, very difficult to come to a deep understanding with them, as they are always first of all sorry for themselves." Meitner finally wrote a summation on June 27, 1945, and gave it to Moe Berg, who promised to take it to the interned Hahn, but never delivered it:

> That is indeed the misfortune of Germany, that all of you lost your standards of justice and fairness. You yourself told me in March 1938 that Hörlein had told you that horrible things would be done to the Jews. He knew about all the crimes that had been planned and would later be carried out; in spite of that he was a member of the Party and you still regarded him—in spite of it—as a very respectable man, and let him guide you in your behavior toward your best friend [meaning Meitner]. . . . All of you also worked for Nazi Germany, and never even attempted passive resistance. Of course, to save your troubled

> consciences, you occasionally helped an oppressed person; still, you let millions of innocent people be murdered, and there was never a sound of protest. I must write you this because so much of what happens to you and the third Reich now depends on your recognizing what all of you allowed to happen. . . . I believe, as do many others, that one possibility would be for you to make a statement, namely that you know you bear responsibility for the occurrences as a result of your passiveness, that you feel it is necessary to help out in making reparations for the occurrences as far as that is even possible. . . . You really can't expect the world to pity Germany. We have heard recently about the unfathomable atrocities of the concentration camps that exceeds everything we had feared. When I heard a very objective report prepared by the British and Americans for British radio about Bergen-Belsen and Buchenwald, I began to wail aloud and couldn't sleep all night. If only you had seen the people who came here from the camps. They should force a man like Heisenberg, and millions of others with him, to see the tortured people. His performance in Denmark in 1941 cannot be forgotten.
>
> Perhaps you remember that when I was still in Germany (and I know today that it was not only stupid, but a great injustice that I didn't leave immediately), I often said to you, "As long as just we [the Jewish people] and not you have sleepless nights, it won't get any better in Germany." But you never had any sleepless nights: you didn't want to see—it was too disturbing. I could prove it to you with many examples, large and small. Please believe me that everything I write here is an attempt to help you.

In his memoirs, Hahn would call Meitner "a bitter, disappointed woman."

After the war, Lise retired to Cambridge, joining Otto Robert. She died at the age of eighty-nine on October 27, 1968, never receiving the Nobel she so obviously deserved. But in 1997, element 109 was named meitnerium for her, and she is memorialized with a lunar crater. Frisch oversaw her England headstone as a rejoinder to both sides of the war. Apart from her equation for fission, it read, "A physicist who never lost her humanity."

The arrival of Niels Bohr in the United States triggered what would become the great political battle in American nuclear physics: Do you believe in arms race, or arms control? Robert Oppenheimer: "Bohr at Los Alamos was marvelous. He took a very lively technical interest. But his real function, I think for almost all of us, was not the technical one. He made the enterprise seem hopeful, when many were not free of misgiving. Bohr spoke with contempt of Hitler, who had tried to enslave Europe for a millennium. His own high hope that the outcome would be good, that the objectivity, the cooperation, of the sciences would play a helpful part, we all wanted to believe." In his office, Oppenheimer kept an April 2, 1944, letter Bohr had written to President Roosevelt about the conversations he'd had with Frankfurter and Halifax in Washington on his way to Los Alamos, that when it came to nuclear weapons research, the Soviets would have to be included, or the Anglo-Americans would risk a postwar global nuclear arms race: "It is already evident that we are presented with one of the greatest triumphs of science and technique, destined deeply to influence the future of mankind. . . . A weapon of unparalleled power is being created which will completely change all future conditions of warfare. . . . Unless, indeed, some agreement about the control of the use of the new active materials can be obtained in due time, any temporary advantage, however great, may be outweighed by a perpetual menace to human security. . . . Knowledge is itself the basis of civilization [though] any widening of the borders of our knowledge imposes an increased responsibility on individuals and nations to the possibilities it gives for shaping the conditions of human life."

Historian Jim Baggott: "Atomic weapons have similar complementary properties, Bohr now realized. Under one set of circumstances, atomic weapons heralded an arms race leading, perhaps inevitably, to nothing less than the destruction of human civilization. At the same time, under different circumstances, atomic weapons heralded the end of war, because in a war fought with atomic weapons there could be no victor. If political, cultural or religious differences were to be settled without an end of the world, no-win scenario, then the advent of atomic weapons meant that recourse to war would no longer be thinkable. Differences would have to be settled in other, less violent, ways. The choice was plain. Arms race or international arms control?"

Physicist David Hawkins: "The implication was that Roosevelt had fully understood. And this was a great source of joy and optimism. . . . We all lived under this illusion, you see, for the rest of our time at Los Alamos, that Roosevelt had understood." In fact, at the same time that Bohr was traveling to Los Alamos, Igor Kurchatov arrived in Moscow to begin the Soviet atomic weapons program. Within a year, on August 20, 1945, the USSR had begun a Special State Committee on Problem Number One, headed by the NKVD's Beria. Problem Number One was the creation of a Soviet nuclear arsenal to match the weaponry of a postwar United States.

From their first encounters with Joseph Stalin to the fall of the USSR and the end of the Cold War, it seemed as if most American leaders considered their counterparts at the Union of Soviet Socialist Republics simultaneously powerful, menacing, backward, and unsophisticated. In fact, in the fields of physics, mathematics, and chemistry, for almost the whole of the twentieth century Russia was as advanced as any country besides Germany, and often far more sophisticated than the United States. In the wake of the Great War, Lenin proclaimed, "The war taught us much, not only that people suffered, but especially the fact that those who have the best technology, organization, and discipline, and the best machines, emerge on top. . . . It is necessary to master the highest technology or be crushed." A mineralogist trained under Curie's Radium Institute, Vladimir I. Vernadski, told the Russian Academy of Sciences as early as 1910 that radioelements meant "new sources of atomic energy . . . exceeding by millions of times all the sources of energy that the human imagination has envisaged."

Immediately after the Joliot-Curies demonstrated artificial irradiation, Petrograd's Institute of Physics and Technology (Fiztekh) began a nuclear physics department, selecting as its leader the thirty-one-year-old Igor Vasilievich Kurchatov. Like Fermi, Kurchatov had a chance meeting with a physics textbook that set the course of his life, but in his case it was *Accomplishments of Modern Engineering*, written by Orso Corbino, who played such a signature role in Fermi's career. Also like Enrico, Igor "worked harder than anyone else. He never gave himself airs, never let his accomplishments go to his head," as one colleague described him, and by 1934 Kurchatov had built the only cyclotron in the world outside Lawrence's Republic. He was nicknamed the Beard because he had stopped shaving until Russia was victorious over Germany, which was distinctive as only old men had beards in post-Peter-the-Great Russia.

Soviet scientists were just as fearful of an atomic Hitler as their American émigré counterparts. Flerov: "It seemed to us that if someone could make a nuclear bomb, it would be neither Americans, English, or French but Germans. The Germans had brilliant chemistry; they had technology for the production of metallic uranium; they were involved in experiments on the centrifugal separation of uranium isotopes. And, finally, the Germans possessed heavy water and reserves of uranium. Our first impression was that

Germans were capable of making the thing. It was obvious what the consequences would be if they succeeded."

Beria wanted to hire one of Russia's most famous scientists to head an atomic program, but Stalin disagreed, saying (as Groves might have about Oppenheimer) "that it was necessary to promote a young, not well-known scientist for whom such a post would be . . . his life work." On February 11, 1943, Kurchatov became the Soviet Oppie, and in March 1943, as part of Lend-Lease, the Soviet Purchasing Commission placed orders for uranium oxide and uranium nitrate, which Groves authorized, but only after being pressured by the Lend-Lease Administration. "Where that influence came from," Groves told a congressional committee after the war, "you can guess as well as I can. It was certainly prevalent in Washington, and it was prevalent throughout the country, and the only spot I know of that was distinctly anti-Russian at an early period was the Manhattan Project. . . . There was never any doubt about [our attitude] from sometime along about October 1942." Then in early 1945, the Russians cleared Czechoslovakia of Germans and began buying pitchblende from Joachimsthal—the same Bohemian source used by the Curies. On August 12, 1945, Henry D. Smyth published "Atomic Energy for Military Purposes," which deliberately left out crucial details, but the Soviets were able to use the materials supplied by Fuchs, Greenglass, Hall, and their other agents to reinstate those details. When they were finished, World War II was essentially over and the Cold War, with its nuclear arms race so long predicted and feared by Leo Szilard and Niels Bohr, had begun.

In early 1945, German surrender seemed inevitable, yet in the same period, 110,000 Japanese died defending Okinawa. Day after day, the mesa was rocked with explosions from the canyons, rattling windows and filling the air with the smell of pine and ordnance.

When Franklin Roosevelt died on April 12, Oppenheimer held a memorial service for the community and said in his speech about hearing that the president was gone, "Many of us looked with deep trouble to the future; many of us felt less certain that our works would be to a good end; all of us were reminded of how precious a thing human greatness is. We have been living through years of great evil, and of great terror. Roosevelt has been our president, our commander in chief, and, in an old and unperverted sense, our leader. All over the world men have looked to him for guidance and have seen symbolized in him their hope that the evils of this time would not be repeated; that the terrible sacrifices which have been made, and those that still have to be made, would lead to a world more fit for human habitation. . . . It is right that we should dedicate ourselves to the hope that his good works will not have ended with his death."

Three weeks later, on May 2, 1945, Berlin surrendered.

7

The First Cry of a Newborn World

As El Camino Real followed the Rio Grande across the American Southwest, the river curved in a 120-mile bend, lengthening and complicating the journey with deep canyons, Apache assaults, and patches of quicksand—the original badlands. A well-known shortcut, well-known for being bleak and harsh, required at least three days of forced twenty-four-hour marching with no water. After Pueblo Indians, revolting against the Spanish, lost over five hundred souls in nine days traveling the route in 1680, the shortcut was named Jornada del Muerto—the Journey of Death. Here in July of 1945, Oppenheimer would test the Nagasaki plutonium gadget that inspired him to quote Hindu scripture (*I am become Death*), building a base camp that meant dust in the lungs, scorpions in the bed, and eighteen-hour workdays of 100°F. Twice, night-flying B-29 pilots mistook the camp for a practice target and bombed it.

Frank Oppenheimer: "We spent several days finding escape routes through the desert, and making little maps so everybody could be evacuated."

Soldier Val Fitch: "In May, one hundred tons of TNT were exploded near the tower site as a calibration of some of the instrumentation. In view of what was to come later I doubt if the exercise was of any value, but at the time I thought that one hundred tons made an incredible explosion."

George Kistiakowsky deliberately transferred the bomb to the Jornada camp on Friday the thirteenth, and Hans Bethe filed his report "Expected Damage of the Gadget":

> Comparison with TNT: The most striking difference between the gadget and a TNT charge is in the temperatures generated. The lat-

ter yields temperatures of a few thousand degrees whereas the former pushes the temperature as high as [tens of millions of degrees]. . . .

The actual damage depends much on the objective. Houses begin to be smashed under shocks of 1/10 to 1/5 of an atmosphere. For objects such as steel supported buildings and machinery, greater pressures are required and the duration of the shock is very important. If the duration of the pressure pulse is smaller than the natural vibration period of the structure, the integral of the pressure over the duration T of the impulse is significant for the damage. If the pulse lasts for several vibration periods, the peak pressure is the important quantity. . . .

Other Damage: The neutrons emitted from the gadget will diffuse through the air over a distance of 1 to 2 km, nearly independent of the energy release. Over this region, their intensity will be sufficient to kill a person.

The effect of the radioactive fission products depends entirely on the distance to which they are carried by the wind. If 1 kg of fission products is distributed uniformly over an area of about 100 square miles, the radioactivity during the first day will represent a lethal dose (= 500 R units): after a few days, only about 10 R units per day are emitted. If the material is more widely distributed by the wind, the effects of the radioactivity will be relatively minor.

Originally Groves planned to detonate Fat Man inside a 240-ton steel canister so that, in case of disaster, the plutonium could be rescued. But it was impossible to construct the bomb inside this containment shell, so instead they sealed the windows of a ranch bedroom with tape to reduce the dust and assembled it there.

The Fat Man gadget, grungy and cobbled together, was composed of thirty-two implosion lenses weighing fifty-three hundred pounds, with an outer layer of 60 percent RDX (also known as cyclonite or hexogen, RDX was a popular industrial and military explosive in this era as it was more powerful than TNT), 39 percent TNT, and 1 percent wax, and an inner layer of 70 percent barium nitrate and 30 percent TNT, surrounding the split sides of the plutonium orb. Each lens was tethered by cloth-insulated wires that arched every which way to a Y-1773 detonator, which could be triggered so that all thirty-two would ignite precisely at the same time. Within the orb was the initiator of beryllium and polonium, already generating power and warm to the touch. When they tried aligning the orb's halves inside Kisty's lenses, however, they wouldn't fit. No one could understand it. Then someone suggested that the plutonium had gotten warm in the car ride over and expanded, and if they waited a bit, it would contract back to its design specifications. This turned out to be the case.

To raise Fat Man to its tower, Groves brought in a $20,000 winch, but everyone was so nervous that the Bomb's five-ton weight would break the winch's cable that soldiers used an untold number of mattresses to mound a fifteen-foot-high cushion on the linoleum-covered tower base. Here was the future of warfare, ominous in its bulging TNT and hurricane of ignition wires, protected by a hillock of beds.

The desert floor was scattered with instruments, to measure as many of the results as possible: Light. Sound. Force. Radiation. Timing. Density. Blowback.

As a much younger man, Oppenheimer had been introduced to poet John Donne by Jean Tatlock, the great love of his life who had committed suicide, and Trinity was named for two Donne poems, "Holy Sonnets XIV" and "Hymn to God, My God, in My Sickness":

> *Batter my heart, three-person'd God; for you*
> *As yet but knock; breathe, shine, and seek to mend;*
> *That I may rise, and stand, o'erthrow me, and bend*
> *Your force, to break, blow, burn, and make me new.*
>
> *As west and east*
> *In all flat maps—and I am one—are one,*
> *So death doth touch the resurrection.*

That day, Kitty gave Robert a four-leaf clover she'd found in their garden.

A thunderstorm eased in on Saturday the fourteenth and lingered. Everyone looked at the high metal tower with its giant metal gadget in its nest of cabling and saw a perfect target for a lightning strike. "Oppenheimer was really terribly worried about the fact that the thing was so complicated, and so many people know exactly how it was put together that it would be easy to sabotage," Dan Hornig, the electric trigger's designer, said. "So he thought someone had better babysit it right up until the moment it was fired. They asked for volunteers, and as the youngest guy present, I was selected. . . . Little metal shack, open on one side, no windows on the other three, and a sixty-watt bulb with just a folding chair for me to sit on beside the bomb, and there I was! All I had was a telephone. I wasn't equipped to defend myself.

I don't know what I was supposed to do. The possibility of lightning striking the tower was very much on my mind." Guarding the tower's base was control-room operator Joe McKibben, who fell asleep on the linoleum floor: "I started dreaming Kistiakowsky had gotten a garden hose and was sprinkling the bomb. Then I woke up and realized there was rain in my face."

Oppenheimer got the results of a dress rehearsal that had been done with a replica bomb and no plutonium in another canyon. It had failed, and he emotionally fell apart. An emergency meeting of Oppenheimer, Groves, Conant, and Kistiakowsky led to the three yelling at Kisty that his design was the problem. The Cossack insisted that his lenses would work.

On Sunday the fifteenth, Hans Bethe called to say that the dummy test failed due to a calculation error that wasn't applicable to Fat Man, and Trinity was rescheduled for July 16, at 4:00 a.m. Joe McKibben was working in the control room: "I was told that [Oppie] came in the door and observed me at the controls and went away. Just to see that I was sane."

The bad weather continued, and besides the lightning, the meteorologist warned that the wet air might short out the electrical triggers, and that winds might carry radiant fallout to nearby towns, including the mesa itself. Groves was so worried about sabotage that when the weatherman suggested another postponement, the general suggested he be hanged.

Dan Hornig: "All the senior scientists who weren't actually involved in the test had a betting pool. The betting ran from a complete dud to little explosions to middle-sized explosions. Just a few people were willing to bet that it would produce what it was supposed to produce with something like twenty thousand tons of TNT's worth. There was a lot of skepticism."

Test director Kenneth Bainbridge was incensed that Enrico Fermi was taking bets "on whether or not the bomb would ignite the atmosphere, and if so, whether it would merely destroy New Mexico or destroy the world," talk that terrified the enlisted men. Ed Teller, meanwhile, was standing with Ernest Lawrence, his face covered in suntan lotion, his hands shielded with thick gloves, and his welder goggles in place. This, too, scared the nearby soldiers, who after all had no idea what to expect. Bainbridge, meanwhile, would become famous at this moment for telling Oppenheimer, "Now we're all sons of bitches."

At 5:03 a.m., the arming party unlocked the switches and started the timer. Everyone was instructed to lie facedown in the sand, turning away from the bomb and burying their faces into their arms. No one did this.

Across the base at 5:10 a.m., speakers and shortwave radios broadcast the voice of physicist Sam Allison announcing for the first time in history what is now known as a countdown. Allison: "I think I'm the first person to count backward."

At T-minus forty-five seconds, an automatic timing drum turned once a second, with a chime that struck at each turn. So there would be forty-four chimes before Allison bellowed, "Zero!"

One of the photographers was Berlyn Brixner, who had been told to be ready with his sixteen-millimeter black-and-white movie camera to film something that had never before been seen, which would begin with the brightest light that had ever reached the earth in human history. Brixner at least saw the irony: "The theoretical people had calculated a ten-sun brightness. So that was easy. All I had to do was go out and point my camera at the sun and take some pictures. Ten times that was easy to calculate."

Civilian tech Jack Aeby had helped Emilio Segrè set up radiation instruments hanging on barrage balloons eight hundred yards from Fat Man; immediately after transmitting their results, they would be vaporized. Aeby had brought his own camera with him, filled with the new Anscochrome color slide film; his boss had gotten it through security. Jack set up a folding chair in the dirt and put on his government-issued welding goggles. He didn't see that one of the lenses had a crack.

At T-minus thirty seconds, a voltmeter indicated that the detonation circuit had achieved full charge. Those standing by Oppenheimer say he seemed to have completely stopped breathing. He held himself up by holding on to a wood fence post and stared at the tower.

Then on July 16, 1945, at 05:29:45 mountain time, in a predawn desert, pitch-black and silent, a light exploded. That light, first white, then red, then purple, was visible on the horizon for 150 miles. As it reddened, it reduced in intensity enough to reveal a fireball raising a self-illuminated mushroom cloud of pulverized debris. The cloud rose 7.5 miles into the air, while the sky around it boiled purple from the ionization of the atmosphere. It smelled first like the desert, and then like a waterfall.

Robert Serber: "At the instant of the explosion I was looking directly at it, with no eye protection of any kind. I saw first a yellow glow, which grew almost instantly into an overwhelming white flash, so intense that I was completely blinded. There was a definite sensation of heat. The brilliant illumination seemed to last for about three to five seconds, changing to yellow and then to red; at this stage it appeared to have a radius of about twenty degrees. The first thing I succeeded in seeing after being blinded by the flash looked like a dark violet column several thousand feet high. This column must actually have been quite bright, or I would not have been able to distinguish it. By twenty or thirty seconds after the explosion I was regaining normal vision. At a height of perhaps twenty thousand feet, two or three thin horizontal layers of shimmering white cloud were formed, perhaps due to condensation in the negative phase of the shock wave. Some time later, the noise of the explosion reached us. It had the quality of distant thunder, but was louder. The sound, due to reflections from nearby hills, returned and repeated and reverberated for several seconds, very much like thunder. A column of white smoke appeared over the point of the explosion, rising very rapidly, and spreading slightly as it rose. In a few seconds it reached cloud level, and the clouds in the immediate neighborhood seemed to evaporate and disappear. The column continued to rise and spread to a height of about twice the cloud level. There was no appearance of mushrooming at any height. A smoke cloud also was spreading near ground level. The grandeur and magnitude of the phenomenon were completely breathtaking."

Edwin M. McMillan: "At about thirty seconds, the general appearance was similar to a goblet; the ball I estimated to be about a mile in diameter and about four miles above the ground, glowing with a dull red; a dark stem connected it with the ground, and spread out in a thin dust layer that extended to a radius of about six miles. When the red glow faded out, a most remarkable effect made its appearance. The whole surface of the ball was covered with a purple luminescence, like that produced by the electrical excitation of air, and caused undoubtedly by the radioactivity of the material in the ball. This was visible for about five seconds; by this time the sunlight was becoming bright enough to obscure luminous effects. At some time near the end of the luminescence (I am not sure whether it was before or after) a great cloud broke out of the top of the ball and rose very rapidly to a height of about eight miles, expanding to a rather irregular shape several times as large as the ball. The whole spectacle was so tremendous and one might almost say fantastic that the immediate reaction of the watchers was one of awe rather than excitement. After some minutes of silence, a few people made remarks like 'Well, it worked,' and then conversation and discussion became general. I am sure that all who witnessed this test went away with a profound feeling that they had seen one of the great events of history."

Enrico Fermi: "After a few seconds the rising flames lost their brightness and appeared as a huge pillar of smoke with an expanded head like a gigantic mushroom that rose rapidly beyond the clouds probably to a height of thirty thousand feet. After reaching its full height, the smoke stayed stationary for a while before the wind started dissipating it. About forty seconds after the explosion the air blast reached me. I tried to estimate its strength by dropping from about six feet small pieces of paper before, during, and after the passage of the blast wave. Since, at the time, there was no wind, I could observe very distinctly and actually measure the displacement of the pieces of paper that were in the process of falling while the blast was passing. The shift was about two and a half meters, which, at the time, I estimated to correspond to the blast that would be produced by ten thousand tons of T.N.T."

Philip Morrison: "The column looked rather like smoke and flame rising from an oil fire. This turbulent red column rose straight up several thousand feet in a few seconds growing a mushroom-like head of the same kind. I noticed two deep thuds which sounded rather like a kettle drum rhythm being played some distance away. I remember the sound as being without any important high frequency components as cracks, etc."

James Conant: "The enormity of the light and its length quite stunned me. My instantaneous reaction was that something had gone wrong and that the thermal nuclear transformation of the atmosphere, once discussed as a possibility and only jokingly referred to a few minutes earlier, had actually occurred. . . . It looked like an enormous pyrotechnic display with great boiling of luminous vapors, some spots being brighter than others. Very shortly this began to fade and without thinking the [welder's] glass was lowered and the scene viewed with the naked eye. The ball of gas was enlarging rapidly and turning into a [ten-thousand-foot-high] mushroom. It was reddish purple, and against the early dawn very luminous."

Next to him was Groves, grumbling as ever: "Well, there must be something in nucleonics after all."

General Farrell's War Department report on the cataclysm in the American desert was released to the press the day after Hiroshima: "The whole country was lighted by searching light with the intensity many times that of the midday sun. It was golden, purple, violet, gray, and blue. It lighted every peak, crevasse, and ridge of the nearby mountain range with a clarity and beauty that cannot be described. . . . Thirty seconds after the explosion came first the air blast, pressing hard against the people and things; to be followed almost immediately by the strong, sustained, awesome roar which warned of doomsday. . . ."

Kisty forgot to take cover when the ten suns exploded. He was knocked flat into the sand.

Dick Feynman was sitting in a military truck and thought the windshield would protect his eyes, so he didn't use any goggles. He guessed wrong and was temporarily blinded.

One man at base refused to use the protective glass and burned his corneas. He was given morphine and was not expected to be permanently blinded.

Jack Aeby had put his Perfex 44 camera on "bulb" and, in the dark before Zero, opened up the shutter, figuring that way he'd get a good image of the flash. Suddenly the light hit the cut in Aeby's glasses and he saw a brilliant electric line. "I could see that crack for some time afterward. . . . I released the shutter, cranked the diaphragm down, changed the shutter speed and fired three times in succession. I quit at three because I was out of film." Berlyn Brixner: "I was temporarily blinded. I looked to the side. The Oscura mountains were as bright as day. I saw this tremendous ball of fire, and it was rising. I was just spellbound! I followed it as it rose. There was no sound! It all took place in absolute silence. Then it dawned on me. I'm the photographer! I've gotta get that ball of fire."

McKibben: "Then an amazing thing: it was followed by echoes from the mountains. There was one echo after another. A real symphony of echoes. Too bad nobody had a recorder on that. It would have been played many times since then." Brixner: "I was looking up, and I noticed there was a red haze up there, and it seemed to be coming down on us. Pretty soon the radiation monitors said, 'The radiation is rising! We've got to evacuate!' I said, 'That's fine, but not until I get all the film from my cameras.' "

Hans Bethe remembered his first thought was "We've done it!" His second was "What a terrible weapon have we fashioned."

Joan Hinton: "It was like being at the bottom of an ocean of light. We were bathed in it from all directions. The light withdrew into the bomb as if the bomb sucked it up. Then it turned purple and blue and went up and up and up."

Frank Oppenheimer: "When one first looked up, one saw the fireball, and then almost immediately afterwards, this unearthly hovering cloud. It was very bright and very purple. . . . I think the most terrifying thing was this really brilliant purple cloud, black with radioactive dust, that hung there, and you had no feeling of whether it would go up or drift towards you." About the conversation with his brother at the time: "I think we just said, 'It worked.' "

Soldier Val Fitch: "Apparently no one had told the military policeman, stationed at the door of the bunker to control access, what to expect. He was absolutely pale and a look of incredible alarm was on his face as he came away from the bunker door to stand beside me and view the sight. I simply said what was on my mind: 'The war will soon be over.' "

The yield confounded all cynics—twenty-one thousand tons of TNT—84,000,000,000,000 joules. The terrain below the tower was pounded down into a crater eleven hundred feet wide; sand melted into a light green glass. When the bomb exploded, its tower was vaporized, the iron staining the silicon of the melted desert sand a startling bloodred.

The shock wave, traveling at the speed of sound, rattled windows at a distance of two hundred miles. It reached the observation point forty seconds after the light, literally rolling the ground under the observers' feet. Kenneth Greisen: "A tremendous cloud of smoke was pouring upwards, some parts having brilliant red and yellow colors, like clouds at sunset. These parts kept folding over and over like dough in a mixing bowl. . . . At about this time I noticed a blue color surrounding the smoke cloud. Then someone shouted that we should observe the shock wave traveling along the ground. The appearance of this was a brightly lighted circular area, near the ground, slowly spreading out towards us. The color was yellow."

Hans Bethe: "Practically everybody at the Trinity test was a scientist except one person, a journalist with the *New York Times* by the name of William Laurence. We were quite far away, twenty kilometers on Compania Hill, so that long after the fireball, the shock wave followed and made a tremendous rumble. Laurence was terribly afraid and cried out, 'WHAT WAS THAT?' So I explained to him that sound takes some time to propagate as compared to light." Actually there was one other nonscientist, nonmilitary of the 240 in attendance—the legendary J. E. Miera. He was there to make the cheeseburgers.

William Laurence: "The big boom came about 100 seconds after the Great Flash—the first cry of a newborn world. It brought the silent, motionless silhouettes to life, gave them a voice. A loud cry filled the air. The little groups that hitherto had stood rooted to the earth like desert plants broke into dance."

To calm area residents, the army immediately notified local radio stations and newspapers that Alamogordo Air Base's munitions dump had exploded, but that there were no fatalities. But many New Mexicans knew that this was no armory mishap. Rowena Baca: "My grandmother shoved me and my cousin under a bed because she thought it was the end of the world." Jim Madrid: "We saw this huge, huge light coming in from the north. It rose from the heavens, so bright, so extremely bright. It was the biggest thing I had ever seen in my life. It was rolling, getting fatter and bigger and taller. My mother said, 'The sun is coming close. The world is coming to an end.' She told me to drop to my knees, but I kept looking. If it was the end of the world, I wanted to see it. I was waiting for God to come out from around the ball of fire."

William Wrye came across some soldiers on his property, twenty miles from Los Alamos: "I went out there and asked what they were doing, and they said they were looking for radioactivity. Well, we had no idea what radioactivity was back then. I told them we didn't even have the radio on." A few months later, Wrye noticed that half of his black cat's fur had turned white.

After the truth came out, a *Santa Fe New Mexican* article reported what locals had imagined was going on at Los Alamos—everything from a manufacturer of submarine windshield wipers to a home for the army's unwed mothers.

Emilio Segrè: " About one hour after the explosion, Fermi donned a protective suit and carrying a radiation meter climbed into a specially shielded tank and cautiously proceeded to the site of the explosion to collect materials to be analyzed for fission products. He was much impressed when he found the sand under the detonation point melted to glass. The remainder of the day was spent collecting the records of the various instruments and other necessary work."

Scientists later discovered a downwind herd of cattle, their hides splotched in gray radiation burns. Soon, they would die.

On the mesa there was joy, for about twenty-four hours—parties, parades, drinking, singing, dancing—a celebration of triumph. Oppenheimer's assistant Anne Wilson: "Feynman got his bongo drums out and led a snake dance through the whole Tech Area." But a number woke up the following morning with hangovers, from both liquor and remorse. When Robert Wilson asked Oppenheimer why he seemed distressed, he replied, "I just keep thinking about all those poor little people." Richard Feynman: "One man I remember, Bob Wilson, was just sitting there moping." Bob Wilson told Feynman, "It's a terrible thing that we made."

8

My God, What Have We Done?

ON May 25, 1945—two weeks after Germany's surrender—Leo Szilard had an appointment to discuss the future of nuclear weapons with President Harry Truman at the White House. Instead, he was redirected to meet with former South Carolina senator and Secretary of State–designate James Byrnes to have a conversation that would define the postwar, Cold War Atomic Age. Szilard explained to Byrnes how the political makeup of the postwar world needed to be taken into account before using this astonishing weapon. Byrnes explained to Szilard that Congress needed to be shown what it had spent so much money on, and that the Soviet Union, which had recently invaded Hungary and Romania, needed to be reminded of its place in the world. Szilard said afterward he was "flabbergasted by the assumption that rattling the bomb might make Russia more manageable" and "was rarely as depressed as when we left Byrnes' house." Byrnes said that Szilard's "general demeanor and his desire to participate in policy-making made an unfavorable impression on me."

These points would be further discussed on May 31, 1945, when Henry Stimson called together a meeting of his Interim Committee—Oppenheimer, Fermi, Lawrence, Groves, Bush, Conant, and Compton, among others—to advise Truman on the use of the atomic bomb. Stimson began by saying that he believed what they had to consider was not "a new weapon merely but as a revolutionary change in the relations of man to the universe." Oppenheimer suggested, "If we were to offer to exchange information before the bomb was actually used, our moral position would be greatly strengthened," and none other than General George Marshall agreed. Marshall said that at least two Russian scientists should be invited to the Trinity test. Secretary of State to-be Jimmy Byrnes aligned with Ernest Lawrence, arguing that if the Soviets were brought aboard, they would insist on a full partnership, and that instead America needed to play two cards at once, "to push ahead as fast as possible in production and research to make certain that we stay ahead and at the same time make every effort to better our political relations with Russia." When asked what the difference was between atomic bombs and firebombs, Oppenheimer said that all living creatures within two-thirds of a mile would be irradiated, and that the look of the explosion—the immense flash of light and the shock wave; the ionized air and boiling flames; the mushroom cloud—was unforgettable. Leslie Groves's sole contribution to the meeting was to insist that, after the weapon was proven a success, "steps should be taken to sever [certain] scientists of doubtful discretion and uncertain loyalty from the program," by which he meant one scientist: Leo Szilard. Finally, Stimson and Conant decided, "The most desirable target would be a vital war plant employing a large number of workers and closely surrounded by workers' houses"—the workers being civilians.

The committee's science panel had a month to offer its report. Back at Met Lab in Chicago, James Franck, Leo Szilard, and Glenn Seaborg submitted a letter suggesting the weapon be demonstrated before members of the United Nations, instead of being used for battle. But Stimson's panel ignored this point, specifically noting that nuclear physicists "have no claim to special competence in solving the political, social, and military problems which are presented by the advent of atomic power."

By July 17, Szilard was able to get fifty-six of his Manhattan Project colleagues to sign "A Petition to the President of the United States," which asked Truman to ponder his "moral responsibilities" in "opening the door to an era of devastation on an unimaginable scale." If the USA cavalierly dropped atomic bombs, "our moral position would be weakened in the eyes of the world and in our own eyes. It would then be more difficult for us to live up to our responsibility of bringing the unloosened forces of destruction under control." Oppenheimer, Compton, Fermi, and a number of others consulting for the Oval Office had been told that the United States could either invade Japan with a loss of hundreds of thousands of American lives or drop the Bomb. They did not sign Szilard's petition. Another who did not sign was Szilard's lifelong friend Ed Teller, who said, "The things we are working on are so terrible that no amount of protesting or fiddling with politics will save our souls. The accident that we worked out this dreadful thing should not give us the responsibility of having a voice in how it is to be used."

To avoid getting even higher on Leslie Groves's enemies list than he already was, Leo submitted his petition through the chain of command. Groves, however, had it classified "secret" so it could not be made public and held on to it until he learned that Little Boy was ready to be dropped on August 1. The petition would not reach Stimson's desk until a week after Fat Man exploded over Nagasaki.

When, during the Potsdam Conference at the end of July, the Bomb was explained to General Eisenhower, he said that it wouldn't be needed since the Japanese were about to surrender: "I was one of those who felt that there were a number of cogent reasons to question the wisdom of such an act." Truman's diary of July 17, 18, and 25 recorded his thoughts on the debate: "Even if the Japs are savages, ruthless, merciless and fanatic, we as the leader of the world for the common welfare cannot drop this terrible bomb on the old Capitol or the new. . . . The target will be a purely military one and we will issue a warning statement asking the Japs to surrender and save lives. I'm sure they will not do that, but we will have given them the chance. It is certainly a good thing for the world that Hitler's crowd or Stalin's did not discover this atomic bomb. It seems to be the most terrible thing ever discovered, but it can be made the most useful."

The great myth of nuclear arms is that they are different from conventional weapons in some magical way beyond their radioactive poisons. This was proved to be a fantasy at the dawn of the Atomic Age, both with Oppenheimer's lunchtime comment at Stimson's Interim Committee—that fission's distinction rests solely with irradiating living creatures for two-thirds of a mile, and its detonation's memorable beauty—and with the history of the B-29 Superfortress under the command of General Curtis Emerson LeMay. On the drawing boards since 1939 and manufactured by Boeing, North American, Bell Aviation, and General Motors' Fisher Body Division, the state-of-the-art B-29 had four engines running a blazing 8,800 horsepower, a pressurized cabin, remote-controlled machine-gun turrets, propellers that could be reversed for a backup taxi, an eleven-man crew, a ceiling of over thirty-five thousand feet, a combat range of more than four thousand miles, pneumatic bomb-bay doors, and a carriage load of up to twenty thousand pounds—the only craft in the entire Army Air Corps with the strength to ferry nuclear bombs. For the Manhattan Project, the Pentagon spent $2 billion. Developing the B-29 cost $3 billion.

Afflicted with Bell's palsy, which turned his normal demeanor into a scowl and inspired the nickname Iron Ass, General Curtis LeMay loved his men, and they loved him back. "I'll tell you what war is about," LeMay said. "You've got to kill people, and when you've killed enough, they stop fighting." A brilliant World War II innovator, LeMay oversaw the campaign of attacking Japan from the Marianas. He abandoned the Army Air Corps scripture of high altitude, precision bombing in daylight, since at the necessary heights the jet stream kept blowing his Superfortresses off course, while after studying flak and strike photos, LeMay's staff determined that Japan had no night fighters, and no low-altitude defense.

Thinking of how Japanese cities are made of paper and wood, LeMay was reminded of the great military innovation of 400 BCE, that mix of pitch, sulfur, granulated frankincense, and pine sawdust known as Greek fire. Stuffed into sacks, set alight, and thrown by catapults into enemy towns, Greek fire ignited everything it touched. Trying to stop its blaze with water only pushed the oil around, taking the fire with it, burning even more of the defender's village. Then in 673 CE, the Byzantines, requiring a method of defending Constantinople from naval assault, developed a liquid form of Greek fire—petroleum mixed with lime, bones, and urine—which could be shot at wooden ships from a pump.

LeMay ordered his B-29s to come in at night, at a mere five thousand feet, ferrying M-69 cluster bombs loaded with napalm, the modern liquid Greek fire that was a gelatin emulsion of gasoline developed by the chemists of Standard Oil, which stuck to anything it touched, and which could not be extinguished by conventional means. On March 9, 1945, 334 Superfortress flights dropped two thousand tons of incendiary bombs on Japan, crisscrossing their targets, creating a tidal wave of conflagration that became hurricanes of annihilating fire, slaughtering by suffocation and heat. Some residents even boiled to death trying to find shelter in ponds and canals.

The Japanese called the planes *B-san*—"Mr. B"—and the attacks were known as the Raid of the Fire Wind and the Raid of the Dancing Flames. Tokyo's Sophia University rector Father Gustav Bitter: "I heard the huzzle-huzzle of something falling, and I ducked and crouched in a corner. It struck beside me, with a noise like a house falling, and I leaped a fine leap into the air. I must have shut my eyes, for when I opened them again I was in a world of fairyland. On every tree in the garden below, and on every tree so far as the eyes could see, some sort of blazing oil had fallen, and it was dancing on the twigs and branches with a million little red and yellow candle-flames. On the ground in between the trees and in all the open spaces, white balls of fire had fallen, and these were bouncing like tennis balls. . . . [Watching the firebombs from a distance, it] was like a silver curtain falling, like . . .

the silver tinsel we hung from Christmas trees in Germany long ago. And where these silver streamers would touch the earth, red fires would spring up . . . and the big fire in the center sent up a rising column of air which drew in toward the center the outer circle of flame, and a hot, swift wind began to blow from the rim toward the center, a twisting wind which spread the flames between all the ribs of the fan, very quickly. Thus, everywhere the people ran there was fire, in front of them and in back of them, and closing in on them from the sides. So that there were only a very few who escaped."

At midnight on March 11, 1945, reconnaissance photographs from the first Raid of the Dancing Flames were finally available, and they shocked everyone in the Army Air Corps. Over sixteen square miles of Tokyo had been destroyed . . . more than Hiroshima and Nagasaki combined. According to Tokyo police records, one-quarter of the capital was gone, with 267,171 buildings demolished and over 1 million left homeless. It would take almost a month to excavate the dead—almost eighty-four thousand men, women, and children—more death than either atomic bomb would yield. On the twelfth, LeMay did the same to Nagoya, on the thirteenth to Osaka, on the sixteenth to Kobe; on the nineteenth, again on Nagoya. His team had run out of incendiaries and had to spend most of April helping in the invasion of Okinawa. Then in May it began again: Nagoya and Tokyo were hit twice and considered so destroyed they were taken off the target list. Osaka, Kobe, and Yokohama took five more raids, and by mid-June the industrial center of Japan had been completely annihilated. The summer meant "mopping up," with certain key targets precision-bombed beyond recognition, sixty smaller cities firebombed into oblivion, and the Japanese coastal waters so extensively filled with mines dropped by LeMay's B-29s that Japan's naval traffic stopped entirely.

One particularly ironic victim was the Riken Institute in Saitama, just north of Tokyo, where one of Bohr's longtime associates, Yoshio Nishina, had a sixty-inch cyclotron, chaired meetings of the Committee on Research in the Application of Nuclear Physics, and was trying to produce fissile uranium with thermal diffusion. In April 1945, Japan's own quest for nuclear power and atomic bombs came to an end when the Riken building was leveled.

The firebombing of sixty-three cities, LeMay claimed, "scorched, boiled, and baked to death" over eight hundred thousand people. The heat was so immense that buildings, and people, would spontaneously erupt in flames. Radio Tokyo called it "slaughter bombing," and Secretary of War Henry Stimson told Oppenheimer he thought it was appalling that no one in

America protested this gruesome carnage. At the time of his Interim Committee, Stimson wrote in his diary, "I was a little fearful that before we could get ready the Air Force might have Japan so thoroughly bombed out that the new weapon would not have a fair background to show its strength."

Comparing Tokyo's fire with Hiroshima's uranium proves nuclear's myth. Besides the iconic multicolored mushroom cloud and the aftereffects of fallout, there is no significant difference except for grandeur. LeMay could've inflicted the same damage to Hiroshima as Little Boy with 210 conventional firebomb strikes, and to Nagasaki as Fat Man with 120 . . . and if LeMay didn't need to show off the country's revolutionary, powerful, and expensive new weapon, that is just what he would have done.

One of those involved in incendiary planning, Robert McNamara—who would become secretary of defense for both the Kennedy and Johnson administrations—remembered, "LeMay said if we'd lost the war, we'd all have been prosecuted as war criminals. And I think he's right. He, and I'd say I, were behaving as war criminals. LeMay recognized that what he was doing would be thought immoral if his side had lost. But what makes it immoral if you lose and not if you win?" After the birth of thermonuclear arms, nuclear physicists from Leo Szilard to Andrei Sakharov would ask the same questions of themselves.

On July 26, 1945, in a Hohenzollern palace outside Berlin, Harry Truman, Winston Churchill, and Chiang Kai-shek demanded "the unconditional surrender of all Japanese armed forces." Japan answered with *mokusatsu*, which means both "no comment" and "silent contempt." That same day, the USS *Indianapolis* arrived at Tinian Island in the Marianas (due east of the Philippines and south of Tokyo) with a three-hundred-pound lead bucket welded to its deck—the heart of Little Boy.

From the air, Tinian looks like Manhattan Island in New York, so the Seabees who built the biggest air base in the world there for 265 of LeMay's B-29s (with six eighty-five-hundred-foot runways and forty thousand employees) gave the roads such names as Forty-Second Street, Wall Street, and Broadway. A herd of cattle for fresh meat was grazed in Central Park, and the headquarters for the 509th Composite Group was on the corner of Eighth Avenue and 112th—in honor of Columbia University. The 509th were given the best of everything—housing, food, liquor, a thousand-seat movie theater, and tubs of ice cream that had been flown to thirty thousand feet for a quick deep-freeze at a cost of $25,000. The only man who knew why they were here and what they were going to do was Colonel Paul Warfield Tibbets: "The people from Trinity had arrived in the Marianas, and they had with them at that particular time color photographs of the Trinity explosion. So we got the gang together and we showed them. We did not use the word 'atomic bomb,' we did not use that, but we said, 'OK, this is the bomb, this is what will happen when we make our flight tomorrow and release it. This is what we're gonna see.'" Tibbets's mother had an unusual name—Enola Gay—which her son's loving intentions would make infamous on August 6 when Superfortress *Enola Gay* rendezvoused at dawn with the squadron's instrument and camera planes. Bombardier Deke Parsons crawled into the bomb bay, pulled out a green plug in Little Boy's innards, and replaced it with a red one. Now the uranium gun was armed.

One of those attending the drop from Los Alamos was Luis Alvarez, who had designed an instrument to measure Little Boy's yield when it was detonated two thousand feet over Hiroshima—the altitude that maximizes the destructive force of a uranium bomb. A set of these instruments were parachuted in advance of the drop; they then radioed their results back to Luis while he flew nearby in one of the instrument planes, the *Great Artiste*.

Enola Gay rose to thirty-one thousand feet. A coded message arrived from the weather plane with the final report: less than three-tenths of a mile of cloud cover obscured the primary, meaning it should be targeted. Tibbets told his men the mission was go. A fan-shaped town nesting on the six estuarial islands of the Ota River, Hiroshima was the home of Mitsubishi Heavy Industries, shipyards, the Second Army headquarters, an ordnance depot, and around 245,000 people. Paul Tibbets: "As we came in from our initial point to the bomb-release point, it was again routine. We were bothered not in the least by any type of fighter opposition, no flak, we didn't see anything to cause us any concern so we were able to concentrate strictly on the bombing problem."

At 8:11 a.m., bombardier Tom Ferraby flipped back his baseball cap so he could peer into his state-of-the-art Norden bombsight. He focused on the designated landmark—a T-shaped river bridge—and the Norden's analog computer directed the plane and opened its bay. The radio emitted a constant whine to alert nearby American aircraft that a bomb was about to fall.

At 8:15:17, Little Boy was dropped. Tibbets sheared across a 158-degree dive-turn to pull away. Just like at Trinity, the crew had been given welder's goggles to protect their eyes, but Tibbets discovered they were so dark he couldn't read his craft's instruments and dropped them to the floor.

On August 6, 1945, at 8:16 a.m., forty-three seconds after release and 1,850 feet over the town, Little Boy's inner gun fired and detonated.

None of the photographers working the film plane were able to capture the blast. Hiroshima's survivors called it the *pika-don*—the "flash-boom."

At the moment of explosion, the air registered 100 million degrees F. In the city below, 5400°F temperatures melted granite, clay roofing tiles, and even the mica of gravestones, for three-fourths of a mile. The flash burned paint but was deflected by foreground objects, leaving atomic shadows on the walls. A blast wave of eleven hundred feet per second (the speed of sound) erupted. Within a one-mile radius, the only structures that survived were of reinforced concrete—rare in Japan. Hiroshima had seventy-six thousand buildings. Now, seventy thousand were gone.

At seventeen thousand spots the water mains cracked apart, leaving firemen helpless against the hurricane of conflagration. There were 150 doctors, but most were dead or injured, and 1,780 nurses, but 1,654 of these were either dead or injured. Then the fire in the sky of ten suns was replaced by an ever-growing darkness, as dust thrown up by the blast combined with smoke from the hurricane of fire. Journalist William Langewiesche: "There is a moment of calm. The fireball is no longer visible, but it is still extremely hot, and it is vigorously rising into the atmosphere. [From this] displacement of air, a result of its rise, the winds now reverse and begin to flow back towards the epicenter at speeds up to 200 miles an hour, ripping apart damaged structures that somehow so far remained standing. These 'afterwinds' raise dirt and debris into the base of the telltale mushroom cloud now beginning to form. The broken city lies like kindling, and whether because of electrical shorts or gas pilot lights, it begins to burn."

Paul Tibbets: "The bomb blast hit us. It hit us in two different shock waves, the first being the stronger. This as I say was a perfectly unexciting and routine thing up to the point of taking a look at the damage that had been done, and then it was a little bit hard to realize, it was kind of inconceivable, as to what we were looking at there. We passed comments back and forth in the airplane, we took pictures, and by the time we had done that I became concerned that we better quit being sightseers and get out of there." He felt a fizziness on his tongue from the radiation, and then a taste of lead.

Enola Gay's copilot, Robert Lewis, wrote in his journal, "My God, what have we done?"

Haruko Ogasawara: "I found that there was nothing around me. My house, the next-door neighbor's house, and the next had all vanished. I was standing amid the ruins of my house. No one was around. It was quiet, very quiet—an eerie moment. . . . I wonder how much time had passed when there were cries of searches. Children were calling their parents' names, and parents were calling the names of their children. . . . A mother, driven half-mad while looking for her child, was calling his name. At last she found him. His head look like a boiled octopus. His eyes were half-closed, and his mouth was white, pursed, and swollen."

Dr. Hiroshi Sawachika: "We heard the strange noise. It sounded as if a large flock of mosquitoes were coming from a distance. We looked out of the window to find out what was happening. We saw that citizens from the town were marching towards us. They looked unusual. . . . Soon afterwards, we learned that many of them had been badly burned. As they came to us, they held their hands aloft. They looked like they were ghosts. . . . When I stepped inside, I found the room filled with the smell that was quite similar to the smell of dried squid when it has been grilled. The smell was quite strong. It's a sad reality that the smell human beings produce when they are burned is the same as that of the dried squid when it is grilled."

Kikuno Segawa: "A woman who looked like an expectant mother was dead. At her side, a girl of about three years of age brought some water in an empty can she had found. She was trying to let her mother drink from it."

Journalist Wilfred Graham Burchett: "Hundreds upon hundreds of the dead were so badly burned by the terrific heat generated by the bomb that it was not even possible to tell whether they were men or women, old or young. Of thousands of others, nearer the center of the explosion, there was no trace. The theory in Hiroshima is that the atomic heat was so great that they burned instantly to ashes—except that there were no ashes."

The immediate aftermath: 78,150 people killed, 13,983 missing, and 37,425 injured. Most of them died from the 85 percent of a nuclear weapon's power: blast and heat. One particularly horrifying group of victims looked directly at the blast. The heat melted their eyeballs, and they survived for a dozen hours or so with hollow sockets.

The doctors and nurses of Hiroshima still alive and able to work now found they were dealing with a strange new disease in those who did not die of injuries from blast or fire. Wilfred Graham Burchett: "I found people who, when the bomb fell, suffered absolutely no injuries, but now are dying from the uncanny aftereffects. For no apparent reason their health began to fail. They lost appetite. Their hair fell out. Bluish spots appeared on their bodies. And then bleeding began from the ears, nose, and mouth. At first, the doctors told me, they thought these were the symptoms of general debility. They gave their patients vitamin A injections. The results were horrible. The flesh started rotting away from the hole caused by the injection of the needle. And in every case the victim died."

Wilhelm Röntgen's cathode ray discovery was so mysterious he called it X—the unknown.

1

2

Outcasts from the inbred society of Parisian science, Pierre and Marie Curie made Alfred Nobel's medal significant, while his prize made them famous across the globe.

3

Marie Curie's daughter Irène and her husband, Frédéric Joliot, created the foundation of nuclear medicine. Marie was the first woman to win the Nobel and Irène, the second.

4

The greatest physicists in history at the 1927 Solvay Conference. In the first row, Max Planck sits to the left of Curie, with Paul Langevin to the right of Einstein. Niels Bohr is the last man in the second row, while Werner Heisenberg is third to the last in the rear.

5

Enrico Fermi and Niels Bohr on the Appian Way, circa 1931. Bohr did not think a bomb made from uranium was technically feasible, and Fermi started the process that proved him wrong.

6

One of the greatest figures in science—Lise Meitner, discoverer of fission—is no longer remembered, as she was written out of history by the Nazis for being a Jew, and by the Germans for being a woman.

7

On December 2, 1942, Fermi and Szilard's nuclear reactor burned to life in a squash court, which Soviet intelligence translated as "pumpkin patch." One of the most dangerous experiments in the history of physics, it was flawlessly executed.

8

Norman Hilberry and Leo Szilard outside the University of Chicago's abandoned football stadium, where nuclear power was born. Szilard convinced Einstein to write Roosevelt letters warning of Hitler's nuclear intentions; these ignited the Manhattan Project and the Atomic Age.

9

Robert Oppenheimer and Ernest Lawrence in a 184-inch cyclotron in 1946. They began the era as the closest of friends, but ended it as enemies over arms control and Robert's affairs.

The Trinity gadget—covered in electrical explosive charges that had to be executed with perfect timing and geometry to compress its inner plutonium core and achieve fission.

10

11

At sixteen milliseconds after ignition on July 16, 1945, the first nuclear bomb's combination of plasma, ionized air, and debris created a "skin" that made it look like a biological creature rising from the deserts of New Mexico.

Edward Teller was the Richard Nixon of physics, testifying against Robert Oppenheimer and leading Ronald Reagan down the path of Star Wars. But many believe his invention of the hydrogen bomb made Alfred Nobel's dream come true, keeping the Cold War cold and the world at peace.

12

Los Alamos's Little Boy uranium bomb was so simple, and its fuel so rare, that its first test detonation would be in the skies over Hiroshima.

Fat Man—the "gadget" as weapon—trundled across Tinian Island on its way to Japan.

15

Over Nagasaki on August 9, 1945: the second and last time an atomic weapon was ever used in war.

16

On July 1, 1946, Able exploded over Bikini, creating revolutions in both weaponry and swimwear.

17

Thousands of atomic bombs were test-detonated in Nevada; May 23, 1953, marked the first time a nuclear warhead was shot from a cannon.

18

This mannequin family lived a typical US suburban lifestyle until a nuclear bomb exploded fifty-five hundred feet from their Nevada test home on May 5, 1955.

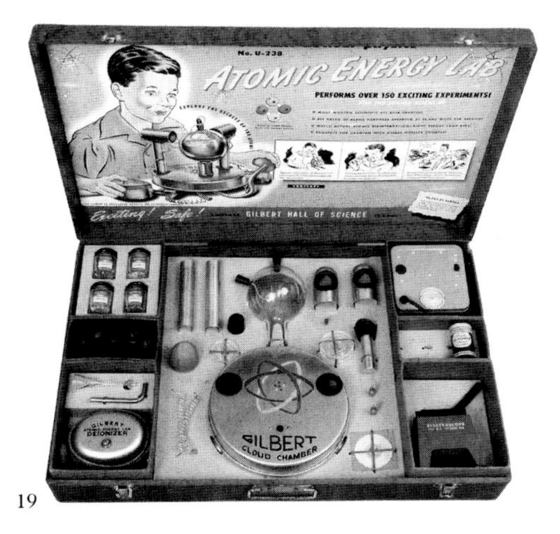

19

During the heyday of atomic euphoria in the United States, Gilbert produced a 1950 children's toy with a Geiger counter, cloud chamber, spinthariscope, electroscope, and five radioactive elements.

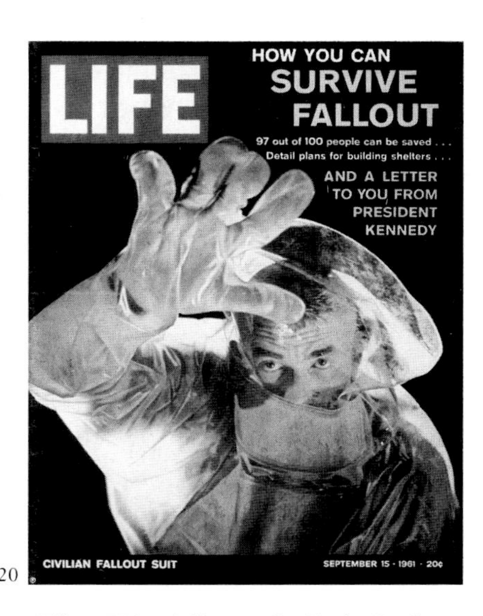

20

The US Office of Civil Defense believed Americans could be trained to survive nuclear attack with shelters dug out of backyards and terrifying exhortations from popular magazines.

22

First a comic book designed to educate children on the wondrous benefits of nuclear medicine, power, and bombs, September 1953's *Picture Parade* became a chilling symbol of nuclear holocaust in the 1960s. Today it is a comical mouse pad.

21

The John Amos Power Plant in Poca, West Virginia, August 1973. Our false sense of alarm in looking at this image—which is not even nuclear, but coal powered—reveals both our beliefs in the atomic myth, and the downfall of the Atomic Age.

23

24

Both the 1986 Chernobyl explosion (pictured on the left) and 2011 Fukushima meldown (pictured below) were dramatic, with Chernobyl releasing a cloud across Europe equal to four hundred Hiroshimas. But the real danger to human health was in core meltdowns through the floor, contaminating the water supply.

25

The "uncanny" had been struck with alpha particles (helium nuclei), beta particles (electrons), and gamma rays (similar to but more powerful than X-rays)—the three forms of radiation produced by nuclear weapons and power-plant meltdowns. The first two do little damage to human bodies—beta particles can burn the skin if they remain in contact for an extended period, and can be dangerous if swallowed, such as with iodine-131, which afflicts the thyroids of children—but gamma rays, the most energetic waves in the electromagnetic spectrum, ionize everything they touch, ripping off electrons and leaving behind ions. A strong gamma dose means radiation sickness (nausea, diarrhea, fever, headache, hair falling out, bleeding gums, and dropping white and red blood cell counts), malnutrition from intestinal damage, immune-system collapse from bone marrow damage, lack of blood flow to the brain from vascular damage, and cancer. Today, acute radiation syndrome can be treated with blood transfusions and antibiotics, but these remedies were unknown in Hiroshima and Nagasaki. Women who were pregnant at the time of the attack found some of their babies had heads that were smaller than normal . . . an outcome that Ukrainian naturalists would discover, in recent years, with the birds of Chernobyl.

Luis Alvarez wrote to his son: "What regrets I have about being a party to killing and maiming thousands of Japanese civilians this morning are tempered with the hope that this terrible weapon we have created may bring the countries of the world together and prevent further wars. Alfred Nobel thought that his invention of high explosives would have that effect, by making wars too terrible, but unfortunately it had just the opposite reaction. Our new destructive force is so many thousands of times worse that it may realize Nobel's dreams."

Hearing about the mission's success, Harry Truman said, "This is the greatest thing in history." On August 6, the president gave a radio broadcast: "The force from which the sun draws its power has been loosed against those who brought war to the Far East. . . . We have spent two billion dollars on the greatest scientific gamble in history—and won. . . . [The enemy] may expect a rain of ruin from the air, the like of which has never been seen on this earth. . . . It is an awful responsibility that has come to us. We thank God that it has come to us instead of our enemies, and we pray that He may guide us to use it in His ways and for His purposes."

Author John Hersey: "A surprising number of the people of Hiroshima remained more or less indifferent about the ethics of using the bomb. Possibly they were too terrified by it to want to think about it at all. Not many of them even bothered to find out much about what it was like. Mrs. Nakamura's conception of it—and awe of it—was typical. 'The atom bomb,' she would say when asked about it, 'is the size of a matchbox. The heat of it is six thousand times that of the sun. It exploded in the air. There is some radium in it. I don't know just how it works, but when the radium is put together, it explodes.' As for the use of the bomb, she would say, 'It was war and we had to expect it.' And then she would add, '*Shikata ga nai*,' a Japanese expression as common as, and corresponding to, the Russian word *nichevo*: 'It can't be helped. Oh, well. Too bad.' . . . Many citizens of Hiroshima, however, continued to feel a hatred for Americans which nothing could possibly erase. 'I see,' Dr. Sasaki once said, 'that they are holding a trial for war criminals in Tokyo just now. I think they ought to try the men who decided to use the bomb and they should hang them all.' "

Survivor Kiyoshi Tanimoto, however, had a revolutionary idea for him and his fellow A-bomb victims to consider. Tanimoto traveled across the United States giving a lecture, "The Faith That Grew out of the Ashes," to American church groups: "The people of Hiroshima, aroused from the daze that followed the atomic bombing of their city on August 6, 1945, know themselves to have been part of a laboratory experiment which proved the longtime thesis of peacemakers. Almost to a man, they have accepted as a compelling responsibility their mission to help in preventing further similar destruction anywhere in the world. . . . The people of Hiroshima . . . earnestly desire that out of their experience there may develop some permanent contribution to the cause of world peace. Towards this end, we propose the establishment of a World Peace Center, international and nonsectarian, which will serve as a laboratory of research and planning for peace education throughout the world." In 1949, the Japanese legislature passed a law calling Hiroshima a Peace Memorial City, leaving the ruins of the Hiroshima Industrial Promotion Hall as they were—it is now called the A-Bomb Dome—and set aside funds for a park designed by Kenzo Tange, with its signature sculpture a *haniwa* cenotaph—a house of the dead.

When after Hiroshima, the Japanese did not unconditionally surrender, Truman approved the dropping of Fat Man on the city of Kokura. The date was set for August 11, but weather predictions made that a problem, so it was moved up to the ninth. That switch meant the plane set for the drop, the *Great Artiste*, couldn't be converted from its role as instrument plane at Hiroshima in time, so instead its crew and pilot, Chuck Sweeney, had to fly *Bockscar* instead.

On August 9, 1945, William Laurence, the writer Leslie Groves had hired from the *New York Times* to explain the Manhattan Project to the general public, joined the squadron, again on the *Great Artiste*:

> As I peered through the dark all around us, I saw a startling phenomenon. The whirling giant propellers had somehow become great luminous discs of blue flame. The same luminous blue flame appeared on the Plexiglas windows in the nose of the ship, and on the tips of the giant wings it looked as though we were riding the whirlwind through space on a chariot of blue fire. . . . I express my fears to Captain Bock, who seems nonchalant and imperturbed at the controls. He quickly reassures me: "It is a familiar phenomenon seen often on ships. I have seen it many times on bombing missions. It is known as St. Elmo's Fire." . . .
>
> Based on earlier weather reports, the crew of Bockscar flew to Kokura fully expecting to drop its bomb on the city and return quickly to Okinawa. Upon arrival, however, the military arsenal at Kokura was obscured by industrial haze and smoke from a nearby fire. The bombardier had specific orders not to drop the bomb unless he could see the target. Three times Sweeney passed overhead, but without success. With the fuel supply now an even greater concern and enemy flak becoming a problem, Sweeney took Bockscar on the most direct route to Nagasaki.
>
> Conditions at Nagasaki were even worse than they had been at Kokura, with cloud cover now as great as nine-tenths. With no possibility of reaching Okinawa with its heavy bomb aboard, a decision had to be made. Ashworth decided that rather than "waste" the multimillion dollar bomb by dumping it into the ocean, the "Fat Man" should be dropped by radar over the Nagasaki target. Less than thirty seconds before the bomb was due to be dropped by radar, an opening appeared in the clouds and Beahan shouted that he could make a visual drop. . . .
>
> Despite the fact that it was broad daylight in our cabin, all of us became aware of a giant flash that broke through the dark barrier of our ARC welder's lenses and flooded our cabin with an intense light.
>
> We removed our glasses after the first flash but the light still lingered on, a bluish-green light that illuminated the entire sky all around. A tremendous blast wave struck our ship and made it tremble

from nose to tail. This was followed by four more blasts in rapid succession, each resounding like the boom of cannon fire hitting our plane from all directions.

Observers in the tail of our ship saw a giant ball of fire rise as though from the bowels of the earth, belching forth enormous white smoke rings. Next they saw a giant pillar of purple fire, 10,000 feet high, shooting skyward with enormous speed.

By the time our ship had made another turn in the direction of the atomic explosion the pillar of purple fire had reached the level of our altitude. Only about 45 seconds had passed. Awestruck, we watched it shoot upward like a meteor coming from the earth instead of from outer space, becoming ever more alive as it climbed skyward through the white clouds. It was no longer smoke, or dust, or even a cloud of fire. It was a living thing, a new species of being, born right before our incredulous eyes.

At one stage of its evolution, covering millions of years in terms of seconds, the entity assumed the form of a giant square totem pole, with its base about three miles long, tapering off to about a mile at the top. Its bottom was brown, its center was amber, its top white. But it was a living totem pole, carved with many grotesque masks grimacing at the earth.

Then, just when it appeared as though the thing had settled down into a state of permanence, there came shooting out of the top a giant mushroom that increased the height of the pillar to a total of 45,000 feet. The mushroom top was even more alive than the pillar, seething and boiling in a white fury of creamy foam, sizzling upwards and then descending earthward, a thousand old faithful geysers rolled into one.

It kept struggling in an elemental fury, like a creature in the act of breaking the bonds that held it down. In a few seconds it had freed itself from its gigantic stem and floated upward with tremendous speed, its momentum carrying into the stratosphere to a height of about 60,000 feet.

But no sooner did this happen when another mushroom, smaller in size than the first one, began emerging out of the pillar. It was as though the decapitated monster was growing a new head.

As the first mushroom floated off into the blue it changed its shape into a flower-like form, its giant petal curving downward, creamy white outside, rose-colored inside. It still retained that shape when we last gazed at it from a distance of about 200 miles.

Nagasaki resident Sumiteru Taniguchi:

> At night, the town, the mountain, and the factories were all on fire, and it was as light as day. Amidst it all, people still searched for families and relatives. I saw an American plane coming down low to shoot these people. When that plane went up again, one stray bullet hit a rock, making a sharp sound. That rock was next to where I was lying.
>
> In the early hours, it started to rain, so I could swallow some water from the leaves. When the morning came, no one lying with me was still alive. And when the rescue team arrived, they thought I was dead like the others. I tried asking for help, but I couldn't muster the strength, so I was left there for two more nights. . . .
>
> Most victims of the A-bomb said that they became infested with maggots, but it took me over a year to have flies lay eggs on me. Even a small fly could not dare to come near my body. A professor of biochemistry said that maybe my body exerted a kind of smell that repelled the flies.

The Japanese would come up with a special term to describe those who survived the American attacks on Hiroshima and Nagasaki—*hibakusha*—"explosion-affected persons." Shunned as filthy and contagious by the cleanliness-is-godliness culture of Japan, for over a decade the government did nothing to help these people.

The actual use of their creation would now split the paradise of Los Alamos into battling camps for the whole of the Cold War. Physicist Robert Wilson: "The day after Trinity. We had done our job, and now the questions had become 'What had we done, and what did it mean?' Most of us stopped the physics that we were doing and began to think hard about that meaning. Three weeks later, the bomb was used at Hiroshima, then we knew, existentially, I suppose, what we had done, and we knew that it should not happen again. We knew that we, also, had not done our job, as perhaps we had thought before. We knew that we, not the army, not the government, should do our best to bring about a general understanding of the mysteries and implications of nuclear energy. We began thinking anew, as social beings and as citizens. We had many arguments. The arguments became furious at times on the hill. Some were agonizing, some were furious, and the wives joined in, all the people on the hill joined in. Five hundred people were involved, and in another three weeks, we had organized the Association of Los Alamos Scientists to help us with what we had appointed ourselves to do: to tell other people about what we would do to have it not happen again."

A few months later, Robert Oppenheimer remarked that the physicists involved in the Manhattan Project had "known sin." Johnny von Neumann's response: "Sometimes, someone confesses a sin in order to take credit for it." Oppie then began his acceptance speech of the army-navy's Excellence Award on November 16, 1945, with "It is with appreciation and gratefulness that I accept from you this scroll for the Los Alamos Laboratory, and for the men and women whose work and whose hearts have made it. It is our hope that in years to come we may look at the scroll and all that it signifies, with pride. Today that pride must be tempered by a profound concern. If atomic bombs are to be added as new weapons to the arsenals of a warring world, or to the arsenals of the nations preparing for war, then the time will come when mankind will curse the names of Los Alamos and Hiroshima."

Ed Teller: "Oppenheimer seemed to lose his sense of balance, his perspective. After seeing the pictures from Hiroshima, he determined that Los Alamos, the unique and outstanding laboratory he'd created, should vanish. When asked about its future, he responded, 'Give it back to the Indians.'" Robert Wilson: "I have to explain about Oppie: About every five years, he would have a personality crisis. He would change his personality. I mean, when I knew him at Berkeley, he was the romantic, radical-bohemian sort of person, a thorough scholar. Then at Los Alamos, he was the responsible, passionate person that we all knew so well there and who was so effective. Later on then, he had another metamorphosis, becoming the high-level statesman who could call [Secretary of State Dean] Acheson by his first name (and such other high-level people), but as a result of that was able to put forward the international plan for controlling atomic energy through the United Nations that we had all agreed was the necessary ingredient for continued survival."

Ernest Lawrence, meanwhile, hosted a V-J day celebration for Rad Labbers at Trader Vic's, where the bartender created an A-Bomb cocktail—rum, blue curaçao, and dry ice to make it bubble and smoke. Lawrence's wife, Molly, remembered it well: *ghastly*.

In the autumn of 1946, Leo Szilard visited Albert Einstein, and they reminisced over the correspondence with Roosevelt that had started it all. Einstein insisted that this was a lesson to be learned: "You see now that the ancient Chinese were right. It is not possible to foresee the results of what you do. The only wise thing to do is to take no action—to take absolutely no action." He later declared, "Had I known that the Germans would not succeed in producing an atomic bomb, I would not have lifted a finger."

In the ensuing decades, Americans would ask themselves: Was the Bomb necessary? After Nagasaki and Japan's surrender, 85 percent of the American public supported Truman's decision for a simple reason: it had ended the war. In April 1945, the US Joint Chiefs had approved Operation Downfall, a November 1 invasion of Japan, beginning with 700,000 Americans landing on Kyushu and eventually an invasion force of 1,532,000 Allied soldiers—bigger than D-day. George Marshall estimated that forty thousand would die; Secretary of War Stimson thought the mortality would fall between half a million and a million. So in this telling, dropping two atomic bombs saved the lives of hundreds of thousands, and many believe to this day that a scientific breakthrough closed that chapter of human suffering. Luis Alvarez: "What would Harry Truman have told the nation in 1946 if we had invaded the Japanese home islands and defeated their tenacious, dedicated people and sustained most probably some hundreds of thousands of casualties and if the *New York Times* had broken the story of a stockpile of powerful secret weapons that cost $2,000,000,000 to build but was not used, for whatever reasons of strategy or morality?"

Others did not see nuclear weapons as being any more immoral than other methods of slaughter. If not short-listed for nuclear war, Hiroshima and Nagasaki would have been firebombed by LeMay into oblivion just as had been done with every other Japanese metropolis of their size. Bombing civilians was a fact of this war, as anyone living in London or Dresden could attest. Philip Anderson: "The firebombing of Tokyo was so close to genocide, killed so many people, that it seemed to me much more of a horror than the atom bombs. Another thing I was conscious of, and I don't know why so few Americans are conscious of it, is Nanking [where in 1937, the Japanese massacred three hundred thousand Chinese]. Nanking and the Japanese behavior in China and Korea was a horrible thing, unbelievably savage. I don't think I have any complaint whatsoever about the atom bombs. And I'm not sympathetic to the Germans about Dresden. The old saying is absolutely right: 'He that soweth the wind shall reap the whirlwind.'" Others at the time asked, what if Stalin or Hitler had been first with nuclear arms? The carnage would have been unimaginable.

Then it was revealed that before Hiroshima, the Army Air Forces chief, General Hap Arnold, said that conventional bombing would have ended the war without requiring an invasion, and Chief of Naval Operations Admiral Ernest King thought Japan could be forced to surrender through a naval blockade, while General Eisenhower called using nuclear weapons "completely unnecessary" and "no longer mandatory as a measure to save American lives," and George Marshall thought the Japanese should have been warned ahead of time with the Potsdam call for unconditional surrender. In a postwar analysis, the United States Strategic Bombing Command determined, "Japan would have surrendered even if the atomic bombs had not been dropped, even if Russia had not entered the war, and even if no invasion had been planned or contemplated." Truman's aide Admiral William Leahy told the British chiefs of staff that it was "because of the vast sums that had been spent on the project," though it meant the United States "had adopted an ethical standard common to the barbarians of the Dark Ages." Yet in truth, even with both Hiroshima and Nagasaki devastated and much of their country in ruins from the napalm of Mr. B, the Japanese military chiefs still refused to surrender. They had lost sixty cities; Hiroshima and Nagasaki were just numbers sixty-one and sixty-two. If they hadn't given up after losing Tokyo, after all, they certainly wouldn't because of Nagasaki. It required the USSR's declaring war against the country, and the sacred intervention of Emperor Hirohito, to change their minds to an unconditional surrender on September 2.

So a new theory arose, based on what the man who would become secretary of state, James Byrnes, mentioned in his conversation with Leo Szilard. Using an atomic bomb against Japan was a diplomatic signal to the USSR, an attempt to make "Russia more manageable in Europe." British physicist Joseph Rotblat: "In March 1944, I experienced a disagreeable shock. [General] Groves said that, of course, the real purpose in making the bomb was to subdue the Soviets. . . . Remember, this was said at a time when thousands of Russians were dying every day on the Eastern Front, tying down the Germans and giving the Allies time to prepare for the landing on the continent of Europe. Until then I had thought that our work was to prevent a Nazi victory, and now I was told that the weapon we were preparing was intended for use against the people who were making extreme sacrifices for that very aim."

With hindsight, Hiroshima and Nagasaki prove that nuclear bombs are ineffective as weapons of war and, with insight, that their place in history as the only nuclear attacks in the world to date did not signal the end of World War II.

They signaled the start of the Cold War.

PART THREE

WORLD'S END

9

How Do You Keep a Cold War Cold?

BEFORE they left Los Alamos, Fermi, Oppenheimer, Bohr, Lawrence, Groves, Chadwick, and Compton had dinner to discuss the future. Some were hopeful about fission-generated nuclear energy; Groves fretted about a decline in American military power with the end of the war. "And Fermi," Oppenheimer remembered, "said, thoughtfully, 'I think it would be nice if we could find a cure for the common cold.' "

Enrico and Laura went home on New Year's Eve 1945, to the University of Chicago's newly created Institute for Nuclear Studies, where Fermi hoped to re-create the intellectual paradise of Weimar Germany with eleven laureates and future laureates—including Urey, Franck, Mayer, Anderson, Segrè, Teller, Dyson, Garwin, and Agnew. As Valentine Telegdi said, "It was a place where you could be proud to be the dumbest one."

Harold Agnew: "Just to show you what a straight shooter and a modest individual Fermi was: When Laura came back from Italy [after the war], she said she'd really like to have a dishwasher and a washing machine. Now, she had a Bendix washing machine, screwed to the floor, and it rotated parallel to the floor, and when it ran—there was no automatic balancing—the whole house sort of shook. It was quite a thing. But she wanted a new one. We were at dinner, and Enrico had just come back from Hanford, I guess. I asked him, wasn't he working for General Electric? Didn't he know the boss? And Laura said before that, that she had gone down to the local hardware store and put her name on a list to get a washing machine and a dishwasher, which was what you did after the war—it wasn't like today, you had to get on a list and wait. And I said, 'Enrico, gosh, you could call your friend, the president of General Electric, and they'd bring it by helicopter, and you'd get it for free, I bet!' Laura was intrigued with this idea. Enrico would have no part of it. No way. He would not use his influence, or whatever you want to call it, to get ahead in line."

Believing nuclear physics had become a mature field with little left to discover, Enrico turned his attention to subatomic particles composed of quarks and antiquarks—mesons. He explained this dramatic change by quoting Mussolini: "Either renew oneself, or perish." Though a complete novice in the high-energy field, Fermi became so engrossed with it for the remaining years of his life that he coined the terms *pion* and *muon*.

In 1951, Chicago's new cyclotron was inaugurated, and as at Oak Ridge, the lab workers had to be careful. One day, Herb Anderson picked up a piece of reinforced concrete, forgetting about the metal in its innards. The cyclotron's magnet yanked it so hard that his hand was crushed.

Valentine Telegdi: "At the yearly Christmas parties, the physics students would compete with the faculty in various tests (always loaded in favor of the students!) and put on theatrical sketches. In some of these an electronic computer able to provide instantly order-of-magnitude estimates, aptly named the ENRIAC, was displayed. This computer consisted of a large box, complete with blinking lights, and contained a junior faculty member who could imitate Fermi's voice and accent."

Nella Fermi: "One day my father brought home a strange substance which was soft like well-chewed chewing gum, yet could be shattered like glass. He told Giulio and me that he had been given a sample of this new material so that he could suggest possible applications for it. We were fascinated. He showed us how we could pull it into a long, thin string like chewing gum if you pulled slowly, but as soon as you jerked, it cracked! You could shape it into a hump or scratch designs on it, but leave it alone, and it melted into a blob. A blow with a hammer shattered it like glass and sent it flying all over the room. My father wouldn't demonstrate that one (I had to take it on faith), because if he did that it would be all over the room and we'd never get it back. I asked a lot of questions and got a physics lesson: this stuff was basically like glass, my father said, it was a liquid. Glass is not a liquid, I said. It is, said my father. I thought he was pulling my leg, but he convinced me. Glass had the molecular structure of a liquid and, given sufficient time, would melt into a blob, but it would take ages. I wouldn't be around to see it, he wouldn't be around to see it, none of us would be around to see it. We spent a happy afternoon with the odd material. My father was puzzling about possible applications but also taking boyish delight in the strange properties of the material. We thought about using it to patch up cracks on windows, but that would be no good, it would only drip down into a blob. He asked us for suggestions for possible uses, but we could come up with none, and neither could he. In spite of the fun that we had with it, we missed the obvious use. It was a great toy. Later it was marketed as Silly Putty."

In the autumn of 1945, Curtis LeMay joined Eighth Air Force chief Jimmy Doolittle to lead a squadron of B-29s flying nonstop from Tokyo to Washington. The trip's "only significance," as reported in the *Chicago Tribune*, was "that it is going to be possible very soon to fly from here to Tokyo in 24 hours by commercial airliner." That this operation had nothing whatsoever to do with the future of business travel became evident in an August 30, 1945, Army Air Forces memorandum, "A Strategic Chart of Certain Russian and Manchurian Urban Areas," identifying the Soviet Union's most important cities, their size, population, and industry, the number of nuclear bombs needed to destroy each, as well as the necessary B-29 flight paths over the north pole from Alaska, Germany, Norway, Italy, Crete, India, and Okinawa to their Red targets.

A more detailed AAF assessment of September 1945 decided that the Soviet Union would surrender after sixty-six "cities of strategic importance" were destroyed, with the field of combat narrowed by the atomic bombing of both the Suez Canal and the Dardanelles—the Middle East gateway to the Mediterranean and the Turkish gateway from the Black Sea to the Aegean—in sum requiring 466 Fat Men. Even Leslie Groves was taken aback: "My general conclusion would be that the number of bombs indicated as required, is excessive." In fact, the B-29's range wasn't anywhere close to accomplishing this absurdly belligerent scheme; the United States did not have enough Fat Men on hand or enough Superfortress-capable airfields at these varied locations; and the necessary aerial refueling technique and technology were in their infancies. So these weren't plans of attack for the moment they were written . . . but nightmares of the future.

At Los Alamos, Norris Bradbury succeeded Robert Oppenheimer and invited Edward Teller to replace Hans Bethe as Theoretical Division chief. Ever since relinquishing implosion calculations to Klaus Fuchs and inspiring Oppenheimer to move him from Bethe to Fermi, Teller had continuously been working on his Super, the great thermonuclear fission-fusion design that would obsess him for decades, which would be, as he saw it, the weapon giving the United States of America the ultimate defense against dictators like Hitler and Stalin.

Following the success of Trinity, Teller believed his Super would be put into fast-track development as the next evolution in atomic weaponry. In their last months in New Mexico together, Teller and Fermi had begun theoretical analysis of thermonuclear ignition, predicting that one cubic meter of liquid deuterium would explode with the force of 10 million tons of TNT. But when Ed asked Oppie to support postwar hydrogen bomb research, Robert coldly replied, "I neither can nor will do so." Then, instead of joining Teller to create the next generation of nuclear weapons, almost all of Los Alamos's most significant physicists returned to civilian life—Wigner going home to Princeton; Alvarez, Seaborg, and Segrè to Berkeley; Kistiakowsky to Harvard; and Bethe to Cornell. As Teller saw it, they had lost "their appetites for weapons work," and only he held fast to the faith. It was the beginnings of a conflict that would rend the scientific community for the next four decades.

On September 28, 1945, the Scientific Advisory Panel of Stimson's Interim Committee—Oppenheimer, Compton, Lawrence, and Fermi—recommended "that no such effort [comparable to the Manhattan Project] should be invested in [the fusion] problem at the present time, but that the existence of the possibility should not be forgotten, and that interest in the fundamental questions involved should be maintained." Oppie reported at the meeting, "General Groves told me very briefly . . . that, with things as they were, the work at Los Alamos ought to continue, but this did not apply to the Super," while Conant, who had been a staunch supporter of weapons development before Hiroshima, said that they had built one Frankenstein and didn't need another, that continuing thermonuclear R&D would be "over my dead body." Compton explained their fusion decision to Vice President Henry Wallace: "We should prefer defeat in war to victory obtained at the expense of the enormous human disaster that would be caused by its determined use."

Fermi knew that Teller had a very different opinion and asked him to write to the committee with his thoughts; Teller responded on October 31, proving himself a pioneer in the Washington tradition of threat inflation: "The time needed [for another country to produce thermonuclear weapons] may not be much longer than the time needed by them to produce an atomic bomb. . . . There is among my scientific colleagues some hesitancy as to the advisability of this development on the grounds that it might make the international problems even more difficult than they are now. My opinion is that this is a fallacy. If the development is possible, it is out of our powers to prevent it." This brief did not overturn the committee's overwhelmingly negative view. At around this time, Oppenheimer wrote Ernest Lawrence that nuclear weapons should be outlawed and shunned, "just like poison gases after the last war," but Lawrence, along with Luis Alvarez, still supported fusion research wholeheartedly, and this was the moment that would split the alumni of Los Alamos into two camps: arms race or arms control.

With support coming from nowhere, Teller gave up and joined Fermi in Chicago, where he was reunited with many old friends and colleagues, coauthored thirteen scientific articles, and appeared regularly in the *Bulletin of the Atomic Scientists*. His life was so good that his friend and colleague Gene Wigner insisted, "The years after Los Alamos and until the renewal of his preoccupation with national security were perhaps Teller's most fruitful years scientifically." Leo Szilard very much also wanted to return to Chicago, but Enrico let colleagues know that there would be no room for him.

In time, Szilard agreed with Fermi's take that nuclear science had matured and looked for another field with the potential for world-changing breakthroughs. By 1947, he was studying molecular biology at Cold Spring Harbor. His most significant work probed the relationship of brain chemistry, electricity, and memory, but as before in Berlin and then Chicago, he was more useful serving as a sounding board and a spark of inspiration for others than as a working lab scientist. About creator's remorse, he joked that having to watch MGM's 1947 *The Beginning or the End*—with John Gallaudet as Szilard, Joseph Calleia as Fermi, Hume Cronyn as Oppenheimer, and Leslie Groves as primary consultant—was more than enough atonement for the physicists who'd crafted the Bomb.

In October 1946, Congress was considering a bill that would keep atomic arms development under the Pentagon's command in an agency headed by Leslie Groves. After learning this, Leo Szilard and Edward Condon arranged for scientists at Chicago and Oak Ridge to send hundreds of telegrams to their congressmen and went to Washington to wage legislative war. Everyone in DC would meet with Ed and Leo since they were "Los Alamos scientists"—and, assuming their phone calls were being monitored, they ended each conversation with "And God bless General Groves!" Staggeringly, they won. On January 1, 1947, Truman signed a bill that transferred the nation's atomic powers from the War Department to a new civilian agency, the Atomic Energy Commission (AEC), headed by David Lilienthal, who'd previously overseen the Tennessee Valley Authority; Oppenheimer was named chairman of the agency's General Advisory Committee (GAC), and began his new life as a man-about-Washington, welcomed into the highest echelons of political power.

Bush, Conant, Oppenheimer, and Lilienthal then met constantly to forge an international policy that would prevent a global arms race, and some of their ideas were included in an American presentation by Bernard Baruch (who called it "the last, best hope of earth") to the just-created United Nations Atomic Energy Commission on June 14, 1946. Secretary of State Jimmy Byrnes, however, was so outraged by one of their ideas—a foreign body (a UN committee) controlling an avenue of American defense—that he included subclauses to maintain the American nuclear monopoly for years to come, something that the Soviets could never accept. Without Byrnes's machinations, the arms race could have been stopped before it had even begun.

Since America was the only nation with nuclear weapons, and since she expected to maintain this status indefinitely because her scientists knew "atomic secrets," it was difficult for most Americans to even understand the point of arms control. The United States was the richest country in the world and, being the only one with the Bomb, the most powerful, and to many, it seemed that this would be the way things would remain for decades to come. Oppenheimer, fearful of the threat of a global outbreak of nuclear arms, and Teller, worrying about a nefarious Uncle Joe Stalin wielding incomprehensible hydrogen weapons, seemed obscure or old-fashioned or even neurotic.

Instead, in the wakes of Hiroshima and Nagasaki and the revelation of their new status as a global superpower that had won World War II, the American public adopted nuclear weapons with the same popular enthusiasm that the world had greeted radium three decades before and X-rays at the start of the epoch. Fat Man and Little Boy became popular salt and pepper shakers; fluoroscopes were used to reveal a proper shoe size; uranium clinics offered cures for bursitis; and budding preteen chemists could experiment in the basement with their own Atomic Energy Lab while reading *Atomic Bunny* comics. Mailing in Kix cereal box tops and fifteen cents would get you 1947's Lone Ranger Atomic Bomb Ring, a plastic spinthariscope in the shape of an atom bomb that held polonium sparking a screen—"genuine atoms split to smithereens!" The AEC organized a popular fair spread over New York's Central Park, "Man and the Atom," sponsored by such nuclear contractors as Westinghouse, which gave youngsters a comic book, *Dagwood Splits the Atom*, whose hero had been chosen by Leslie Groves himself. Eventually one AEC chief, Lewis Strauss, would predict that in the next decade "it is not too much to expect that our children will enjoy in their homes electrical energy too cheap to meter; will know of great famines in the world only as a matter of history; will travel effortlessly over the seas and under them and through the air with a minimum of danger and at great speeds; and will experience a life span far longer than ours, as disease yields and man comes to understand what causes him to age. This is the forecast of an age of peace."

By the summer of 1948, Edward Teller's Chicago idyll was upended by news of the Soviet invasions of Hungary, his birthplace, and Czechoslovakia, with its uranium mother lode at St. Joachimsthal. Communists were victorious in China, and soon enough, they would successfully blockade Berlin. It appeared to many that America's foes were taking over the world, that the United States was in real danger. "Russia was traditionally the enemy," John von Neumann said of his countrymen. "I think you will find, generally speaking, among Hungarians an emotional fear and dislike of Russia." Had Edward Teller been certain that a hydrogen bomb was impossible, that nobody could make it, he would have set his sights elsewhere. But like Leo Szilard's thinking of Hitler, Ed was tormented by what might happen if the Americans failed to create such a mighty weapon, and the totalitarians succeeded.

Then came the call from the mesa: Los Alamos wanted Teller back. Director Norris Bradbury "was rather diffident in his approach to the scientists who had left," Stan Ulam said, explaining the ironic origins of what would turn out to be an epic clash of Hungarian physicist and Polish mathematician. "He felt that they should recognize by themselves how important for the country and the world it was for them to come back. As a result, although he wanted to, he did not like to ask people like . . . Teller to visit. It was actually left to me, with his consent, to write such invitations. . . . Thus, in a way I was instrumental in bringing Teller back to Los Alamos." Teller asked Oppenheimer what he thought; Oppie encouraged him to return; and in August 1948, Teller replied to Bradbury, "[I am] giving most serious consideration to this possibility. . . . The main reason that attracts me is the great importance of the work on the atomic bomb. I fully realize the menacing international situation and I believe that the United States must develop its military strength to the utmost if we are not to succumb to the danger of communism."

In the wake of his return to New Mexico, Ed Teller would so turn against Stan Ulam that Ulam would give him a photograph and inscribe it *To My Enemy*, while a student of Ed's, Freeman Dyson, would conclude that Teller was "a good example of the saying that no man is so dangerous as an idealist."

At a Council of Foreign Ministers meeting in London in September of 1945, Secretary of State Jimmy Byrnes decided to try using America's atomic monopoly to cow Soviet representative Vyacheslav Molotov: "You don't know Southerners. We carry our artillery in our pocket. If you don't cut out all this stalling and let us get down to work, I'm going to pull an atomic bomb out of my hip pocket and let you have it." Molotov seemed unconcerned. Later that night, he in turn was joshing with Britain's delegate and suddenly blurted out, "You know, we have the atomic bomb." No one knew what to think; Byrnes assumed Molotov was bluffing. In fact he was, but only by degree.

The previous July 24, at Potsdam, Truman remembered that he "casually mentioned to Stalin that we had a new weapon of unusual destructive force. The Russian premier showed no special interest. All he said was that he was glad to hear it and hoped we would make 'good use of it against the Japanese.'" "I was sure," Churchill said, "that [Stalin] had no idea of the significance of what he was being told . . . his face remained gay and genial and the talk between these two potentates soon came to an end. As we were waiting for our cars, I found myself near Truman. 'How did it go?' I asked. 'He never asked a question,' he replied." Jimmy Byrnes later explained that this was a deliberate feint, that the American president didn't want the Soviet premier to rush in against Japan when that war had already been won to claim spoils, as he was doing so furiously across Eastern Europe. But Stalin's apparent indifference at this momentous news had a good explanation—he knew about America's nuclear weapons program long before Truman did, since the Georgian had the world's best espionage agency, and the American didn't have a high enough security clearance before inheriting the Oval Office.

The day after *Enola Gay* struck Hiroshima, a Soviet nuclear physicist remembered, "Stalin summoned [atomic bomb chief Igor] Kurchatov and accused him of not demanding enough for maximum acceleration of the work. Kurchatov answered, 'So much is destroyed, so many people perished. The country is on starvation rations and everything is in shortage.'" He barely knew the full horror: Over 20 million Russians had died in World War II, and another 10 million had been executed by Beria's NKVD. Twenty-five million more were homeless; one hundred thousand farms, seventy-two thousand villages, and thirty-one thousand factories had been demolished; the Russian people were indeed starving. Yet, "Stalin said irritably to his Oppenheimer, 'If the baby doesn't cry, the mother doesn't know what he needs. Ask for anything you need. There will be no refusals.' "

During 1946, Igor Kurchatov created—with help from his NKVD spy network and Stalin's urgent priority of no refusals—a Soviet copy of Fermi's Chicago pumpkin patch at Arzamas-16, the Russian Los Alamos. Following Fermi's technique, Kurchatov built a series of ever-larger subcritical piles, and just as Anderson and Zinn had seen on the evening of their creation, it became clear to the Soviets on December 24, 1946, that adding one more layer would trigger fission.

Physicists B. G. Dubovsky and I. S. Panasyuk were there for Christmas, and their memories are almost a perfect echo of Chicago Pile-1's birth, with shutdown rods slowly removed from the pile, but in this case, the winching was done by Kurchatov himself: "We were all very anxious, of course. . . . Everyone was silent. The only sounds were the clicks of neutron counts from the loudspeakers and Kurchatov's brisk orders. As the seconds passed, the graph showed an almost linear growth of reactor power. For the first time the sound turned into a roar. The indicator lamps no longer blinked but burned with a reddish-yellow light. . . . The tension became extreme when the second boron trifluoride counter, which was located within the underground control room, began producing more frequent clicks than its background of two or three per minute—an increase which meant that neutrons from the reactor had penetrated the thick layers of earth and cement and reached the room."

A few days later, Beria came to see his new miracle. Stalin's whip was disappointed that the only sign of magical weaponry was the clicks and flashes of a Geiger-like counter. When Kurchatov told his boss he couldn't go into the room with a burning reactor since it was too dangerous, Beria became suspicious that he was being tricked, that this whole nuclear operation was some kind of fraud.

Two and a half years later in the summer of 1949, a coal-powered locomotive carried First Lightning, its scientists, its technicians, and its soldiers across two thousand miles to the northeast of Kazakhstan, to an isolated desert steppe swept in feather grass and wormwood. A hundred-foot tower had been built for the test and, beside it, a concrete hull for assembling the first Soviet atomic bomb. Nearby were recently built tunnels, bridges, train cars, brick and wooden houses, tanks, and livestock—targets to test the weapon's destructive power.

The blast was scheduled for 6:00 a.m. on August 29, 1949, but, as at Trinity, desert thunderstorms delayed history. At the countdown, Beria told

Kurchatov he was certain it wouldn't work; the scientist tried to appear confident. Then, after witnessing the explosion of ten suns, Kurchatov merely said, exactly as Oppenheimer had, that it worked. A technician later insisted that if it had been a fizzle, everyone there would have been shot.

In New Mexico, the bomb had fused the sand into a green glass; here, a scientist reported the array field as a glass plate "sparkling bluish black." The still-suspicious Beria was attended by two "journalists" who had witnessed an American Pacific test, so they could confirm that Kurchatov's explosion was authentic, that the Kremlin indeed had her own nuclear device. Beria hugged Kurchatov and then immediately called his confirmation eyewitnesses: "Did it look like the American one? How much? Haven't we slipped up? Did Kurchatov humbug us? Quite the same? Good! Good! So may I report [to] Stalin that the experiment was a success? Good! Good!"

Beria insisted that Stalin be woken up to hear this great news. But the dictator was angry: "What do you want? Why are you calling?"

"Everything went right."

"I know already." Stalin hung up.

Lewis Strauss, the financier who'd worked with Leo Szilard in the Hungarian's first years in America, was now an AEC commissioner who firmly believed that America needed a detection system to discover if any other country detonated a nuclear weapon. Private company Tracerlab believed that a test's atmospheric fallout would agglomerate into particles that could be detected in some manner, but at a meeting with Oppenheimer and Teller, the physicists insisted that such a thing was impossible, that only sonic and seismic detection would work. With Strauss's encouragement Tracerlab technicians set up an experiment to test their thesis anyway at a series of American detonations called Sandstone on the Pacific island of Eniwetok, where sniffer planes with fuselage ducts lined with paper filters flew across the jet stream, collecting atomic fallout. By measuring isotopes captured in the filters, Tracerlab could use half-life calculations to determine when the isotopes had been born. If all the birthdays were the same, than they must have been released in the aftermath of a nuclear device. The test results were so spectacular that in July 1948 the air force told the AEC that it had the technology and could detect nuclear detonations.

Outfitted with Tracerlab's sniffer technology, the 375th Weather Reconnaissance Squadron, three teams of B-29s based in Alaska, regularly patrolled from their home base across the Arctic and down to Japan in an arc downwind from the coastal mass of the USSR. On September 3, 1949, a member of the 375th was flying east of Kamchatka and came home with paper filters registering three hundred times normal. The analysis was so clear that, combined with rainwater tests by the navy, a Joint Committee on Atomic Energy staffer could track First Lightning's fallout cloud drifting across the Pacific, then splitting over the midwestern United States and Canada, with the southern half floating to Washington, where it hovered for two or three days, showering the district with rain and radioactive debris. Named Joe One in honor of Stalin by officials in the United States, Tracerlab calculated that the explosion had taken place at 6:00 a.m. on August 29 in Semipalatinsk, Kazakhstan. They were off by an hour. The Americans were shocked at how quickly they had lost their nuclear monopoly—Truman just couldn't believe "those Asiatics" had built a nuclear weapon—and Stalin in turn was flabbergasted that Washington could detect his Bomb set off in the middle of nowhere.

After a meeting on October 29, 1949, with the Joint Chiefs, where General Omar Bradley said he was unimpressed with the military potential of Teller's thermonuclear warheads as such an enormous weapon would have only "psychological" value, the Atomic Energy Commission's General Advisory Committee voted, again, against the Super. At the start of the meetings that ended in this decision, Fermi and Rabi were in favor of supporting Teller as an interim step, but both changed their minds, writing that Truman should "invite the nations of the world to join us in a solemn pledge" never to build thermonuclear arms and continued, "Necessarily such a weapon goes far beyond any military objective and enters the range of very great natural catastrophes. By its very nature it cannot be confined to a military objective but becomes a weapon which in practical effect is almost one of genocide. . . . It is necessarily an evil thing considered in any light." Oppenheimer at an AEC meeting on January 30, 1950, insisted of hydrogen bombs, "If the Russians have the weapon and we don't, we will be badly off. And if the Russians have the weapon and we do, we will still be badly off. . . . Going down this path ourselves, we are doing the one thing that will accelerate and ensure their [thermonuclear] development." In the middle of the terrifying discovery of Joe One and Russian nuclear power, though, Harry Truman was shocked to learn he could have an even bigger threat in his arsenal but had never been told about it. On January 31, Lilienthal tried to explain the AEC's antifusion thinking to Truman, and the president asked, "Can the Russians do it?" Lilienthal had to admit they could. Truman: "In that case, we have no choice. We'll go ahead." He rationalized, "We're going to use this for peace and never use it for war—I've always said this, and you'll see. It'll be like poison gas [never used again]."

On September 23, 1949, Truman announced to the American public that the Soviets had the Bomb, and at the end of January 1950 that America would develop thermonuclear weapons. Isidor Rabi: "I never forgave Truman. . . . For him to have alerted the world that we were going to make a hydrogen bomb at a time when we didn't even know how to make one was one of the worst things he could have done." An added wrinkle to this policy was that any scientist working with the government was told he or she should not make any public comments about the nation's nuclear strategies, as these would be "contrary to the national interest."

In a meeting at Cornell, Edward Teller was able to convince Hans Bethe, the Los Alamos boss he had spurned, to return to New Mexico to work on the Super. But after that meeting, Bethe took a walk with a colleague, theoretician Victor Weisskopf, and remembered how "Weisskopf vividly described to me a war with hydrogen bombs—what it would mean to destroy a whole city like New York with one bomb, and how hydrogen bombs would change the military balance by making the attack still more powerful and the defense still less powerful. . . . We both had to agree that after such a war, even if we were to win it, the world would not be . . . like the world we want to preserve. We would lose the things we were fighting for. This was a very long conversation and a very difficult one for both of us." A few days later, Bethe told Teller he'd changed his mind and wouldn't come back to the mesa. When Teller then tried to enlist Emilio Segrè at Berkeley, Teller's aggressive enthusiasm was countereffective. Segrè: "I soon realized . . . that he was dominated by irresistible passions much stronger than even his powerful rational intellect." Segrè, too, passed. But Teller's "irresistible passions" and fears of foreign menace would spread across Washington. Physicist Herbert York: "You have to recall that in 1948 was the Berlin blockade, in 1948 was the coup in Czechoslovakia, and the expansion that these things represented seemed quite real. And then there was the fall of China, as reported, to the Communists, the creation of the Sino-Soviet bloc, the Korean War in the early fifties. So looking back in the late fifties, what we saw was a lot of successes or what seemed to be successes on the part of the Russians, including territorial expansion. That was the high-water mark, but we didn't know it at the time."

Facing an ever-growing global Communist menace, Washington now reversed course. Those at the AEC long committed to arms control became a peculiar and disdained minority, while instead of Bradley's "psychological" dismissal, the January 13, 1950, Joint Chiefs' report decided that it was "necessary to have within the arsenal of the United States a weapon of the greatest capability, in this case the super bomb. Such a weapon would improve our defense in its broadest sense, as a potential offensive weapon, a possible deterrent to war, a potential retaliatory weapon, as well as a defensive weapon against enemy forces. [It was better] that such a possibility be at the will and control of the United States rather than of an enemy." White House assistant press secretary Eben Ayers said that three weeks later, on February 4, Truman told him "that we had to do it—make the [H-]bomb—though no one wants to use it. But, he said, we have got to have it if only for bargaining purposes with the Russians."

As historian Richard Rhodes pointed out, "When the GAC argued that building the Super might unleash unlimited destruction . . . it unwittingly enlarged the scope of its opponents' fears and encouraged them to pursue the project with even greater urgency, because they immediately translated the weapon's destructive potential into a threat and imagined the consequences if the enemy should acquire it first. An arms race is a hall of mirrors." The same confused strategic thinking that had led to Hiroshima and Nagasaki would now produce a weapon suitable only for genocide.

In 1948, Beria had insisted Kurchatov and his team faithfully reproduce America's Fat Man plutonium implosion bomb from the designs provided by Fuchs, Greenglass, and Hall to create First Lightning–Joe One. They knew that Trinity had worked, and they had no time to waste on fundamental research. But there was one significant exception to Beria's command. Andrei Sakharov had twice turned down offers to join the Soviet nuclear program, "but the third time, nobody bothered to ask my consent," and he was ordered to become part of the team. But then, just as Teller had worked on fusion while everyone else in Los Alamos was developing fission, in a few months at Arzamas-16, Sakharov created a thermonuclear design known as Layer Cake—*sloika*—a fission bomb surrounded by alternating layers of uranium and deuterium so that, when the fission ignited, it would in turn ignite fusion, becoming a hydrogen bomb. The Soviets had their Oppenheimer in Kurchatov, and now they would have their Teller in Sakharov. After Truman's announcement that America would build the Super, Beria immediately went forward with *sloika*, and the following day, on February 1, 1950, those frightened by the knowledge of vast Communist conspiracies running clandestine operations throughout the government of the United States had their paranoia confirmed when it was made public that Klaus Fuchs had been captured, and that he had confessed to being an atomic spy.

The revelation that America had been infested by a nest of Communist agents was the result of backbreaking intelligence work by two cryptanalysts, the FBI's Robert Lamphere and the Army Security Agency's Meredith Gardner, who spent twenty-seven years at Virginia's Arlington Hall clawing through Venona, an archive of World War II–era cables sent from the Soviet consulate in New York to Moscow Center. On December 20, 1946, Mr. Gardner broke a 1944 cable, which contained the name of every significant Manhattan Project scientist. The pair's next decrypt was "that someone (designated by the code name LIBERAL) had approached a man named Max Elitcher and had requested that Elitcher provide information to him on his current work at the Navy's Department of Ordnance," as Lamphere later said. Then they uncovered LIBERAL's wife, ETHEL, who "acted as an intermediary between [a] person or persons who were working on wartime nuclear fission research and for KGB agents" (LIBERAL and ETHEL would in time be revealed as Julius and Ethel Rosenberg). "Then, in mid-September [1949]," Lamphere continued, "still before the President's announcement [that the Russians had the Bomb], I found a startling bit of information in a newly deciphered 1944 KGB message." The cable was a summary of Harold Urey's gaseous diffusion process, and it meant that a Soviet agent, part of the British mission, had been a member of both the US and Canadian atomic programs. Within two weeks, the AEC pinpointed the author: Klaus Fuchs.

Fuchs had left Los Alamos on June 14, 1946, to begin his next assignment, developing nuclear weapons for the United Kingdom at Harwell, a former air force base just to the south of Oxford, where he was still currently employed. Lamphere informed his FBI higher-ups as well as British intelligence, MI5, with whom he'd had a strong professional relationship, since he had discovered in 1948 "that someone in the British embassy in Washington in 1944–45 had been providing the KGB with high-level cable traffic between the United States and Great Britain." That someone was Donald Maclean, who turned out to be key to the most humiliating espionage betrayal in British history, and now the same Robert Lamphere would present MI5 with the second-most humiliating betrayal in its history. The British, however, were not the only ones to feel the sting of embarrassment, for Maclean's colleague Kim Philby regularly received copies of Venona decrypts, while KGB agent Bill Weisband had worked within Arlington Hall (the signal intelligence unit's operation in what was once a Virginia girls' school) for five years.

Since the FBI had originally stolen the Russian cables decrypted by the Americans, under English law these could not be used by the prosecution as evidence at trial, and as the Soviets were allies during Fuchs's years of betrayal, his acts could not be called treason. The harshest sentence he could get would be fourteen years. Additionally, without the FBI documents, MI5 would have to get Fuchs to confess, a job assigned to traitor specialist William Skardon, who was, according to Lamphere, "sort of a British Columbo character, complete with disheveled appearance and an intellect that was sometimes hidden until the moment came to use it to point to incongruities in a suspect's story."

On December 21, Skardon began his interrogation with a friendly conversation about Fuchs's childhood. Then he asked, "Were you not in touch with a Soviet official or a Soviet representative while you were in New York?" Fuchs replied vaguely, "I don't think so," and when Skardon announced that they had "precise information which shows that you have been guilty of espionage on behalf of the Soviet Union," Fuchs again said, "I don't think so." The conversation continued, with Skardon edging Fuchs further and further away from ambiguity, until, on January 24, 1950, Fuchs admitted what he had done. The trial then lasted less than two months, since the government avoided using many witnesses needed to convict as they would in turn reveal the incompetence that had allowed so much to slip to the Kremlin, and that those who knew the technical details of Fuchs's work were seriously alarmed. That he was spying for Moscow wasn't just a revelation of how the Soviets had created First Lightning so quickly—Fuchs and von Neumann's thermonuclear patent of 1944 included much of Teller's Super design. Klaus Fuchs had done so much to help the Russians with fission, and now, through him, they could be well on their way to fusion.

It was a turning point in American history. Though the United States "came out of World War II the most powerful nation on earth—perhaps, briefly, the paramount nation of all time," as Richard Barnet, founder of the Institute for Policy Studies, remarked in 1985, "it has not won a decisive military victory since 1945 despite the trillions spent on the military and the frequent engagement of its military forces. What the United States got instead of victory was a national-security state with a permanent war economy maintained by a military-industrial complex—much like the Soviet Union in those departments, but with a far greater reserve of resources to squander. . . . It is one of history's great ironies that, at the very moment when the United States has a monopoly on nuclear weapons, possessed most of the world's gold, produced half the world's goods on its own territory, and laid down the rules for allies and adversaries alike, it was afraid." Almost seven decades on . . . and we are still afraid, and it all began with the concomitant revelations of Soviet bombs and atomic spies.

Six days after Fuchs's arraignment, on February 9, 1950, Wisconsin senator Joseph McCarthy announced from Wheeling, West Virginia, that he had a list of 205 Communists employed by the State Department, and two weeks after McCarthy's announcement, Robert Lamphere deciphered another 1944 Soviet cable that gave "reason to believe that someone in a lower-level position at Los Alamos, who had had furlough plans in late 1944 and early 1945, was a KGB agent." Lamphere informed the FBI in Albuquerque, who determined that the "most logical suspect for [another] Soviet agent" was none other than Edward Teller, as he was a "close associate of . . . Fuchs at Los Alamos [and] Dr. Teller had considerable contact with Fuchs in England in the summer of 1949," and he "made frequent trips away from the Los Alamos Project and could have furnished information to the Russians on a regular basis." The level of absurdity was institutional. An NKVD officer familiar with the FBI's operations in World War II said that the Bureau's agents were "like children lost in the woods."

With information from Klaus Fuchs's confession, Philadelphia FBI agents visited Harry Gold in March 1950 and found, hidden behind a bookcase, a Chamber of Commerce map of Santa Fe. They told him the "jig was up," and one reported, "After about one minute and at 10:15 a.m., Gold stated, 'I am the man to whom Klaus Fuchs gave his information.'" Fuchs was given a photograph of Gold on May 24 and admitted, "Yes, that is my American contact."

The unveiling of Klaus Fuchs then led to the Rosenbergs. At the end of 1949, Julius Rosenberg told his wife's brother, David Greenglass, that he was "hot . . . something is happening which will cause you to leave the United States." David's immediate reaction: "I'll never be able to read *Li'l Abner* again." Greenglass had been honorably discharged from the army in February of 1946 and had returned to live in New York City, where he was regularly extorting his brother-in-law: "Julius had money. I went to Julius, 'Look, I need money,' and he would give me money . . . about a thousand dollars all told." After Fuchs's arrest, Julius "said to me I would have to get out of the country with my family. . . . You remember that man out in Albuquerque. . . . This man knew me and that when Fuchs was taken . . . he would tell about Gold and he would lead them to me. . . . He wanted me to go with my whole family: pouf, disappear! . . . I figured I might [really disappear], so I better not go."

On the morning of February 14, 1950, Ruth Greenglass's nightgown caught fire from the open-gas heater in their Lower East Side apartment, and she spent nearly a month at Gouverneur Hospital getting skin grafts. They could not flee, as Julius was planning, to Czechoslovakia. David Greenglass: "The day after my wife came home from the hospital [Julius arrived with] the *Herald Tribune* or *Times*. Anyway, there was a picture of Gold on the front page. And he said, 'That's your man, look at the picture.' I said, 'You're silly, that's not the fellow; my wife said it was not him.' He said, 'That's the man.' . . . He feared he would be arrested; they would pick me up, I would lead them to him. . . . [I] said, we can't go anywhere, we have an infant here; we can't just up and leave. . . . He said your baby won't die; babies are born in the air and on trains, and she will survive."

On June 2, Harry Gold said his other contact in New Mexico was in the "US Army . . . twenty-five years of age, perhaps even younger [whose wife] may have been Ruth." On June 15, two FBI agents visited David and Ruth Greenglass at 265 Rivington Street, found a picture of the couple back in Albuquerque, took it to Philadelphia, had it verified by Gold, and confronted Greenglass. David immediately confessed, taking Julius down with him, but not Ruth, later saying, "I told . . . the FBI right from the start that if my wife was indicted, I would not testify. I told [them] I would commit suicide and [they] would have no case. . . . I got two children. If the choice was between [Ruth] and my sister, I'll take [Ruth] any day. That was the choice that I thought I had." When Ruth was interrogated, she then implicated Ethel. Julius was arrested on July 17, Ethel on August 11, and Ruth was never indicted.

David Greenglass pled guilty on October 18, 1950, then testified at the Rosenbergs' trial in March 1951, as did Ruth Greenglass and Harry Gold. The Rosenbergs insisted, throughout, that they were innocent, and many in the global audience transfixed by this dramatic story of atomic spies believed them. Even though both Hoover and his Justice Department colleagues did not want to pursue the death penalty—especially against Ethel, a mother of two young children—Judge Irving Kaufman believed his trial could "make people realize that this country is engaged in a life-and-death struggle with a completely different system." At their sentencing on April 5, he told the Rosenbergs, "I believe your conduct in putting into the hands of the Russians the A-bomb, years before our best scientists predicted Russia would perfect the bomb, has already caused, in my opinion, the Communist aggression in Korea, with the resultant casualties exceeding fifty thousand, and who knows but what that millions more innocent people may pay the price of your treason. Indeed, by your betrayal, you undoubtedly have altered the course of history to the disadvantage of our country."

Ethel Rosenberg publicly replied, "And what of our children, noble testament to our sacred union, fruit of our deep and enduring love; what manner of 'mercy' is it that would slay their adored father and deliver up their devoted mother to everlasting emptiness? Know then, you warped, gross eaters of dust, you abominations upon this beauteous earth, I should far rather embrace my husband in death than live on ingloriously upon your execrable bounty."

In time, Venona would reveal 108 Soviet spies working within the United States and Great Britain. Meredith Gardner's discovery of a November 1944 message listing the key scientists of Los Alamos was tracked to the nineteen-year-old Ted Hall, who was found to be now working with Ed Teller in Chicago. When the FBI interviewed him there, though, Hall revealed nothing. That night, he and his wife, Joan, "took all the left-wing stuff and packed it in boxes, and put it in the car, and put [our daughter] Ruthie into her snowsuit," Joan remembered. "She was just, then, just over a year. Put her in her car seat and got in the car and drove to the bridge that crosses over the Chicago drainage canal. We dumped all the stuff into the canal. . . . [Ted] told me that he had done it because he was afraid that the United States might become 'a very reactionary power' after the war—those were his words—and that this would give the Soviet Union a better chance of standing up to them. . . . He certainly broke the law. He certainly broke his security oath. But he did not betray his country. He didn't betray the people. Everything that he did was done because of his concern for the people. It was a humanitarian act. His motive was a humanitarian motive. Now, if you want to call that sort of thing treason, go right ahead."

The FBI was never able to amass enough evidence to bring Ted Hall to trial. Instead, Harry Gold was sentenced to thirty years; David Greenglass to fifteen years; and the Rosenbergs were sentenced to death. President Eisenhower, wanting to show American fortitude in the morning hours of the Cold War, refused clemency.

Lewisburg Federal Penitentiary inmate Alger Hiss: "The June evening of the executions was calm and cloudless. We were all aware that . . . the executions had been scheduled for just before sunset. As the sun sank, silence spread over the recreation yard. Men stopped their games of baseball, *boccie*, handball, their exercises with weights, their trotting about the cinder track, their endless conversation. We sat or stood in an eerie quiet until after the sun had disappeared. . . . We felt we were honoring the very moments of death. . . . We had all been aware of the worldwide demonstrations and protests, which at those moments were proved to be futile. In all the months I spent at Lewisburg this occasion was unique—the inmates transcending their own unhappiness and self-involvement and joining in a mood of universal sadness at an act of inhumanity."

Julius Rosenberg died in the electric chair at Sing Sing on June 19, 1953, at 8:06 p.m.; Ethel at 8:15.

In early 1950, North Korea's Kim Il Sung told his ally Joseph Stalin that he was going to reunite the Koreas, that "the attack will be swift, and the war will be won in three days." Stalin replied, "According to information coming from the United States . . . the prevailing mood is not to interfere." As the American army was next door occupying Japan and as the skirmish reminded Pentagon chiefs of the early days of World War II, however, the United States immediately moved to defend its ally—unlike what Britain and France did for Poland—and made great advances into North Korean territory. At his first postintervention meeting with the Joint Chiefs, Truman asked for a study on the use of the Bomb if the Russians entered the conflict. Then in response to a request from Douglas MacArthur two weeks later, the Joint Chiefs decided that ten to twenty atomic bombs could be made available for his use, which the general thought could stop either the Russians or the Chinese from intervening: "I would cut them off in North Korea . . . I visualize a cul-de-sac. The only passages leading from Manchuria and Vladivostok have many tunnels and bridges. I see here a unique use for the atomic bomb—to strike a blocking blow—which would require a six months' repair job. Sweeten up my B-29 force." The Joint Chiefs, however, vetoed this strategy.

This moment was the start of the great debate in military and political circles that would continue for six decades: How, beyond the original demonstration of their power, could atomic bombs be used as weapons in war? A few months after America entered the conflict, Oppenheimer gave a speech to the National War College: "Are [atomic bombs] useful in ground combat? Are they useful in preventing the delivery of atomic bombs? What can we do with them? . . . It is a job that calls for a great deal of imagination to think what is the atom good for in war." Then for the summer of 1953's

Foreign Affairs, he wrote "Atomic Weapons and American Policy," which concluded, "The very least we can say is that, looking ten years ahead, it is likely to be small comfort that the Soviet Union is four years behind us, and small comfort that they are only about half as big as we are. The very least we can conclude is that our twenty-thousandth bomb, useful as it may be in filling the vast munitions pipelines of a great war, will not in any deep strategic sense offset their two-thousandth." Hearing this, physicist John Wheeler complained to a congressman, "Anybody who says twenty thousand weapons are no better than two thousand ought to read the history of wars." Sharing Wheeler's perspective was Hungarian mathematician John von Neumann, who announced in 1950, "If you say why not bomb them tomorrow, I say why not today? If you say today at five o'clock, I say why not one o'clock?" Von Neumann's promotion of a preemptory nuclear strike was one of the many oddities that inspired Einstein to nickname his Princeton colleague *Denktier*, "think animal."

Washington military and civilian policymakers during the Cold War would be baffled by the conundrum of "What is the atom good for in war?"—to the point of making them angry and confused. This cognitive dissonance, this inscrutable puzzle, was demonstrated in full force by the president of the United States when, after Mao bolstered North Korea with Chinese troops in the autumn of 1950, Truman threatened nuclear retaliation at a notorious November 30 news conference:

THE PRESIDENT: We will take whatever steps are necessary to meet the military situation, just as we always have.
QUESTION: Will that include the atomic bomb?
THE PRESIDENT: That includes every weapon that we have.
QUESTION: Mr. President, you said "every weapon that we have." Does that mean that there is active consideration of the use of the atomic bomb?
THE PRESIDENT: There has always been active consideration of its use. I don't want to see it used. It is a terrible weapon, and it should not be used on innocent men, women, and children who have nothing whatever to do with this military aggression. That happens when it is used.

In December, MacArthur again requested permission to employ at his discretion twenty-six atomic bombs in a strategy he insisted would end the war in ten days, explaining later, "I would have dropped thirty or so atomic bombs . . . strung across the neck of Manchuria [and] spread behind us—from the Sea of Japan to the Yellow Sea—a belt of radioactive cobalt. . . . It has an active life of between 60 and 120 years. For at least 60 years there could have been no land invasion of Korea from the North. . . . My plan was a cinch." The Joint Chiefs again said no. After Truman replaced MacArthur with Matthew Ridgway, the JCS also turned down Ridgway's December 24, 1951, request for thirty-eight atomic bombs, though one historian's theory as to why MacArthur was removed (a theory as yet unproved, as the sixty-year-old documents are still classified) is that Truman wanted a less volatile commander in the field in case nuclear weapons were in fact unleashed.

Even after 36,568 Americans, about 600,000 Chinese, and 2 million Koreans died during the three-year Korean War, the conflict could not be resolved into victory and defeat, as both sides were nuclear armed. Instead, it fizzled out in stalemate. For a brief part of the struggle, Soviet and American fighters had a couple of dogfights, and this is the full extent that the two nations would directly battle each other over the entire history of the Cold War. Contrary to what everyone believed in Washington during those three years, Nikita Khrushchev later admitted, "America had a powerful air force and, most important, America had atomic bombs, while we had only just developed the mechanism and had a negligible number of finished bombs. Under Stalin we had no means of delivery. . . . This situation weighed heavily on Stalin. He understood that he had to be careful not to be dragged into a war."

Stalin's death on March 5, 1953, led to renewed negotiations to end the conflict, but when these bogged down in August, President Eisenhower decided a theatrical reminder of LeMay's nuclear strike force might move things forward, so when his boys landed at Okinawa, LeMay called a press conference just to make sure everyone knew exactly what was standing by. Lieutenant General James Edmundson: "Our mission called for me to take twenty B-36s with nuclear weapons on board and go to Okinawa and sit on the alert. The B-36 really wasn't much fun to fly. It's a gigantic thing. It's like—they used to say it was like sitting on your front porch and flying your house around. . . . We stayed at Kadena and sat on the alert, crews at the airplanes, while the hostility-cessation papers were being signed. And the B-36s being there, with atomic weapons and ready to go, was a warning to the North Koreans and the Russians and the Chinese not to try anything funny when we were sitting around the peace table." After his men came home to the United States, though, General LeMay never returned his bombs to the Atomic Energy Commission, as he was required to do, and this would not be the last of LeMay's nose-thumbing of civilian oversight.

During the Korean War, the AEC grew from eight sites with 55,000 employees to twenty with 142,000. While running the country during this vast and nonsensical expansion, Harry Truman said, "The war of the future would be one in which man could extinguish millions of lives at one blow, demolish the great cities of the world, wipe out the cultural achievements of the past—and destroy the very structure of a civilization that has been slowly and painfully built up through hundreds of generations. Such a war is not a possible policy for rational men." Variations on this two-faced strategy would be heard in speeches from leaders of both the United States and the USSR over the decades to come, while at the same time both would always find new reasons to be belligerent, and afraid. Sergei Khrushchev: "When America elected General Dwight D. Eisenhower, a hero of the Second World War, as president in November 1952, we had no doubt what it meant: *The USA has decided to fight, otherwise, why would they need a general?*"

10

A Totally Different Scheme, and It Will Change the Course of History

In the first years after Truman approved going forward with the Super, Edward Teller made absolutely no progress. Physicist Lee DuBridge said that at meetings of the Atomic Energy Commission's General Advisory Committee "every time he reported, we thought he'd taken a step backwards." Finally Enrico Fermi returned to Los Alamos to help. Some explained his reversal in opinion as that, since Truman had given the go-ahead, Fermi thought he should contribute, while others said he was there to prove that hydrogen bombs would never work, even to the point of saying to Teller directly that he hoped they would fail. But perhaps since Enrico had originally given the idea of thermonuclear to Teller so many years before, he wanted to wrestle with the scientific puzzle and see how it all turned out.

Instead of actual lab work producing physical, measurable results, the fusion group was forced to calculate phenomena arising from the intersection of fission and fusion, including how an immense cascade of neutrons would behave (neutronics); where the heat would go and what it would do (thermodynamics); and the effect of particles and radiation released in the fluid flow of explosion (hydrodynamics). What was needed for thermonuclear's immense calculations of all these forces, combined with the complicated interactions when they were arrayed against each other, were counting machines. A group around Dick Feynman had tried working with IBM punch-card units, but the hardware just wasn't powerful enough. Instead, most of the work began with raw human labor. Physicist Richard Garwin: "The computers of [Polish mathematician Stanislaw] Ulam and Fermi in those days were young women, who would come in the morning to present the results of the previous day's run. The run was the use of Marchant mechanical calculators, to fill in successive boxes on a spreadsheet, where various differential equations had been reduced to first-order differential equations, so there was only adding, subtracting, and multiplying, as one crawled one's way across the spreadsheet."

One night in 1945, John von Neumann admitted to his dear wife, Klari, "What we are creating now is a monster whose influence is going to change history, provided there is any history left. Yet it would be impossible not to see it through." The mathematician was not speaking of hydrogen bombs, but of computers. During the first meeting of von Neumann with Herman Goldstine, one of the engineers at the University of Pennsylvania who had created the best computing machine at the time, Goldstine recalled, "When it became clear to von Neumann that I was concerned with the development of an electronic computer capable of 333 multiplications per second, the whole atmosphere of our conversation changed from one of relaxed good humor to one more like the oral examination for the doctor's degree in mathematics. Soon thereafter the two of us went to Philadelphia so that von Neumann could see the ENIAC. . . . The story used to be told about him at Princeton that while he was indeed a demigod, he had made a detailed study of humans and could imitate them perfectly. Actually he had great social presence, a very warm, human personality, and a wonderful sense of humor."

When the ENIAC proved erratic and too weak for the calculations the Super required, von Neumann expanded on the ideas of British mathematician Alan Turing to create a more powerful calculating machine that used binary code for both its content and its programming—wife Klari wrote the code—with memory provided by oscilloscope tubes. He called it the Mathematical Analyzer, Numerical Integrator and Computer—MANIAC. As he and Princeton did not patent the design, MANIAC became the original open source—it was copied by universities across the United States and is still the foundation of modern computer architecture. Besides calculating the contrary effects of fission on fusion to create the most enormous weapon in human history, MANIAC was also used to forecast the weather.

Before von Neumann could wield his MANIAC, Teller, Ulam, and Russian physicist George Gamow were regularly meeting to brainstorm and search for solutions. Stan Ulam: "Both Gamow and I showed a lot of independence of thought in our meetings, and Teller did not like this very much. . . . I wrote [Gamow], prophetically it seems, that great troubles would follow because of Edward's obstinacy, his single-mindedness and his overwhelming ambition." Teller, in turn, said that Ulam "originally came with my friend Johnny von Neumann's recommendation, but I found him difficult company. He seemed to think very highly of himself and expended much effort in demonstrating his cleverness (which was strange because his ingenuity was obvious). Although we had limited contact with each other during the war and postwar, I developed an allergy to him. His demeanor made it clear that his feeling about me was even stronger."

The universally acclaimed von Neumann told Françoise, Stan's wife, that he'd never met anyone so self-confident as her husband, "adding that perhaps it was somewhat justified." For one example, Ulam had noticed that he could predict the outcome of a solitaire card game by noting what happened in the first plays, instead of having to memorize every possible outcome. He and von Neumann developed the details into a form of statistics called the Monte Carlo method, and these mathematical principles of probability would produce insights into the calculations of fission-induced fusion and help create the hydrogen bomb.

In June 1950, following von Neumann's suggestion, Ulam was able to use ENIAC to test Teller's original math. He proved that the 1946 calculations that were the Super's basis, Teller's claim to genius, were all wrong, and von Neumann, one of Teller's lifelong friends and most ardent supporters, used his own computer prototype at Princeton to confirm Ulam's numbers. The handsome mathematician had undone the brooding physicist, and Teller felt his life work collapse. Stan's wife, Françoise: "I was well placed to watch how personally Teller took the fact that Stan and [Wisconsin mathematician C. J.] Everett were the first to blow the whistle with their crude calculations. Every day Stan would come into the office, look at our computations, and come back with new 'guesstimates,' while Teller objected loudly and cajoled everyone around into disbelieving the results. What should have been the common examination of difficult problems became an unpleasant confrontation." Hans Bethe: "Nobody will blame Teller because the calculations of 1946 were wrong, especially because adequate computing machines were not then available. But he was blamed at Los Alamos for leading the laboratory, and indeed the whole country, into an adventurous program on the basis of calculations which he himself must have known to have been very incomplete."

Then, in a story as old as myth itself, Teller's great foe became his savior. Françoise Ulam: "Engraved on my memory is the day when I found [Stan] at noon staring intensely out of a window in our living room with a very strange expression on his face. He said, 'I found a way to make it work.'

'What work?' I asked. 'The Super,' he replied. 'It's a totally different scheme, and it will change the course of history.'" Françoise, who "had rejoiced that the 'Super' had not seemed feasible," was "appalled by this news."

Inspired by von Neumann's Fat Man ordnance lenses, Ulam's idea was to implode a fission shell into a fusion core, compressing and igniting it, creating a real Super bomb magnitudes larger than Teller's Alarm Clock or Sakharov's Layer Cake. Ed Teller: "[Ulam's] suggestion was far from original; compression has been suggested by various people on innumerable occasions in the past. But this was the first time that I did not object to it." Ulam thought neutrons would be the key force, but Teller insisted X-ray radiation would symmetrically compress the inner layers of tritium, deuterium, plutonium, and uranium and be the answer.

After writing a joint report with Ulam on March 9, 1951, Teller went to Washington to present their new concept to the AEC in June. The man who had so significantly opposed Teller and his Super over the previous six years, J. Robert Oppenheimer, now said that this new direction had to be fully supported as the science was so "sweet." Los Alamos would develop a bomb based on the Teller-Ulam design, and it would be tested on the Elugelab atoll in Eniwetok, a Marshall island three thousand miles west of Hawaii, on November 1, 1952. Understanding the change in thinking of atomic bureaucrats during this period can be seen with the simple evidence that, instead of the lyrical Trinity, this first significant thermonuclear test would be called Mike.

Now, just as Edward Teller after decades of struggle and failure was to finally achieve the great triumph of his life's work, the cancerous element in his nature erupted. Françoise Ulam: "From then on, Teller pushed Stan aside and refused anything to do with him any longer. He never met or talked with Stan meaningfully again. Stan was, I felt, more wounded than he knew by this unfriendly reception, although I never heard him express ill feelings toward Teller. He rather pitied him instead." Herb York: "What Ulam did was not a thermonuclear device. It was a general idea. What Teller did was convert that into something which was a sketch of a Super that would work. Teller sketched out a Super bomb. Ulam simply presented a fairly general idea in dealing with that topic. I think Teller has slighted Ulam, but I think also Teller does deserve 51 percent of the credit."

Hans Bethe: "Before the end of the summer of 1951, the Los Alamos Laboratory was putting full force behind attempts to realize the new concept. However, the continued friction of 1950 and early 1951 had strained a number of personal relations between Teller and others at Los Alamos. . . .

There was further disagreement between Teller and [director] Bradbury on personalities, in particular on the person who was to direct the actual development of hardware. Bradbury had great experience in administrative matters like these. Teller had no experience and had in the past shown no talent for administration. He had given countless examples of not completing the work he had started; he was inclined to inject constantly new modifications into an already going program, which becomes intolerable in an engineering development beyond a certain stage; and he had shown poor technical judgment. Everybody recognizes that Teller more than anyone else contributed ideas at every stage of the H-bomb program, and this fact should never be obscured. However: . . . Nine out of ten of Teller's ideas are useless. He needs men with more judgment, even if they be less gifted, to select the tenth idea which often is a stroke of genius." Then "Teller accused the leadership of Los Alamos of not working wholeheartedly on the hydrogen bomb, on the Mike test," Dick Garwin said, "and he walked out."

With the full backing of Ernest Lawrence, Teller insisted that, as fusion development was haphazard and erratic on the mesa, there needed to be a new lab competing with Los Alamos to create his revolutionary weapons. Los Alamos director Norris Bradbury: "Lawrence believed Edward. Simple as that. Why not? Edward could sell refrigerators to Eskimos." Just as Szilard was able to use his Los Alamos credentials to talk Congress out of giving nuclear oversight to the Pentagon, Lawrence and Teller campaigned across Washington to build another weapons lab. At meetings with the Department of Defense, the physicists insisted that the Russians, through their just-revealed American spy nest of Fuchs, Gold, Greenglass, and the Rosenbergs, would soon achieve atomic dominance if things did not change immediately. They were so convincing that the air force warned the AEC that if the agency didn't accede to Teller and Lawrence's plan, the USAF would build its own weapons lab, regardless of the civilian mandate.

After being expelled from the paradises of prewar Budapest and Weimar Berlin, the once-shy foreign émigrés with then-comical-to-American-ears glottal accents were now éminences grises, invited into the private chambers of Washington to consult with the most powerful men on earth. Attending Pentagon advisory boards and AEC committees, or making appearances before Congress and writing memos to the White House, Fermi, Szilard, Teller, and von Neumann had all achieved a remarkable version of the American dream. Men long famous for being solitary eccentrics now learned to mount political campaigns, and the intersection of science, business, and government pioneered by Marie Curie reached a new pinnacle under the rise of Eisenhower's noted military-industrial complex and Lawrence's big science.

A few years before, the AEC had awarded Lawrence $7 million to turn a defunct naval air station into a neutron factory—its swimming pool would be a coolant tank—to generate fuel for the Super. On September 2, 1952, the civilian agency then gave in to the Pentagon's threats, and Lawrence's Livermore was transformed into a second federal nuclear weapons lab. Instead of Edward Teller as its director, though, Lawrence appointed Herbert York, and by now it was too late for Teller to take Mike with him to Livermore; the Teller-Ulam design was instead birthed at Los Alamos. Hans Bethe: "Once Teller left Los Alamos, even though they were working on 'his' weapon, he found all sorts of reasons why it wouldn't work." Then, the first of Livermore's designs were tested. They were spectacular failures. Ed Teller: "Our first computer at the Livermore site had a glass case standing on top of it, complete with a small hammer and the instructions *In case of emergency, break glass.* Inside the case was an abacus."

Vannevar Bush, the brilliant World War II OSRD chief, was now working on arms control, and he tried to get Mike postponed, as he was developing a breakthrough treaty with Moscow: "I felt strongly that that test ended the possibility of the only type of agreement that I thought was possible with Russia at that time, namely, an agreement to make no more tests. For that kind of an agreement would have been self-policing in the sense that if it was violated, the violation would be immediately known. . . . I think history will show that was a turning point . . . [and] that those who pushed that thing through to a conclusion without making that attempt have a great deal to answer for." Mike had another party interested in its delay, for when he learned it was set for three days before the November presidential election pitting Adlai Stevenson against Dwight Eisenhower, Harry Truman sent word to the AEC "that the President would not change the date, but he would certainly be pleased if technical reasons cause a postponement."

All of these concerns were completely ignored as Mike became a physical manifestation of Cold War hysteria. By October 1952, over nine thousand enlisted men and two thousand civilians were in tents or on ships in the Eniwetok vicinity, supported by eighty aircraft and a full navy task force of ships. Five hundred scientific bases were on thirty islands; the control room was on Estes atoll, while the core scientists and technicians were on Parry. Troops bullied Elugelab's coral rock and sand into a platform to improve the explosion's visibility, then built atop that a six-story-high shot cab. Seven mirrors reflected the explosion to streak cameras housed in a bunker, one of which could attain a speed of 3.5 million frames per second. A nine-thousand-foot plywood tunnel with a shell of concrete filled with helium balloons would carry neutrons and gamma rays to instruments on the island of Bogon, almost two miles away. C-54 photography planes, F-84 sniffer jets, and B-47 and B-36 technical planes would orbit between ten thousand and forty thousand feet overhead. Serviceman Michael Harris: "We called the place the Rock. Like the federal prisoners did on Alcatraz. The local bar was called the Snakepit in honor of Olivia de Havilland's insane asylum."

It was popularly known as the hydrogen bomb . . . but hydrogen was the least of it. On November 1, 1952, at 7:15 a.m., ninety-two detonators triggered Mike's outer high-explosive shell, which created a shock wave compressing the uranium layer onto its plutonium ball, collapsing the centerpiece—the urchin initiator, which had beryllium and polonium to infest the supercritical mass of surrounding uranium and plutonium—creating a fission fire hotter than the sun. This in turn began to heat and compress a tank of liquid deuterium (which until then had been kept motile at -417°F) into a gas of tritium, which additionally activated a fission spark-plug, the plug's X-rays irradiating the compressed deuterium, mingling with the radiation coming from the outside shells, pushing the atoms of the heavy water over the electrostatic barrier and triggering thermonuclear fusion—the power of starlight; the energy of our sun—with a neutron density 10 million times that of a supernova.

With the Teller-Ulam design, Los Alamos achieved fission-fusion-fission and created a weapon of unimaginable force. Even those who'd attended a prior atomic blast were flabbergasted by the outcome in the Pacific—10.4 megatons, eight times the power of Hiroshima, a three-mile-diameter fireball—Hiroshima's was one-tenth of a mile—surrounded by a hundred-mile-wide and twenty-five-mile-high mushroom cloud. One sailor wrote home, "You would swear that the whole world was on fire." Elugelab was vaporized into a crater, its 80 million tons of obliterated cremains rising into the sky as a cloud of radioactive fallout. After this cloud began to disperse, rains brought it down as a radioactive mud, and within this mud was a new element, number 100. It would be named after the Italian navigator who had overseen the first fission reactor and given Edward Teller the idea for the thermonuclear bomb in the first place: fermium.

Though Mike was the great dream Teller had spent eleven years imagining, designing, and fighting over, he was so insistent that Los Alamos would fail that he refused to attend the test. Instead, after having estimated the time Mike's seismic wave would take to travel across the Pacific, "I went down into the basement of the University of California geology building in Berkeley, to a seismograph that had a little light-point marking on photographic film. A tremor of that point would show when the shock wave, generated thousands of miles away on Eniwetok Island, reached Berkeley. . . . At exactly the scheduled time I saw the light point move. . . . There was the signal, just as predicted. . . . I at once wired to a friend . . . 'It's a boy!' "

One side effect of Mike was to make Fermi's and Bohr's apprehensions about Szilard's wartime strategy of research secrecy come true. Since the AEC and the Pentagon keep their fusion science classified, to this day much of the results of their thermonuclear tests in Nevada and the Pacific remain top secret and unavailable to scientists not working on classified research for the federal government. Secrecy meant that for decades, there was little effort to achieve anything with this revolutionary science except make more, and bigger, and ever-less-useful bombs. Photographs from these tests are about as awe inspiring as anything produced by the hand of man—such power, such strength, such magnificence, emanating from the tiniest of sources . . . it is remarkable to contemplate. Just as remarkable is to think of what all that money, manpower, and scientific endeavor could have done if it had been directed to something worthwhile, instead of hundreds and hundreds of ever more absurd weapons and atmospheric tests that poisoned the whole of the continental United States. Besides the mythic imaginings of thermonuclear annihilation in a cataclysmic mushroom cloud big enough for God to see, there is, in the end, a dispiriting tragedy in this spectacle.

Mike's success did exactly what every arms-race breakthrough did—thrilled those American who thought twenty thousand atomic bombs were better than two thousand, while spurring the Russians to bake Sakharov's *sloika*. On August 8, 1953, after the death of Stalin, Nikita Khrushchev's coup removed Lavrenti Beria from power, and to make sure its great foe would not try to take advantage of the turmoil, the Kremlin announced, "The United States is said to have a monopoly on the hydrogen bomb. Apparently it would be of comfort to them if that were the truth. But it is not. The government considers it necessary to announce that the United States does not hold a monopoly in the production of the hydrogen bomb." Like Molotov with fission, this was a demi-bluff. Layer Cake hadn't yet been proven a success; its first test would be four days after this proclamation. Andrei Sakharov: "Of course, we worried about the success of the test, but for me, anxiety about potential casualties was paramount. Catching a glimpse of myself in a mirror, I was struck by the change—I looked old and gray."

Andrei's *sloika* was tasty. Though having a fraction of Mike's power, unlike

Los Alamos's immense gadget with its tank of liquid deuterium, this was a real weapon that could be ferried in the bay of a Soviet bomber and dropped on American cities. The US Air Force sniffed it, called it Joe 4, and brought back such detailed information on its effluvia that Hans Bethe could tell it was a design similar to Teller's two-year-old Alarm Clock—named for being the bomb that would wake up the world. Sakharov was given the Stalin Prize, the title Hero of Socialist Labor, promoted to chief of theory at Arzamas-16, and went back to working like two dogs in the yard.

Physicist German Goncharov: "An absolutely insane task was set for us not to lag behind the United States by one iota. We had to have everything the Americans had. There couldn't be the slightest gap. And so as soon as new information arrived about the work in this or that direction, we absolutely had to do the same thing." On March 1, 1955, Tesla exploded with seven kilotons; on March 7, Turk did forty kilotons; and Russian scientists began voicing the same regrets as had so many from Los Alamos. Sakharov: "When you see the burned birds who are withering on the scorched steppe, when you see how the shock wave blows away buildings like houses of cards, when you feel the reek of splintered bricks, when you sense melted glass, you immediately think of times of war. . . . All of this triggers an irrational yet very strong emotional impact. How not to start thinking of one's responsibility at this point?" Physicist Nikolay Larionov: "Even if you strike first, you will perish together with the defeated side. That is the paradox of our time."

Just as with First Lightning, the Soviet's fusion tests terrified a coterie of American policymakers. A committee headed by General James Doolittle suggested offering the Kremlin two years to come to an agreement, and to launch a nuclear first strike if it refused, while a Joint Chiefs' study group thought the United States should "deliberately precipitate war with the USSR in the near future . . . before the USSR could achieve a large enough thermonuclear capability to be a real menace to [the] Continental US." Eisenhower passed on both attempts to jump-start global holocaust. Instead, the president now wanted to offer the Russians Vannevar Bush's original test-ban agreement. Eisenhower's AEC director, Lewis Strauss, talked him out of it.

The Doolittle Report paved the way for a new era in American foreign policy as it proposed, "We are facing an implacable enemy whose avowed objective is world domination. There are no rules in such a game. Hitherto acceptable norms of human conduct do not apply. . . . [American citizens need to] be made acquainted with, understand and support this fundamentally repugnant philosophy." Even Eisenhower had to agree: "I have come to the conclusion that some of our traditional ideas of international sportsmanship are scarcely applicable in the morass in which the world now flounders. Truth, honor, justice, consideration for others, liberty for all—the problem is how to preserve them . . . when we are opposed by people who scorn . . . these values. I believe that we can do it, but we must not confuse these values with mere procedures, even though these last may have at one time held almost the status of moral concepts."

The ratcheting levels of American terror in the face of the Soviets' presumably malevolent intentions can readily be seen in Eisenhower's 1958 State of the Union address, as Ike was one of the least fearmongering of Cold War leaders: "What makes the Soviet threat unique in history is its all-inclusiveness. Every human activity is pressed into service as a weapon of expansion. Trade, economic development, military power, arts, science, education, the whole world of ideas—all are harnessed to this same chariot of expansion. The Soviets are, in short, waging total cold war." And Washington's paranoia escalated all over again when Nikita Khrushchev repeatedly threatened to annihilate the West. In one incident during Senator Hubert Humphrey's visit to the Kremlin, Khrushchev asked where the American politician was from, then used a blue pencil to mark Minneapolis, explaining, "That's so I don't forget to order them to spare the city when the rockets fly." In August of 1957 the Soviet Union launched its first intercontinental ballistic missile—ICBM—and then, on October 4, *Sputnik*, the first man-made earth satellite.

But it was all a magnificent bluff. The premier's rocket-scientist son, Sergei Khrushchev, later admitted, "We threatened with missiles we didn't have." Their long-range bombers could attack the United States, but only as one-way suicide missions, and their missiles couldn't accurately strike their targets. By the end of 1959, the Soviets had a total of six long-range-missile sites, and each missile needed twenty hours to prepare for launch, meaning that the total number of Soviet missiles available to attack the United States before retaliation was . . . six.

In the traditional version of Los Alamos's black postscript—the history of Robert Oppenheimer's security-clearance hearings—the founder of the mesa is portrayed as a martyr both to science and to arms control, necessarily atoning for the sin of having created our nuclear plague. Foreign-policy adviser George Kennan summed up this position: "On no one did there ever rest with greater cruelty the dilemmas evoked by the recent conquest by human beings of a power over nature out of all proportion to their moral strength. . . . In the dark days of the early fifties . . . I asked him whether he had not thought of taking residence outside this country. His answer, given to me with tears in his eyes: 'Damn it, I happen to love this country.' "

In 1947, after Oppenheimer was named chairman of the Atomic Energy Commission's General Advisory Committee, the FBI turned over the twelve-pound file of its surveillance to the AEC. Some of the agency's new board members went through these documents and became so alarmed they went to meet directly with J. Edgar Hoover. Hoover said he was convinced that Oppenheimer had turned away from communism and was worthy of a security clearance . . . but the same assurances could not be made for his brother, Frank. Even so, for eight years after World War II, the FBI, at the personal insistence of Hoover, generated another eight thousand pages or so from its spying on J. Robert Oppenheimer.

On June 7, 1949, Oppie appeared before the House Un-American Activities Committee and so charmed the congressmen that they all rose from their seats at the end of the session to shake his hand. Two days later, he testified before Congress's Joint Committee on Atomic Energy, which was deciding whether the AEC should allow the export of radioisotopes to foreign nations. The sole AEC commissioner who thought this was a peril was the balding, moon-faced Lewis Lichtenstein Strauss—financier at Kuhn, Loeb, aide-de-camp to President Herbert Hoover, "troubleshooter" for Secretary of the Navy James Forrestal, cobalt experimenter with Leo Szilard, and board member of Princeton's Institute for Advanced Study. In testimony employing his patented withering, Oppenheimer showed Strauss to be a fool: "No man can force me to say you cannot use these isotopes for atomic energy. You can use a shovel for atomic energy. In fact you do. You can use a bottle of beer for atomic energy. In fact you do. But to get some perspective, the fact is that during the war and after the war these materials have played no significant part and in my knowledge no part at all." David Lilienthal remembered Strauss's reaction: "There was a look of hatred there that you don't see very often in a man's face."

Physicist Max Born said that Oppenheimer "was a man of great talent, and he was conscious of his superiority in a way which was embarrassing and led to trouble. . . . Vast insecurities lay forever barely hidden beneath his charismatic exterior, whence came an arrogance and occasional cruelty befitting neither his age nor his stature." Lewis Strauss was apparently just the kind of man who brought out Robert's arrogance, and cruelty. The year before, Oppenheimer had politically maneuvered to keep Strauss from getting a stronger leadership position at Princeton. The year after, in 1950, at Strauss's fifty-fourth birthday party, when the financier tried to introduce his children to the father of the atomic bomb, the physicist snubbed them. Strauss had a global reputation for being ruthless and vindictive, which Oppenheimer was possibly unaware of, and his great achievement with Trinity and the resulting public hubbub perhaps inspired him to feel invincible. But such behavior explains why Oppenheimer wasn't a wholly innocent bystander—or was, at the very least, suffering from a classic level of hubris—in the drama that would end with his public martyrdom.

During this period, Edward Teller wanted to create a Super measured in megatons, not kilotons, and so did nothing further with his workable Alarm Clock fusion bomb design. Instead, his attempt to make the ultimate thermonuclear weapon, pre-Ulam, was going nowhere, and on top of feeling like a failure, Teller interpreted the lack of support of the AEC's General Advisory Committee and its chief, Oppenheimer, as a brutal rejection. In the spring of 1952, Teller repeatedly went to the Albuquerque FBI office with such information as that "[Oppenheimer] delayed or hindered development of H-bomb from 1945 to 1950 by opposing it on moral grounds." Oppenheimer's opposition, Teller thought, was not due to any subversive intent "but rather to [a] combination of reasons including personal vanity in not desiring to see his work on A-bomb done better on H-bomb, and also because he does not feel H-bomb is politically desirable. Teller also feels [Oppenheimer has] never gotten over the shock of first A-bomb being dropped. . . . A lot of people believe Oppenheimer opposed the development of the H-bomb on direct orders from Moscow. Teller states he would do most anything to see [Oppenheimer] separated from General Advisory Committee because of his poor advice and policies regarding national preparedness and because of his delaying of the development of H-bomb."

At an interview with AEC public information officer Charter Heslep, "Teller feels deeply that [Oppenheimer's] 'unfrocking' must be done or else—regardless of the outcome of the current hearings—scientists may lose their enthusiasm for the [nuclear weapons] program," and Teller told the Joint Committee on Atomic Affairs that "were Robert, by any chance found to be disloyal (in the sense of transmitting information) he could of course do more damage to the program than any other single individual in the country."

In an aria of threat inflation, Teller then told Lewis Strauss that the Soviets were rapidly moving forward with implosion research and that, any day now, the United States would cede nuclear superiority to Moscow. This news enraged and terrified Strauss. Strauss told Teller that Oppenheimer "had been instrumental in bringing to Los Alamos a number of men known to him to be Communists. It would be reasonable to suppose that they were doing what Fuchs and others did, viz., passing on to the Soviets everything they could discover. Oppenheimer's later decision, therefore, to do what he could to prevent the United States from developing the Super [meaning the various times he had voted against Teller along with the rest of the AEC's committee] was a decision reached in the knowledge that such weapon data as we then had were in the hands of men whose leaning to the Soviets he knew. Consequently, if he had been able to block the development of the weapon by the United States, its denial to the Russians was beyond his control. It is hardly conceivable that the consequences of such a condition could have been overlooked by a mind as agile as his."

In a speech given on February 17, 1953, Oppenheimer insisted that government secrecy about nuclear science and weapons did not protect anyone; instead, it led to ignorance, fear, magical thinking, gossip, and paranoia, so that "we may anticipate a state of affairs in which the two Great Powers will each be in a position to put an end to civilization and life of the other, though not without risking its own. . . . We may be likened to two scorpions in a bottle, each capable of killing the other, but only at the risk of his own life." Here was the Atomic Age's creator, the most famous scientist in America after Einstein, calling defense policy ignorant and foolish. Eisenhower was nearly alone in Washington in thinking that candor was a good suggestion and the scorpions a good analogy; others thought the speech treasonous and insane.

To those arrayed against Oppenheimer, there had to be a pattern, there had to be meaning. Why was it that the man at the very center of the nation's atomic weaponry program opposed the ultimate hydrogen bomb, argued against the air force's nuclear-powered bombers, and called civilian nuclear power plants "a dangerous engineering undertaking. I was astonished to know that many people were wishing for this proving ground in their state"? Suffering from what is today called conspiracism, some of the capital's elite began to wonder, was the great hero of Los Alamos actually a traitor, secretly working for the Soviets to ensure they would win the arms race?

On May 25, 1953, when Eisenhower asked Lewis Strauss to replace Gordon Dean as AEC chairman, he accepted on the condition that Oppenheimer would no longer be "connected in any way" to the agency. After Strauss was sworn in, however, the president said, "Lewis, let us be certain about this. My chief concern and your first assignment is to find some new approach to the disarming of atomic energy. . . . The world simply must not go on living in fear of the terrible consequences of nuclear war." To the new chief of the AEC, this liberal directive was more evidence of Oppenheimer's demonic and treasonous influence within the highest levels of Washington.

Lewis Strauss began meeting secretly with the twenty-eight-year-old, square-jawed, all-American William Borden—executive director of Congress's Joint Committee on Atomic Energy, who, while a student at Yale, had written a nuclear horror fantasy, *There Will Be No Time*—to amass evidence against Oppenheimer. Strauss arranged for the FBI to dramatically increase its Oppenheimer surveillance, bugging and wiretapping Oppie's home, his Princeton office, and even his lawyer's office, such a breach that the FBI Newark supervisor wrote to headquarters questioning the phone taps' legality, "in view of the fact that [the taps] might disclose attorney-client relations." Washington responded that it was fine to record Oppenheimer's conversations with his attorneys since, at any moment, he might defect to Moscow. Strauss then ensured that the resulting transcripts were set to a security clearance higher than that held by Oppenheimer, so that only Strauss and his allies would have access to them.

When Strauss presented excerpts of the transcripts along with Borden's research to various government agency chiefs, Hoover was unimpressed, the secretary of defense was alarmed, and the president thought it all material that had been gone over before and cleared up by army intelligence. Still, Eisenhower worried: "The truth is that no matter now what could or should be done, if this man is really a disloyal citizen, then the damage he can do now as compared to what he has done in the past is like comparing a grain of sand to an ocean beach. It would not be a case of merely locking the stable door after the horse is gone; it would be more like trying to find a door for a burned-down stable."

On December 21, 1953, Strauss informed Oppenheimer that his AEC security clearance would have to be reviewed, presented him with a list of accusations, and asked if he would like to gracefully resign. The exceedingly proud Oppenheimer did not want to quietly exit from his life's work and tried to get Strauss to fire him. Neither would accede. Later Oppenheimer wrote Strauss, "I have thought most earnestly of [resignation]. Under the circumstances, this course of action would mean that I accept and concur in the view that I am not fit to serve this Government that I have now served for some twelve years. This I cannot do." When Isidor Rabi then told Strauss that the whole of the General Advisory Committee would testify on behalf of

Oppenheimer, Strauss said that he considered that nothing less than blackmail. An FBI report said that Strauss "felt that if this case is lost, the atomic energy program and all research and development connected thereto will fall into the hands of 'left-wingers.' If this occurs, it will mean another 'Pearl Harbor' as far as atomic energy is concerned. Strauss feels that the scientists will then take over the entire program. Strauss stated that if Oppenheimer is cleared, then 'anyone' can be cleared regardless of the information against them."

Valentine Telegdi: "The day the Oppenheimer case broke, we were having lunch with Fermi at the Quadrangle Club. He said, 'What a pity that they took him and not some nice guy, like Bethe. Now we have to all be on Oppenheimer's side!'" Einstein told Oppenheimer that he "had no obligation to subject himself to the witch hunt, that he had served his country well, and that if this was the reward that she offers, he should turn his back on her," and told Abraham Pais, "The trouble with Oppenheimer is that he loves a woman who doesn't love him—the United States government."

At the beginning of April 1954, American newspaper headlines were dominated by Senator Joseph McCarthy's allegations that secret Communist agents employed by the federal government had stifled for eighteen months "our research on the hydrogen bomb." Then on April 12, in World War II barracks converted to AEC Building T-3, with a table for the security inquisitors, two more for the defense team, a chair for the witnesses, and a couch for the victim, the inquiry known as "In the Matter of J. Robert Oppenheimer" began.

On the very first morning AEC attorney Roger Robb, using FBI and army documents the defendant and his attorneys were not allowed to see, caught Oppenheimer in a series of contradictions about Los Alamos events from ten years before. "I felt sick," Robb said. "That night when I came home, I told my wife, 'I've just seen a man destroy himself.'"

The story was fairly straightforward. In 1942, the Soviet consulate had asked British engineer George Eltenton to contact Robert Oppenheimer, Ernest Lawrence, and Luis Alvarez for any information on atomic bomb research at the University of California's Radiation Laboratory. Eltenton may have asked Oppenheimer's friend Haakon Chevalier to talk to Robert or to Frank Oppenheimer. Nothing more happened, until Oppenheimer started working at Los Alamos and began to wonder if Eltenton constituted a material threat. He wanted army security to be aware of these entreaties, but did not want to implicate his friend Chevalier or, more importantly, his brother, Frank. So Oppenheimer told one version of these events in

1943, and a different one in 1946, but he told the truth to Leslie Groves, who kept the matter quiet until "In the Matter of J. Robert Oppenheimer," when Groves testified, "There was an approach made, that Dr. Oppenheimer knew of this approach, that at some point he was involved, in that the approach was made to him—I don't mean involved in the sense that he gave anything—I mean he just knew about it personally from the fact that he was in the chain, and that he didn't report it in its entirety as he should have done. . . . [He] was doing what he thought was essential, which was to disclose to me the dangers of this particular attempt to enter the project. . . . It was always my impression that he wanted to protect his brother, and that his brother might be involved in having been in this chain, and that his brother didn't behave quite as he should have, or if he did, [Robert Oppenheimer] didn't even want to have the finger of suspicion pointed at his brother, because he always felt a natural loyalty to him and had [a] protective attitude toward him." Robb was able to prod Groves into admitting that, under the agency's more stringent rules, for this fibbing, Oppenheimer did not now deserve a security clearance.

The vast majority of the hearings were then spent analyzing Robert's years of opposition to the hydrogen bomb—before he declared the Teller-Ulam design "sweet" and encouraged its development—and whether that opposition meant he was a traitor to the United States. Rabi, Conant, Bethe, Bush, Groves, and even von Neumann all testified to his loyalty, and to his character. Like Teller, childhood friend Ernest Lawrence had turned on Robert, infuriated that he had opposed the Super, opposed creating the Livermore lab, and had conducted a love affair with the wife of a mutual friend. Lawrence agreed to testify against clearance, but felt too ill with colitis to attend and so provided a transcript saying that a man with such a poor moral sense should obviously not have influence over defense policy.

On April 28 at 4:00 p.m., Edward Teller took the stand. After talking over Oppenheimer's many faults ad nauseam with Lawrence and Alvarez and seeing the damning FBI and army intelligence materials about Haakon Chevalier, it's clear that Teller believed he was following his conscience by testifying. Freeman Dyson: "Teller thought Oppenheimer was somehow a Machiavelli who had far more influence than he really had in the real world. And Teller must have had, somehow, the feeling that if he could once destroy Oppenheimer's political power, that somehow things would be all right." At this moment in history, any number of Americans in Washington were fearfully trying to save the world in their own fashion, perhaps no one with as much fervor as Edward. But Teller couldn't save Teller from Teller:

Q. Is it your intention in anything that you are about to testify to, to suggest that Dr. Oppenheimer is disloyal to the United States?
A. I do not want to suggest anything of the kind. I know Oppenheimer as an intellectually most alert and a very complicated person, and I think it would be presumptuous and wrong on my part if I would try in any way to analyze his motives. But I have always assumed, and I now assume, that he is loyal to the United States.

Q. Now, a question which is the corollary of that. Do you or do you not believe that Dr. Oppenheimer is a security risk?
A. In a great number of cases I have seen Dr. Oppenheimer act—I understood that Dr. Oppenheimer acted—in a way which for me was exceedingly hard to understand. I thoroughly disagreed with him on numerous issues, and his actions frankly appeared to me confused and complicated. To this extent I feel that I would like to see the vital interests of this country in hands which I understand better, and therefore trust more. In this very limited sense I would like to express a feeling that I would feel personally more secure if public matters would rest in other hands.

Q. I would then like to ask you this question: Do you feel that it would endanger the common defense and security to grant clearance to Dr. Oppenheimer?
A. I believe, and that is merely a question of belief and there is no expertness, no real information behind it, that Dr. Oppenheimer's character is such that he would not knowingly and willingly do anything that is designed to endanger the safety of this country. To the extent, therefore, that your question is directed toward intent, I would say I do not see any reason to deny clearance. If it is a question of wisdom and judgment, as demonstrated by actions since 1945, then I would say one would be wiser not to grant clearance.

After Edward finished his testimony, he shook Robert's hand and murmured, "I'm sorry." Oppie replied, "After what you've just said, I don't know what you mean."

For J. Robert Oppenheimer, the Teller damnation would in the long run mean little, since Lewis Strauss was clearly determined Oppenheimer would not get his needed security clearance. A decade later, Oppie would tell the *Washington Post*: "The whole damn thing was a farce." But for Edward Teller, this moment would make of him a pariah in the scientific community. His colleagues in nuclear physics were enraged, not just because of loyalties to Oppenheimer or because Teller was such a remarkably difficult character to work with, but because if *anyone* was responsible for delaying the progress of thermonuclear research, it was not Robert Oppenheimer, but Edward Teller. His Alarm Clock was ready to move forward, but he was so insistent on achieving megatons of destruction that he only pursued the most extreme version of the Bomb, the design that required Stan Ulam's intervention. The eighteen-month delay that so riled up Joe McCarthy and Lewis Strauss was entirely Edward Teller's doing.

His hypocritical Oppenheimer testimony, combined with his relentlessly hawkish views, incinerated most of Teller's professional relationships. "He's a danger to all that's important," Hans Bethe decided. "I really do feel it would have been a better world without Teller." "I've never seen [Teller] take a position where there was the slightest chance in the interest of peace," Isidor Rabi said. "I think he is the enemy of humanity." Twenty years later, Teller told his biographers, "If a person leaves his country, leaves his continent, leaves his relatives, leaves his friends, the only people he knows are his professional colleagues. If more than ninety percent of these then come around to consider him an enemy, an outcast, it is bound to have an effect. The truth is it had a profound effect." In the community of nuclear physicists, he was shunned as harshly as any Old Testament whoremonger and would in time become the Richard Nixon of American science—dark, brooding, rejected, isolated, and alone.

On May 27, the security board decided that while America owed Oppenheimer "a great debt of gratitude for loyal and magnificent service," that he was "a loyal citizen," and that "no man should be tried for the expression of his opinions," his security clearance would not be reinstated: "Dr. Oppenheimer is not entitled to the continued confidence of the Government and of this Commission because of the proof of fundamental defects in his 'character.'" Bob Serber: "I think it broke his spirit, really. He had spent the years after the war being an adviser, being in high places, knowing what was going on. To be in on things gave him a sense of importance. That became his whole life. As Rabi said, he could run the institute with his left hand. And now he really didn't have anything to do."

Isidor Rabi: "I was indignant. Here was a man who had done so greatly for his country. A wonderful representative. He was forgiven the atomic bomb. Crowds followed him. He was a man of peace. And they destroyed this man. A small, mean group. There were scientists among them. One reason for doing it might be envy. Another might be personal dislike. A third, a genuine fear of Communism. He was an aesthete. I don't think he was a security risk. I do think he walked along the edge of a precipice. He didn't pay enough attention to the outward symbols. He was a very American person of a certain kind." But Freeman Dyson had a different view: "The real tragedy of Oppenheimer's life was not the loss of his security clearance bur his failure to be a great scientist." In a 1939 paper, Oppenheimer and his student Hartland Snyder created the concept of black holes—that echo of the death of stars into a continuous free fall of matter, time, and space—with "On Continued Gravitational Contraction." By the 1950s, however, Oppenheimer had completely turned away from the subject, and the great expert in black holes was his Princeton colleague and political antagonist, John Wheeler.

In April 1962, J. Robert Oppenheimer was asked to join Robert Frost, John Glenn, and forty-nine Nobel laureates at the Kennedy White House for dinner. The following spring, the president announced Oppenheimer would be awarded the $50,000 Fermi Prize for his service to the American people. Three years later, in February of 1966, Robert was diagnosed with throat cancer, and the following year, on February 18, 1967, he died at the age of sixty-two.

A few months after Oppenheimer's security-clearance hearings, Enrico Fermi returned to Chicago from summering in Europe, "and all of us were absolutely shocked by his appearance," physicist Maurice Glicksman remembered. "We asked him what was the trouble, and he said that he just couldn't eat. What he said was that food tasted like dirt, and he couldn't get it in."

Fermi soon learned the diagnosis: stomach cancer. Emilio Segrè: "Fermi was resting in the hospital, with his wife in attendance, and was being fed artificially. In typical fashion he was measuring the flux of nutrient by counting drops and timing them with a stopwatch. It seemed as if he were performing one of his usual physics experiments on an extraneous object. He was fully aware of the situation and discussed it with Socratic serenity. . . . He preserved to the last an almost superhuman courage, strength of character, and clarity of thought."

From his deathbed, Enrico begged Edward Teller to repair his broken professional relationships. Segrè: "One of the last times I saw him, at the hospital when he knew he had very little time to live, he said that he wanted to set straight a friend whose testimony he thought had been unethical. He smiled with slight irony and said, 'What nobler thing for a dying man to do than to try to save a soul?' " The following year, Teller published "The Work of Many People" in *Science* (February 1955), giving Ulam the credit he deserved for thermonuclear implosion.

Enrico Fermi died in 1954, at the too-young age of fifty-three. Nella Fermi: "I personally have discussed [whether my father's death was caused by his work with radioactivity] with two doctors. One doctor was one of the doctors that was on the scene at the time, and he said absolutely not. But, that definite 'no' . . . I don't know whether I believe it or not. The other doctor I talked to is my own personal doctor, who was not there at the time, and he said that he had always assumed that, in fact, that was the case. But, he also qualified it and said that there would be no way of knowing, that it would look like cancer in any case, that the only thing you could really say or speculate was, yes, this man has been exposed to a lot of radiation, and therefore, this may have been due to the radiation. He had stomach cancer, but after his death when they did the autopsy, they found that he had another cancer as well, which was apparently unrelated to that. That, to me, strengthens the case for radiation because it would seem that there was damage all over the place." Laura lived until December 28, 1977, when she died of pneumonia at the age of seventy, while the sixty-four-year-old Nella died of lung cancer on March 2, 1995.

Besides Carl Sagan's catalog of Fermi honors, for decades America hosted the world's leading cyclotron—Fermilab's four-mile-circumference Tevatron, housed under sixty-eight hundred acres outside Batavia, Illinois—where, nearby, buffalo roam. The machine's biggest achievement was the discovery of three of seventeen subatomic particles considered to be the building blocks of everything, and its technology led to the birth of MRI medical diagnostic technology. Four other US nuclear reactors are named for Enrico, as is element number 100, fermium. NASA's Fermi Gamma-ray Space Telescope has revealed a previously unknown fifty-thousand-light-year remnant of a black hole eruption at the very center of the Milky Way. Cosmologist Dan Hooper: "We've considered every astronomical source, and nothing we know of, except dark matter, can account for the observations. No other explanation comes anywhere close."

In its obituary, the *New York Times* said, "More than any other man of his time, Enrico Fermi could properly be named 'the father of the atomic bomb.' It was his epoch-making experiments at the University of Rome in 1934 that led directly to the discovery of uranium fission, the basic principle underlying the atomic bomb as well as the atomic power plant. And eight years later, on Dec. 2, 1942, he was the leader of that famous team of scientists who lighted the first atomic fire on earth, on that gloomy squash court underneath the west stands of the University of Chicago's abandoned football stadium. That day has been officially recognized as the birthday of the Atomic Age. Man at last had succeeded in operating an atomic furnace, the energy of which came from the vast cosmic reservoir supplying the sun and the stars with their radiant heat and light—the nucleus of the atoms of which the material universe is constituted."

Though now a molecular biologist, Leo Szilard continued to use his celebrity status as a nuclear pioneer to promote arms control. In February of 1950, he terrified the American public during a radio broadcast by describing how if cobalt was used as a tamper, a nuclear weapon could be designed that would bring an end to all life on earth, inspiring movie director Stanley Kubrick to fashion cobalt bombs for his satire *Dr. Strangelove*. Leo wrote to Stalin in 1947, and then to Khrushchev in 1960; Stalin never answered, but Khrushchev said he would meet him on September 27 at the Soviet mission to the United Nations. Szilard was promised fifteen minutes but unsurprisingly to any of his friends, the talk went on for two hours, with Khrushchev finally agreeing to the possibility of an international agency that would limit arms escalation and a communications hotline between the Soviet premier and the American president in case of nuclear crisis. That hotline would also appear in *Dr. Strangelove*, but would not exist in the real world until the Cuban Missile Crisis and its series of delayed telegrams made it clear to both sides that this was worthwhile.

In April 1958, the USSR unilaterally suspended nuclear testing, and after the AEC's Strauss warned Eisenhower that a reciprocal American test ban would turn Los Alamos and Livermore into "ghost towns," the president growled that he "thought scientists, like other people, have a strong interest in avoiding nuclear war." Later that year after a joint ban was enacted, Freeman Dyson wrote a letter to his parents from Livermore, now under the directorship of his longtime mentor, Ed Teller: "A lot of the talk of Livermore was about cheating the test ban. We've done a lot of ways to cheat which would be quite impossible for any instrument to detect. The point of this is not that the Livermore people themselves intend to cheat, but we are convinced the Russians can cheat as much as they want anytime they want, without being found out."

In the fall of 1958, Leo Szilard was meeting with Soviet and Hungarian delegates at the international Pugwash scientific conference. He asked them for a favor—to arrange for exit visas for Ed Teller's mother and sister—and they did. For the first time in twenty-three years, the family was reunited. Pugwash also turned out to be a fine venue to begin discussing arms control in person, and with friendly demeanors. The US-USSR antiballistic-missile treaty of 1972 was a direct outcome of the conference's work.

When Leo was diagnosed with bladder cancer in November 1959, he moved to Memorial Sloan-Kettering in New York City and designed his own radiation therapy, telling his doctors, "If worse comes to worse, I'll be dead ten years longer." If the program failed, he invented a bag that dispensed cyanide in such an undetectable way that he wouldn't be judged a suicide and his widow, Trude, would not be denied her life insurance benefits. After the cancer went into remission, Trude told Ed Teller, "The hospital was even more relieved to be rid of Szilard than Szilard was to be rid of the hospital."

In 1961, Russian hydrogen bomb inventor Andrei Sakharov stopped by his colleague Victor Adamsky's office to show him a short story, Szilard's "My Trial as a War Criminal." Adamsky: "I'm not strong in English, but I tried to read it through. A number of us discussed it. It was about a war between the USSR and the USA, a very devastating one, which brought victory to the USSR. Szilard and a number of other physicists are put under arrest and then faced the court as war criminals for having created weapons of mass destruction. Neither they nor their lawyers could make up a cogent proof of their innocence. We were amazed by this paradox. You can't get away from the fact that we were developing weapons of mass destruction. We thought it was necessary. Such was our inner conviction. But still the moral aspect of it would not let [Sakharov] and some of us live in peace." Leo's story inspired Andrei Sakharov to become a dissident, and his protests eventually helped end the Cold War's arms race.

In 1962, news of the Cuban Missile Crisis led Szilard to flee America for Geneva; he returned, abashed, in December of that year. Two years later, he moved to La Jolla to work at Jonas Salk's Institute for Biological Studies, where he worked until his death on May 30, 1964, of a heart attack, in his sleep.

On Sunday, February 7, 1960, the Russian Oppenheimer, Igor Kurchatov, visited with his colleague Yuli Khariton and Yuli's wife, Maria, in the suburbs of Moscow. When the radio played a waltz, Igor danced with Maria and then turned to Khariton: "Let's go for a little walk, Yuli Borisovich, and talk some shop." They went to a nearby park where, though below freezing, it was sunny. Kurchatov picked out a bench, and they sat for a moment, Yuli describing the results of his latest work. Then he noticed that his boss's eyes were glazing over. He shouted out, "There's something wrong with Kurchatov!" But it was already too late. At fifty-seven, Kurchatov died of a heart attack. The Russians say it was the strain of working for Beria that shortened his life, but Nella Fermi, among others, may have a different opinion: that like so many of his global peers, Kurchatov succumbed to the side effects of being a nuclear physicist.

With the deaths of Fermi, Szilard, Oppenheimer, Kurchatov, and Niels Bohr (in 1962), the great heroic generation of nuclear science had passed. They would be replaced by a very different group of men and women, a group that rarely had second thoughts about their miracle with two faces.

11

The Origins of Modern Swimwear

In December 1949, the Atomic Energy Commission announced it would buy uranium at a startling inflated price as well as pay $10,000 bonuses for any successful new US ore discovery. The largesse triggered a nuclear gold rush in the Colorado plateau across the American Southwest, where nine hundred mines sprang to life. The public reaction was so explosive that both Popeye and the Bowery Boys had pitchblende adventures, while Milton Bradley amended the Game of Life to include "Discover Uranium! Win $240,000." On July 6, 1952, a down-at-his-heels geologist, Charles Steen, cracked open a record-breaking uranium mine in the Big Indian Wash outside Moab, Utah. He had his boots bronzed and, every week, flew in his private plane to Salt Lake City for rumba lessons. Decades later, federal officials discovered that one Colorado mining town, Uravan, was so radioactive that its buildings, its streets, and its trees had to be torn down, chopped up, and buried, at a cost of $127 million.

If the Cold War meant the USA and the USSR continuously threatened each other with an ever-greater arsenal of annihilation, the real nuclear bomb targets would turn out to be American citizens. From 1951 to 1992 the AEC detonated 928 nuclear devices at the Nellis Air Force Gunnery and Bombing Range Proving Grounds (in time called the Nevada Test Site, or NTS), fifty miles north of Las Vegas—and an additional 126 tests were run elsewhere, notably at the Pacific Proving Grounds in the Marshall Islands. During the same period, over four thousand radiation experiments were conducted by the AEC, the CIA, the Department of Defense, the Centers for Disease Control, the National Institutes of Health, the Veterans Administration, and NASA on (for the most part uninformed) human subjects. As documented by Pulitzer Prize–winning journalist Eileen Welsome, hospital patients at New York's Sloan-Kettering, Cincinnati's General, Houston's Baylor College, and San Francisco's University of California were fully irradiated to produce military data; eight hundred pregnant woman were dosed with radioactive iron and their fetuses monitored by Vanderbilt University; seven newborns and over a hundred Native Americans were injected with radioactive iodine; seventy-three mentally disabled children in Massachusetts were fed radioactive cereal by Quaker Oats and the AEC; two hundred cancer patients were dosed with enormous amounts of cesium and cobalt; and 232 inmates had their testicles irradiated at carcinogenic levels by the University of Washington. The prisoners were paid a hundred bucks and sterilized at the end of the experiment to "keep from contaminating the general population with radiation-induced mutants."

Around two hundred of Nevada's explosions directly irradiated over two hundred thousand site witnesses, most of them servicemen, while around ninety atmospheric demolitions afflicted thousands who would be known as "downwinders"—residents of Nevada, New Mexico, Colorado, and Utah. The tests were so common that Las Vegas began promoting them as tourist attractions. Tom Saffer: "We Marines were brought [to Nevada's Frenchman Lake] at 3:30 in the morning [on June 24, 1957], the trucks disembarked us, we were left here. Hanging from a balloon two miles in that direction was a nuclear test called Priscilla [a part of Operation Plumbbob, which detonated twenty-nine bombs in Nevada from May 28 to October 7, 1957, twenty-seven of which were successful]. This was the spot where a twenty-two-year-old Marine Corps lieutenant, yours truly, was irradiated and became part of the population of 250,000 American veterans who were used in nuclear testing. Approximately half an hour before this test was conducted, a voice from an unseen loudspeaker said, 'Good morning, gentlemen, welcome to the land of the giant mushrooms. You are going to be closer to a nuclear weapon, or an atomic bomb, than anyone since Hiroshima.' We were told to kneel, put our forearms over our eyes and close our eyes tightly, and then the countdown started. Our right shoulder was towards the blast. We were told not to look up and none of us dared look. Then the countdown started—5, 4, 3, 2, 1 . . . We heard a sharp 'click' and this intense heat on the back of the exposed neck. And the most shocking part of this was you could see the two bones in your forearm, and a bright red light. Within a few seconds, shock waves from the bomb hit these trenches and I was immediately thrown from one side of the trench wall to the other. And I was frightened beyond belief."

At the time, the Centers for Disease Control told residents living downwind from the Nevada blasts that the only thing that would make them get cancer was worrying about getting cancer, especially if that worrier was a woman. In fact, to take one memorable example, of 220 cast and crew who worked on the movie *The Conqueror* in 1956 in Utah downwind from the NTS, 91 were diagnosed with cancer, a morbidity rate of 41 percent, with 46 dying of it by 1980, including the film's two stars, John Wayne and Susan Hayward.

Fourteen months after Mike, an American thermonuclear device, called Bravo, small enough to be carried by Curtis LeMay's Strategic Air Command, was ready to be tested. On March 1, 1954, Livermore and Los Alamos joined forces in the Marshall Islands, at the Bikini Atoll. Commander of the test, Vice Admiral W. H. P. Blandy, tried to reassure his men: "The bomb will not start a chain reaction in the water, converting it all to gas, and letting all the ships on all the oceans drop down to the bottom. It will not blow out the bottom of the sea and let all the water run down the hole. It will not destroy gravity."

Serviceman Michael Harris: "The explosion was going to be behind us. The major reminded us that this was essential, as he had done so many times before. Face the blast, he said, and our eyes could be damaged permanently, even if they were closed. And not just our eyes. Although he never did get specific about what other body parts might be affected. The disembodied voice repeated the warning again and again over the loudspeaker. 'Don't turn around before the countdown reaches zero. Don't turn around after the countdown reaches zero. Don't turn around until you are told it is safe to turn around.' We stood at attention. And paid attention. Lined up as instructed. Backs to the ocean. Following orders. Careful to avoid damage. The countdown and the disembodied voice: 'Four, three, two, one, zero.' The flash of light. The low, distant rumble. The shaking of the earth. And guess what? The pilot missed the target, but our eyes hit the target. We didn't have to turn around after thirty seconds. The fireball and the mushroom cloud were right there in front of us. We goggle-less enlisted men were facing ground zero. A result of pilot error. Major Maxwell gathered us together to explain: 'You win a few and you lose a few and sometimes things don't turn out the way you want. When that happens, you take it on the chin like a man and start all over again.' "

Additionally, the Livermore physicists had missed an important part of the equation. They thought it would be a 5-megaton blast, but instead it was a 14.8 Goliath with a four-mile-diameter fireball. Plasma physicist Marshall Rosenbluth: "I was on a ship that was thirty miles away and we had this horrible white stuff raining out on us. I got ten rads of radiation from it. It was pretty frightening. There was a huge fireball with these turbulent rolls going in and out. The thing was glowing. It looked to me like what you might imagine a diseased brain, or a brain of some madman would look like. You know, the surface, with the cortex convolutions, and so on. And it just kept getting bigger and bigger. It spread until the edge of it looked as if it was almost directly overhead. It was a much more awesome sight than a puny little atomic bomb. It was a pretty sobering and shattering experience."

Serviceman Michael Harris: "A mixture of radioactive materials, including pulverized coral, was forced high into the air, dispersed over a wide area by the winds, and showered down on hundreds if not thousands of people, covering them with white, gritty, hail-like 'snow.' Otherwise known as fallout. At least 236 Marshall Islanders, 23 Japanese fishermen, and a minimum of 31 Americans were dangerously exposed. The victims inhaled 'hot' ash, and radioactive particles whitened their hair, clung to their skin, and caused radiation burns. They developed nausea, diarrhea, itching, eyes that smarted and watered, and a significant decrease of white corpuscles in the blood. Eighteen Marshallese children died (after playing in the 'snow') and so did Aikichi Kuboyama, a Japanese fisherman aboard a boat with the unfortunate name of *Lucky Dragon*."

Though Washington denied it was responsible, "as a token of sympathy" a check for 2.5 million yen was sent to the dead fisherman's widow. Unfortunately, *Lucky Dragon*'s irradiated tuna was sold into the market before anyone knew why the crew's hair was falling out.

Bravo was the first of the Castle series, which continued with such designs as Runt, Shrimp, and Nectar. Able and Baker were detonated to see if they could annihilate a flotilla of ninety ships hosting 57 guinea pigs, 109 mice, 146 pigs, 176 goats, and 3,030 rats. They could. Publicity surrounding the tests included the announcement of a revolution in women's swimwear: "Like a bomb, the bikini is small and devastating!"

George Cowan watched Baker from a B-17: "An Air Force photographer was on board. He removed the door on the port side of the plane and looked directly toward the zero point. He carefully strapped himself and his equipment to buckles by the side of the door. We were supposed to be at least two miles from the detonation point but our pilot was obviously creeping closer. The voice of the test manager began the last ten-second countdown. At 'zero' the brilliant flash was dimmed by the overlying water, but the entire bay seemed to rise toward us. Then the shock wave arrived. It would've stripped the wings off a plane less sturdy than a B-17. The photographer hadn't practiced this part of the exercise. He and his equipment tumbled out the door and dangled outside on straps. We pulled him and his cameras back in. [His] work entered the history books. His picture of the huge 'wedding cake' of water and vapor rising from the bay, with ships decorating the cake fringes, was republished countless times."

Fifty years later, a team of biologists returned to Bikini in the submersible *M.Y. Octopus* to study the long-term effects of such massive doses of radiation. They were shocked to discover the only remaining trace of twenty-three thermonuclear detonations were in the Marshall Islanders' tombstones, which were radiant because they were made of absorbing sandstone. The lagoon waters were completely free of taint, with dosimeters not registering a blip above normal even when testing one of the sunken target ships, *Saratoga*. The area's marine life was extravagantly abundant, and as far as the investigators could tell, the ocean had miraculously diluted the massive amounts of cesium left by Castle and washed itself clean.

At the same time that the world was learning about an exciting new swimming suit and weapons beyond human imagination, the era's leading geneticists, Thomas Morgan and Hermann Muller, were irradiating the babies and grandbabies of fruit flies, with alarming results. The simultaneous news of *Lucky Dragon* and fruit-fly mutations would merge into a new story line in popular culture beginning in 1954, when atomic lizard Godzilla rampaged through Tokyo. While the rest of the world found *Godzilla*'s low-budget effects laughable, the Japanese, targets once again of American radioactive poisons, watched in silence, many breaking down in sobs. In 1957, *Incredible Shrinking Man* related the saga of a sailor who found himself engulfed in a mysterious cloud and grew ever smaller until he shrank into nothingness itself. After a New Mexico test site accidentally created twelve-foot ants in *Them!*, a scientist explained that this would become an everyday event in our modern atomic world. On TV screens *The Outer Limits* and *The Twilight Zone* dramatized atomic holocaust, genetic mutations, radioactive powers, and vague, invisible, indefinable terrors. In 1964's *Fail-Safe*, Pentagon consultant Walter Matthau predicts nuclear holocaust would be survived by file clerks and hardened criminals, but one of America's most glamorous socialites tells him that he's all wrong, that no one will survive a nuclear war, and that's the beauty of it:

"I've heard nuclear war called a lot of things . . . never beautiful."

"People are afraid to call it that, but that's what they feel."

Beginning in the 1960s, Marvel Comics created a new America where radiation had nothing to do with terrors about the end of the world and everything to do with the forging of superheroes. In *Fantastic Four*, *Daredevil*, and the *Hulk*, superheroes were all created through some form of atomic mishap, while *Spider-Man* combined a child's fantasy of spiders as aggressive, toxic, and voracious in attacking human beings with radiation, the magical force that in the world of Marvel could make anything happen. It was a brilliant remaking of atomic superstition, and like all myths it worked at the intersection of doubt and faith. Do people really believe irradiation can produce a Spider-Man? No. Do they know for an absolute fact that this is impossible? *Not really*.

The overwhelming power of Bravo pushed the Soviets to transcend *sloika* and create a weapon of equal force. Within a few weeks, Arzamas-16 physicists were doing the final calculations for their own multimegaton implosion, after having uncovered on their own the Teller-Ulam fission-fusion-fission design. On November 22, 1955, their version of Mike was ready, and it was a remarkable success. Physicist German Goncharov: "Immediately, it felt as if you had put your head into an open oven. The heat was unbearable. Then we had this impressive view of the fireball, the mushroom cloud, all of it on a huge scale. What was shocking was that this great scene was unfolding in absolute silence. And when the shock wave approached us, we dropped to the ground. Thunder. Stones were flying. Someone was hit by a large rock. There were several claps of thunder and the ground was shaking. I remember when we arrived back at our hotel, the windows and doors had been blown out. It felt as if the place had been hit by an air raid. But our joy was indescribable. We started celebrating immediately. We took out all our supplies. Someone brought alcohol. There was a sense of fulfillment, of having completed our task. That this beautiful, complex device—and that's what it was from a physicist's point of view—had worked was a triumph of science, of course. We all understood that."

Both Moscow and Washington now possessed the means to rid the world of humankind. They would spend the next five decades menacing each other with thermonuclear annihilation, while internally trying to solve a political riddle: If you already have the biggest weapon in human history, but a military insatiable for growth, what do you do next?

12

The Delicate Balance of Terror

THE rise of thermonuclear arsenals triggered consequences unforeseen by either their biggest supporters or gravest detractors, as the major nations of the earth were now armed with a weapon so grotesquely overpowered that, no matter what the circumstance, only a lunatic would deploy it. Super, indeed. Yet, just as the genocidal devastation of these ever-greater devices defied human reason—bombs, torpedoes, and missiles forever expanding in omnipotence, efficacy, and power—so, too, did overseeing these cataclysmic weapons seem to inflict a type of mental disability on its bureaucrats. Ardent Cold Warriors on both sides of the Iron Curtain found themselves beset with an ordnance version of the dysmorphia seen in some bodybuilders. No matter how many bombs they had or how big their explosives grew, they needed more, and bigger: enough was never enough. No one in charge seemed to grasp the point that Winston Churchill made resonant: "If you go on with this nuclear arms race, all you are going to do is make the rubble bounce." The only strategic logic each side followed was, if Washington had enough Bombs to make the rubble bounce and bounce again, then Moscow needed enough to make it bounce, bounce, and bounce a third time, which incited the USA to then need at least four, but better yet five, bounces, and this math would continue on as infinitely as the half-life of uranium.

Nikita Khrushchev made a joking threat of this: "I remember President Kennedy once stated . . . that the United States had the nuclear missile capacity to wipe out the Soviet Union two times over, while the Soviet Union had enough atomic weapons to wipe out the United States only once. . . . But I'm not complaining. . . . We're satisfied to be able to finish off the United States the first time round. Once is quite enough. What good does it do to annihilate a country twice? We're not a bloodthirsty people." His generals and admirals, however, did not at all share this perspective. They wanted, always, to one-up the Americans with ever more bouncing, bouncing, bouncing.

This aspect in the history of arms escalation reveals a deep paradox in the administration of every American president. After three months in office, Truman's successor, Dwight D. Eisenhower, famously told the American Society of Newspaper Editors:

> The cost of one modern heavy bomber is this: a modern brick school in more than thirty cities.
> It is two electric power plants, each serving a town of sixty thousand population.
> It is two fine, fully equipped hospitals.
> It is some fifty miles of concrete highway.
> We pay for a single fighter plane with a half million bushels of wheat.
> We pay for a single destroyer with new homes that could have housed more than eight thousand people.

Yet, simultaneously under Ike's watch, when the nation was—in practice if not in demeanor—at peace, America made for itself a surfeit of Armageddon. In 1950, the country had around 400 atomic bombs; by 1955, she had 2,280, twenty times more than the Russians and the start of a spiral of warhead escalation:

1957	3,500
1959	7,000
1961	2,305
1963	23,000
1967	32,500

Over the Cold War's four decades, every American president except Nixon publicly spoke of his regret at this state of affairs, yet under the watch of every president, including Nixon, the tools of apocalypse exponentially grew. John Kennedy's science adviser Jerome Wiesner discussed 1963's twenty-three-thousand-bomb arsenal: "I will give you a simple piece of calculus. For most cities it is reasonable to equate one bomb in one city. It would take a bigger bomb for Los Angeles or New York. . . . In any event, it does not take many. And if you ask yourself, 'Where would you put three hundred large nuclear weapons to be most destructive?' You run out of vital cities and towns and railroad junctions and power plants before you get to three hundred. The same thing is true in the United States and the Soviet Union. If I was not trying to be conservative, I would say fifty bombs, properly placed, would probably put a society out of business, and three hundred in each of the two countries leading the arms race would destroy their civilizations. That is a pretty clear-cut fact." Yet, his president did not ask the Pentagon why it had to have those twenty-three thousand Bombs when it only needed three hundred.

Immediately after Nagasaki, Yale political scientists Bernard Brodie and Jacob Viner began to develop theories of nuclear strategy to try to answer Oppenheimer's question of what the atom is good for in battle beyond Bradley's "psychological" use. Viner almost immediately arrived at a revolutionary concept: "The atomic bomb makes surprise an unimportant element of warfare. Retaliation in equal terms is unavoidable and in this sense the atomic bomb is a war deterrent, a peace-making force." It took a year before Brodie was able to see this point, with the 1946 paper that would make him famous, "The Atomic Bomb and American Security": "Thus far the chief purpose of our military establishment has been to win wars. From now on its chief purpose must be to avert them." He also echoed Oppenheimer's math, arguing, "If two thousand bombs in the hand of either party is enough to destroy entirely the economy of the other, the fact that one side has six thousand and the other two thousand will be of relatively small significance."

The State Department was so taken with these ideas that Brodie was made an agency consultant on nuclear control, but Pentagon chiefs did not share his viewpoint. By 1950, Vandenberg's air staff had created a strategy of nuclear bombing three categories of Soviet targets—liquid-fuel refineries, electrical power stations, and nuclear energy plants—a strategy it called Killing a Nation. The Strategic Air Command's Curtis LeMay predicted the next war would be nuclear—"If there is another war, we will be first, instead of last to be attacked, and the war will start with bombs and missiles falling on the United States"—and the American defense would be conducted and her atomic victory won by his global force of nuclear bombers. His idea was to strike the USSR with the entirety of his arsenal, killing over 77 million people in 188 Soviet cities—three-fourths of the population—in thirty days. But instead of Killing a Nation, he called it the Sunday Punch. Now at State, Bernard Brodie countered that the air staff was wrong—since it didn't know enough about the Soviet Union to be sure that it had the full target list to kill a nation—and that SAC was wrong—since why would the United States use all of its nuclear weapons at once; why not hold some in reserve to use as a coercive threat? Again, the military wasn't interested.

LeMay's first major decision on becoming SAC chief in 1948, in fact, was to have the whole Strategic Air Command simulate a Sunday Punch against Dayton, Ohio—"a realistic combat mission, at combat altitudes, for every airplane in SAC that we could get in the air." The exercise was an out-and-out failure, LeMay calling it "just about the darkest night in American military aviation history. Not one airplane finished that mission as briefed. Not one. . . . I'll admit the weather was bad. There were a lot of thunderstorms in the area; that certainly was a factor. But on top of this, our crews were not accustomed to flying at altitude. Neither were the airplanes, far as that goes. Most of the pressurization wouldn't work, and the oxygen wouldn't work. Nobody seemed to know what life was like upstairs." His men practiced again and again, targeting Baltimore as military intelligence insisted she resembled Soviet cities, and dummy-bombing San Francisco over six hundred times in one month. LeMay: "We attacked every good-sized city in the United States. People were down there in their beds, and they didn't know what was going on upstairs. . . . My determination was to put everyone in SAC into this frame of mind: We are at war now! So, if we actually did go to war the very next morning, or even that night, no preliminary motions would be wasted."

As seen when he kept his warheads after threatening America's foes in the Korean War on behalf of Eisenhower, LeMay had little regard for civilian oversight of the armed forces. He decided the USAF's sniffer planes patrolling the Soviet Union's borders weren't enough; he wanted aerial reconnaissance, which he started in early 1950. Almost immediately, Soviet fighters brought down an American PB4Y-2 eavesdropping over Soviet territory, killing ten crewmen. As the Kremlin could interpret these flights as acts of war, Truman had them banned. But after getting approval from the Joint Chiefs of Staff and Winston Churchill in March 1952, in exchange for English crews conducting reconnaissance flights over the USSR and sharing the results with him, LeMay gave the British the best high-altitude American craft currently in production, the B-45. This sidestepping of Truman continued for two years, ending only when the US resumed its own direct surveillance, notably with the infamous and spectacular U-2.

Before the switch to satellites, the Soviets brought down at least twenty US reconnaissance craft, killing an estimated one hundred to two hundred Americans. But these flights weren't just for surveillance, as LeMay revealed after his retirement from the service: "There was a time in the 1950s when we could have won a war against Russia. It would have cost us essentially the accident rate of the flying time, because their defenses were pretty weak. One time in the 1950s we flew all of the reconnaissance aircraft that SAC possessed over Vladivostok at high noon. Two reconnaissance airplanes saw MiGs, but there were no interceptions made. It was well planned, too—crisscrossing paths of all the reconnaissance airplanes. Each target was hit by at least two, and usually three, reconnaissance airplanes to make sure we got pictures of it. We practically mapped the place up there with no resistance at all. We could have launched bombing attacks, planned and executed just as well, at that time."

In January 1956, he scrambled many of his bombers in a simulated nuclear attack. In another exercise, Operation Powerhouse, SAC crews flew nearly a thousand simultaneous sorties from more than thirty bases around the world to intimidate Moscow. A few weeks after, Operation Home Run sent B-47 Stratojets from Thule, Greenland, over the north pole and into Siberia looking for gaps in Soviet radar, as well as a squadron of RB-47 Stratojets flying in attack formation, in daylight, over the USSR. The Soviets had no way of knowing if the bombers were armed and about to launch nuclear strikes. And that was the point.

The most egregious SAC insult to civilian oversight came during the Kennedy era, when Robert McNamara decided that Minutemen ICBMs needed tighter security controls. In 1962 he arranged for an executive order that all silos be outfitted with secret launch codes—Permissive Action Links, or PALs—to keep a local officer from launching, on his own initiative, nuclear war. SAC commanders, however, came to believe that using these PALs might cause a delay in defending the homeland and decided to upend the president's order with a policy they would maintain for two decades. "The Strategic Air Command in Omaha quietly decided to set the 'locks' to all zeros in order to circumvent this safeguard," missile launch officer Bruce G. Blair said. "During the early to mid-1970s during my stint as a Minuteman launch officer, they still had not been changed. Our launch checklist in fact instructed us, the firing crew, to double-check the locking panel in our underground launch bunker to ensure that no digits other than zero had been inadvertently dialed into the panel. . . . So the 'secret unlock codes' during the height of the nuclear crises of the Cold War remained constant at 00000000." At the same time, missile silo operators worried about a fail-safe aspect that required two officers, standing at separate consoles, to turn their keys in unison and hold them for two seconds before their ICBMs would fire. What would happen if an order came down, one of the men refused to obey, or what if only one man was alive in the silo? They practiced with various alternatives and found that a key tied with string could be turned with a spoon from the other console, and the fail-safe of two consenting officers was thwarted.

Today, SAC headquarters outside Omaha, Nebraska, is called the Underground Command Post, with steel-lined corridors and doors immune against a five-megaton warhead and its radioactive breath. Nuclear Armageddon would be started from the Command Balcony, a mezzanine of SAC's two-story war theater. At the theater's center is the swivel chair for CINCSAC, the commander in chief of the Strategic Air Command, and his two telephones. The red phone is for incoming calls from the president and the Joint Chiefs of Staff; the gold phone is for CINCSAC to pass along their commands to his staff. If the bunker is destroyed, command would be assumed by the Looking Glass Plane, presumably secure against any threat, which has been continuously floating over the Midwest since February of 1961.

At the end of World War II, Army Air Forces chief Hap Arnold decided, "We have to keep the scientists on board. It's the most important thing we have to do." On September 31, 1945, he took $10 million of the $30 million in funds he had left over from the war research budget to start a new R&D outfit in California far from the Pentagon. RAND was a think tank investigating science, engineering, psychology, sociology, weapons, tactics—anything that would keep Arnold's the best air force in the world. General LeMay engaged RAND to research what would become, in May 1946, its first report, "Preliminary Design of an Experimental World-Circling Spaceship" (i.e., a satellite), which "would inflame the imagination of mankind, and would probably produce repercussions in the world comparable to the explosion of the atomic bomb. . . . The development of a satellite will be directly applicable to the development of an intercontinental rocket missile." This would, of course, all come true, as the Soviets' *Sputnik*. Herb York: "The reason for having RAND do the study was to get a jump on the navy, which also was studying satellites. LeMay was determined that it wasn't going to be a navy program, it wasn't going to be a joint navy–air force program, it was going to be an air force program."

At various times, employing a cavalcade of American intellects, including John Forbes Nash Jr., Donald Rumsfeld, Daniel Ellsberg, Francis Fukuyama, Condoleezza Rice, Henry Kissinger, and Margaret Mead, RAND followed a state-of-the-art curriculum in assessing the future of war. At its very birth the agency was dead right about *Sputnik*, but over the ensuing decades it would be dead wrong about almost everything else. While the CIA reported during the Eisenhower administration that the Russians had fewer than fifty ICBMs, RAND insisted they had over five hundred and fed enough backing data to presidential candidate John Kennedy's staff on this "missile gap" that it could be used in his campaign against Richard Nixon. After Kennedy won, being president included the security clearance to learn what the U-2 surveillance program had uncovered—that the USSR had, indeed, forty-one ICBMs. RAND continued to get things wrong over the following decades, with many of the agency's misfires rising from RAND's strategy of finding out what Pentagon brass wanted, then recommending that they get exactly that. In regards to nuclear arms, what service chiefs wanted was more, more, and more, so what RAND recommended was bounce, bounce, bounce . . .

Following the Department of Defense's "enough is never enough" policy, the AEC mirrored labs and production facilities so that, if one was destroyed in a nuclear strike, another just like it would continue on the vital work of making more and better Bombs. Hanford was replicated at Aiken, South Carolina, as the Savannah River Plant; Los Alamos was twinned with Livermore; and Oak Ridge was built over again in Paducah, Kentucky. Creating this massive enterprise required more than 11 percent of America's nickel, 34 percent of her stainless steel, 33 percent of her hydrofluoric acid, and by 1957, 6.7 percent of her electrical power. From its opening annual budget of $1.4 billion in 1947, the Atomic Energy Commission's capital budget hit $9 billion in 1955, greater than those of General Motors, Alcoa, DuPont, Goodyear, and Bethlehem and US Steel combined. The AEC's labs turned out atomic weapons for all occasions—gravity bombs, submarine-pen penetration bombs, atomic artillery shells, ship-to-ship and air-to-air rocket-propelled devices, antiaircraft weapons, cruise-missile warheads, antiballistic-missile nuclear explosives, and the Davy Crockett, fired from a recoilless rifle; by 1970, Lawrence Livermore had designed sixty different nuclear weapons on its own, including George Gamow's New Mexico Jumping Bean, which had a fission engine to fly to Russia, where it would turn itself into a bomb and fall from the sky. For the next two decades, widow Molly Lawrence did everything she could to get her husband's name removed so he wouldn't go down in history as the father of a nuclear munitions factory. Her effort failed.

Bernard Brodie left State to become RAND's first great nuclear strategist. When Ed Teller leaked news of the forthcoming Teller-Ulam test to a RAND physicist, a group headed by Brodie calculated that a five-to-ten megaton warhead would annihilate every creature in fifty square miles, that fifty-five twenty-megaton thermonuclears could take out Russia's fifty biggest cities and kill 35 million people. The carnage was so extravagant that any idea of strategy or targeting was removed; the Backyard bomb of Teller's Los Alamos dreams—the one so huge and so lethal you didn't need to take it and drop it on an enemy, you could just set it off in your own backyard—had come true. Brodie: "We no longer need to argue whether the conduct of war is an art or a science—it is neither. The art or science comes in only in finding out, if you're interested, what not to hit."

In 1954, Eisenhower's secretary of state, John Foster Dulles, announced that, as the Soviet army was so vast and could attack anywhere and anytime, America had a new defense policy: Massive Retaliation. If provoked, the United States would attack with everything it had, including the whole of its nuclear arsenal. Inspired by a 1947 fashion magazine, Eisenhower called this approach the New Look, though it was just another Sunday Punch. In the 1960s, Washington with NATO created Flexible Response, a strategy described in 1982 by Reagan's secretary of defense, Caspar Weinberger: "Under this concept, the United States and NATO planned to strengthen general purpose warfare forces in order to better equip them to deal with a Soviet conventional attack; at the same time, US nuclear capabilities were increased in order to provide the President with the option of using nuclear forces both to support our general purpose forces and to respond selectively (on less than an all-out basis) to a limited Soviet nuclear attack. The option of retaliation on a more massive scale was retained in order to deter the possibility of a major Soviet nuclear attack. This concept of flexible response remains as a central principle of our strategy today."

Even though the strategy required ever more and ever bigger weapons bouncing, bouncing, bouncing, Flexible Response's amending of the all-out first attack of Sunday Punch and Massive Retaliation meant that Bernard Brodie's key points were becoming mainstream defense policy. At the same time, his thoughtful stance was being wholly overturned by his successor at RAND, the neoconservative icon Albert Wohlstetter. Wohlstetter made a name for himself in 1951 by showing how the USSR could use a lightning strike to destroy 85 percent of SAC with 120 nuclear bombs—a presentation that terrified many in the air force, even though it was outlandishly implausible considering Soviet military capabilities at the time. Wohlstetter initiated the concept of "fail-safe," ensuring the ability to guard against accidental Armageddon, counseled the air force and SAC to target military installations instead of cities, and promoted the use of surveillance planes and satellites to monitor the enemy's capabilities. LeMay was uninterested, so Wohlstetter had to present his notions directly to the air force's chief of staff, giving over ninety-two briefings from 1952 to 1953, most of which became national defense policy.

On July 21, 1955, Eisenhower unveiled an arms control plan that became informally known as Open Skies, which would have allowed the US and the USSR to monitor each other's military installations to make sure they were treaty-compliant. Khrushchev immediately rejected it as an "espionage plot," which the Americans were expecting; the plan was primarily hatched as a global public relations maneuver portraying Washington as working in favor of arms control, with Moscow secretive and belligerent. But the reason for Khrushchev's refusal was the worry that if the United States found out the truth about Soviet defenses, it would strike and invade. Nikita Khrushchev: "Our missiles were still imperfect in performance and insignificant in number. . . . [So] we couldn't allow the US and its allies to send their inspectors crisscrossing around the Soviet Union. They would have discovered that we were in a relatively weak position, and that realization might have encouraged them to attack us." Twice, in June 1961 and August 1963, John Kennedy proposed that the Americans and the Russians fly to the moon together, and each time Khrushchev refused, citing espionage, but the reason was the same as for Open Skies; the premier worried that a joint program would reveal to the enemy his country's profound weakness.

In 1956, "the USSR had a total of 426 nuclear warheads," rocket scientist (and son of the Soviet premier) Sergei Khrushchev explained. "That year the United States had an overall nuclear superiority 10.8 times greater. To restrain the West from a possible attack on the Soviet Union, Father decided to resort to bluff and intimidation. During his visit to England in April 1956, he casually inquired from time to time—once during an official lunch, once in the course of a five-hour tea at the fireplace of the prime minister's country residence at Chequers—if his hosts knew how many nuclear warheads it would take to wipe their island off the face of the earth. An awkward silence followed. But Father did not drop the subject, and with a broad smile on his face he informed those present that if they didn't know, he could help them, and he mentioned a specific number. Then he added, quite cheerfully, 'And we have lots of those nuclear warheads, as well as the missiles to deliver them.' It was in those years that he used the famous phrase 'We are producing missiles like sausages.' When I asked him how he could say that, since the Soviet Union had no more than half a dozen intercontinental missiles,

Father only laughed: 'We're not planning to start a war, so it doesn't matter how many missiles are deployed. The main thing is that Americans think we have enough for a powerful strike in response. So they'll be wary of attacking us.' Such statements by Father were in fact received with enthusiasm on the other side of the ocean, since they made it easier for the American military to receive additional funds."

The Russians were especially alarming adversaries to American political and military leaders as it seemed impossible to know what they would do next. On the one hand they seemed barbaric to Washington powers-that-be—"Asiatics," as Truman sneered. On the other, they would trump American beliefs that the West had superior science and technology by having fission and fusion bombs long before the United States expected and would shock the world by launching a satellite orbiting the globe before the United States even had a rocket that could lift such a payload into orbit. On October 4, 1957, Americans were horrified to learn they now had a Red Moon orbiting over their heads, as *Sputnik* circumnavigated the globe 1,440 times for ninety days, before reentering the atmosphere and burning up. Here was direct physical evidence that the Soviets could deliver nuclear strikes anywhere in the world. In his memoirs, Khrushchev explained his reasoning for the space and missile race: "Of course we tried to derive the maximum political advantage from the fact that we were first to launch our rockets into space. We wanted to exert pressure on the American militarists—and also influence the minds of more reasonable politicians—so that the United States would start treating us better."

Instead, US military leaders now demanded a dramatic increase in spending for conventional and nuclear weapons, even more than the dramatic increase in spending for conventional and nuclear weapons that had already been budgeted. Three weeks after *Sputnik*'s launch, an October 29 CIA report held that the Soviets were well on their way to having a sophisticated missile program up and running and concluded that "the country is in a period of grave national emergency," while the US Air Force across the early 1960s insisted that the USSR had dramatically eclipsed the USA in the strength of its missile forces. When the U-2 spy plane's recon photos did not reveal either of these dramatic assertions, Curtis LeMay said that he knew the missiles were there, but that the American cameras had missed them. In fact the opposite was true: the Soviets then had four ICBMs, but moved them around on tracks to be photographed again and again by American surveillance to create an illusion of strength. When Kennedy then used this "missile gap" in his presidential campaign against Richard Nixon, Khrushchev believed Kennedy was warmongering and might wage a preemptive American strike. In 1961, his fears were confirmed by the Bay of Pigs invasion of Cuba.

The month after the launch of *Sputnik*, the November 7 Gaither Report prepared for Eisenhower insisted that the USSR would soon surpass the United States in nuclear power, that the Kremlin had already prepared its citizenry for nuclear combat, and urged the government to spend $25 billion on a nationwide system of fallout shelters and a massive campaign instructing Americans on what to do when the Bombs began to fall. A new agency, the Office of Civil Defense, would spend the next two decades promoting the fantasy that everyday Americans could survive nuclear war, creating a cartoon hero, Bert the turtle, who when attacked by monkeys wielding sticks of TNT knew to "duck and cover." Bert then explained that, as a nuclear bomb explosion "would break windows all over town," American schoolchildren should themselves duck and cover under furniture to survive the forthcoming holocaust. But at least this was a sincere effort; when Khrushchev told the West, "It would take really very few multimegaton nuclear bombs to wipe out your small and densely populated countries and kill you instantly in your lairs," Kennedy said there was only one answer: "A fallout shelter for everybody, as rapidly as possible." Kennedy's own fallout shelters were built on Nantucket, Massachusetts, near the family compound in Hyannis Port, and on Peanut Island, Florida, near the Kennedy winter home in Palm Beach.

The OCD informed suburbanites—urban dwellers were excluded from the program as it was assumed cities would be targets, and the majority of their populations immediately slaughtered—that their backyard fallout shelters needed concrete walls at least a foot thick, additionally shielded by five hundred cubic feet of air and three and a half inches of packed dirt or a half inch of lead. Twinkies were a staple of shelter pantries since they were supposed to "stay fresh forever," but supporting human life for the duration of thermonuclear aftereffects required far more sophisticated ventilation than these backyard designs, as well as some kind of waste disposal. Journalist Susan Roy: "One thing that was rarely talked about was going to the bathroom. You were supposed to take a garbage bin and line it with plastic. Imagine. When it was full, you were supposed to close it, run outside really fast, and leave it." Cities such as San Francisco, Seattle, Las Vegas, and Philadelphia provided their children with dog tags so they could be identified in case they were incinerated in a nuclear holocaust—New York City distributed 2.5 million tags between February and April of 1952—and the 1951

Journal of the National Education Association had a discussion on whether kids should be tattooed with their identities. The experts concluded that, as skin could not survive nuclear war, tattooing was not effective.

Besides testing the effects of atomic bombs on various building materials and producing motion pictures so that American citizens could see paint being boiled away and edifices being thrown in one direction by the shock wave and then reversed by the pull of the rising mushroom cloud, the Federal Civil Defense Administration ran Operation Cue at the Nevada Test Site to determine the effects of nuclear attacks on the typical American family. Journalist Laura McEnaney: "They called them mannequin families and they dressed them up in JC Penney clothing and they positioned them in various rooms of the house. And in one room there was a child taking a nap, in another room there was a family dinner party, and one of the things they did, they placed families hiding underneath shelters. And they equipped each of these homes with refrigerators, typical appliances one would find in a home, the kinds of food one would eat, from baby food to adult food, and they exposed them to blast. These were sort of like morality tales. Each film, each brochure, was a morality tale. If you didn't prepare, you have only yourself to blame for being wounded or killed. And it asked American families to think about themselves not just as friends, neighborhoods, family members, but as warriors of a cold war. And this really introduced a military purpose and practice into American family life."

Post-*Sputnik*, Eisenhower's Open Skies would come true through another avenue with Corona, the CIA's surveillance-satellite program, which began to replace the U-2s in 1959, and which, using high-resolution, remarkably detailed 70 mm Eastman Kodak film, provided Washington with information about every aspect of Soviet and Chinese military power. Yet, even though the United States now knew how much weaker the Soviets were than previously believed, it reassured no one. At any minute, after all, the Red Army might surpass the red, white, and blue's forces and achieve nuclear dominance. This never happened, but the Soviets did begin their own successful spy-satellite program, and fairly quickly the superpowers reached eye-in-the-sky reconnaissance parity. But oddly enough, neither deployed antisatellite weapons against the other, perhaps realizing that knowledge made for less frightened and less trigger-happy foes. Leaving each other's satellites alone, it turned out, was a more brilliant strategy than Killing a Nation, Sunday Punch, Massive Retaliation, and Flexible Response combined.

By the time of the Eisenhower administration, LeMay's SAC was a nuclear monopoly that threatened to swallow the whole of the DoD budget. Many advisers, including von Neumann, raised Wohlstetter's fear of the USSR's wiping out all of SAC in one blow. In 1955, Ed Teller suggested developing nuclear warheads small enough to fit on rockets launched from submarines and arranged for Livermore to outbid Los Alamos to research and develop such a system, giving the brilliant Admiral Hyman Rickover a nuclear role for his navy—a fleet of forty-five submarines, twenty-nine of them deployed 24-7, capable of launching atomic-warhead-tipped cruise missiles against 232 Russia targets, with a build cost of $7–$8 billion and annual operating expenses of $350 million. At the same time, von Neumann and the air force developed warhead-bearing rockets, and by 1967, America had a thousand Minutemen intercontinental ballistic missiles in underground silos. The Pentagon now wielded a global force of sub-ferried ballistic missiles, SAC-borne atomic bombs, and bunker-based ICBMs—the Triad Doctrine. Today, over two decades after the fall of the USSR, the United States maintains this Triad Doctrine for conducting a massive and overwhelming nuclear strike against the Soviets, even though there are no more Soviets. As early as 1958, the American nuclear arsenal was already so morbidly obese that, when exercises were conducted to assess operations, they revealed more than two hundred incidents in which bombers and missiles overlapped, meaning that, during nuclear war, they would be blowing up each other instead of the enemy.

In 1961, RAND created the Single Integrated Operational Plan or SIOP-62, a revision of Massive Retaliation/Sunday Punch, as it would release the whole of the American nuclear arsenal if the country was provoked. One billion people would eventually die from fire and radiation, with 285 million in the ellipsoidal target from China to Eastern Europe perishing in the initial blast. From about 1961 to 1968, the concomitant strategy of Furtherance meant that, if the United States was attacked and the president died or disappeared, secret instructions would be sent to military commanders giving them authorization to launch the nation's nuclear arsenal, meaning an automatic "full nuclear response" against both the Soviet Union and China. When Marine Corps chief David Shoup asked SAC chief Thomas Power at a 1960 briefing if, in the event China had nothing to do with the attack, they could be spared, Power replied, "Well, yeah, we could do that, but I hope nobody thinks of it because it would really screw up the plan."

When JFK was briefed on SIOP-62 after taking office, he muttered, "And we call ourselves the human race," while after his first atomic briefing in September 1953, Khrushchev "learned all the facts of nuclear power [and]

I couldn't sleep for several days. Then I became convinced that we could never possibly use these weapons, and when I realized that, I was able to sleep again." Witnesses reported that two of his successors, Brezhnev and Kosygin, after learning that an American attack would annihilate 99.99 percent of the Soviet military, 85 percent of the nation's industry, and kill 80 million Russians, were visibly shaken.

Defense Secretary Robert McNamara announced on September 18, 1967, "The cornerstone of our strategic policy continues to be to deter nuclear attack upon the United States or its allies. We do this by maintaining a highly reliable ability to inflict unacceptable damage upon any single aggressor or combination of aggressors at any time during the course of a strategic nuclear exchange, even after absorbing a surprise first strike. This can be defined as our assured-destruction capability." The strategy of Mutual Assured Destruction—MAD—was accompanied by the Pentagon's Nuclear Utilization Target Selection: NUTS. In the nuclear arms race, "each individual decision along the way seemed rational at the time. But the result was insane," admitted McNamara. "Each of the decisions, taken by itself, appeared rational or inescapable. But the fact is that they were made without reference to any overall master plan or long-term objective. They have led to nuclear arsenals and nuclear war plans that few of the participants either anticipated or would, in retrospect, wish to support. . . . Despite an advantage of seventeen to one in our favor, President Kennedy and I were deterred from even considering a nuclear attack on the USSR by the knowledge that, although such a strike would destroy the Soviet Union, tens of their weapons would survive to be launched against the United States. These would kill millions of Americans. No responsible political leader would expose his nation to such a catastrophe."

Adding to MAD and NUTS was the Nixon administration's Madman Theory, or "the principle of the threat of excessive force." In the wake of so many ex-presidents and premiers admitting that they could never pull the nuclear trigger, Madman Theory was a strategy to convince foreign powers that one of the leaders of the United States was wildly aggressive, ruthlessly cold-blooded, and fearfully unpredictable. In the summer of 1974 when his impeachment was an open topic of conversation, Nixon idly mentioned to a group of congressmen, "I could go into the next room, make a telephone call, and in twenty-five minutes, 70 million people will be dead." Afterward, Senator Alan Cranston asked Defense Secretary James Schlesinger about "the need of keeping a berserk president from plunging us into a holocaust," and Schlesinger initiated a new procedure to have anyone in the nuclear chain contact him upon receipt of unusual commands or requests from President Nixon.

Soviet colonel Viktor Girshfield: "All of us, more or less, know that nuclear war would be the end. All our theoreticians say that there is no way of preventing nuclear war from escalating to the global level, that you cannot win a nuclear war. That is our general theoretical position. But from a professional military point of view, such a position is impossible. Can a professional military man say that nuclear war is inconceivable? No, because some fool of an American president may really start a nuclear war. A professional military man must consider what to do in that event. . . . Consider the point of view of another professional, the doctor who knows that his patient is suffering from an incurable disease. He cannot for that reason abandon further efforts. . . . To make no plans for [a nuclear war] would be openly to proclaim our helplessness. It would be psychologically wrong." Physicist Herbert York: "Let's put it this way, in more understandable terms. All roads in the strategic equation lead to MAD [Mutual Assured Destruction]. All the other ones . . . are games, are window dressings, and they are window dressing for upmanship. . . . But when you take away all these layers of cloth, at the bottom of the thing, basically, is MAD, and no one likes it."

Nikita Khrushchev's son, Sergei: "For thousands of years peoples have resolved their conflicts by armed clashes. There was good reason for Karl von Clausewitz to write that war is a continuation of politics by other means. With the invention of nuclear weapons, politicians suddenly realized that war would no longer lead to victory, that both sides would lose. But they didn't know how to behave differently. So they behaved the same way, but without going to war. War without war was called 'cold war.' . . . Thus the Cold War was a kind of transitional period from a disconnected world that used weapons as its main instrument for resolving world conflicts to some kind of different state of being—to a new world order, if you like."

At RAND in the Kennedy era, Albert Wohlstetter was eclipsed both physically and theoretically by Herman Kahn, who was described as looking "like a prize-winning pear" and "a thermonuclear Zero Mostel." Kahn was renowned for his two-day, twelve-hour lectures, which employed von Neumann's game theory to "think about the unthinkable"—nuclear holocaust as a round of poker. In a world of strategies called MAD and NUTS and evermore-deranged notions of atomic warfare, Kahn was the endgame.

Inspired by Wohlstetter's 1959 *Foreign Affairs* article, "The Delicate Balance of Terror," Kahn's 1960 *On Thermonuclear War* analyzed and compared various apocalyptic scenarios, including hundreds of megadeaths—a unit invented by Kahn, one megadeath equaling 1 million murdered—and entire continents uninhabitable for centuries. It insisted that "despite a widespread belief to the contrary, objective studies indicate that even though the amount of human tragedy would be greatly increased in the postwar world, the increase would not preclude normal and happy lives for the majority of survivors and their descendants." The elderly would be fine eating radioactive food since "most of these people would die of other causes before they got cancer." In crafting defense in the Atomic Age, "any power that can evacuate a high percentage of its urban population to protection is in a much better position to bargain than one which cannot do this. . . . It might well turn out that US decision makers would be willing, among other things, to accept the high risk of an additional 1 percent of our children being born deformed if that meant not giving up Europe to Soviet Russia."

Scientific American called *On Thermonuclear War* "a moral tract on mass murder: how to plan it, how to commit it, how to get away with it, how to justify it." Yet fundamentally, to launch the global annihilation of all-out nuclear assault, any leader from either side pushing the button had to agree with Kahn's theses. National Security Council member Roger C. Molander attended one meeting at the Pentagon to discuss these new realities: "A Navy captain was saying that people here and in Europe were getting much too upset about the consequences of nuclear war. The captain added that people were talking as if nuclear war would be the end of the world when, in fact, only 500 million people would be killed. 'Only 500 million people!' I remember sitting there and repeating that phrase to myself: 'Only 500 million people!' "

On Thermonuclear War was one of the key inspirations for Stanley Kubrick's 1964 *Dr. Strangelove*, with Peter Sellers's title character being a mélange of Kahn, Teller, von Neumann, and Nazi rocketeer Wernher von Braun; his Austrian accent was inspired by Kubrick's photographic consultant, Weegee. So much of Kahn's book reappears in *Dr. Strangelove*—in such preposterous moments as George C. Scott's general explaining, "Mr. President, I'm not saying we wouldn't get our hair mussed. But I do say no more than ten to twenty million killed, tops, uh, depending on the breaks," and the strategy of preserving the best of the American species in mine shafts, with two females for each male—that Kahn complained to Kubrick about being paid royalties. "It doesn't work that way," the director replied.

A key *Strangelove* plot point used Kahn's invention the Doomsday Machine, a computer-run automatic response that was the ultimate in retaliation as it would destroy your foe, even though you, yourself, were already dead. If the only purpose of nuclear arsenals is deterrence, the machine was a great one—"The whole point of the Doomsday Machine is lost if you keep it a secret!" Strangelove screams at one point at the Soviet ambassador. Stanley Kubrick: "The present nuclear situation is so totally new and unique that it is beyond the realm of current semantics; in its actual implications, and its infinite horror, it cannot be clearly or satisfactorily expressed by any ordinary scheme of aesthetics. What we do know is that its one salient and undeniable characteristic is that of the absurd." An extraordinary number of the movie's inventions came true, including the Moscow-Washington hotline, and the Soviet version of the Doomsday Machine. It was called Dead Hand.

On July 25, 1980, Jimmy Carter signed presidential directive PD-59, which replaced the Sunday Punch–styled SIOP—which had been in effect since Kennedy was disgusted by it—with a strategy of nuclear combat that would directly target and attack Soviet leaders. A sequel to Killing a Nation, this was known as Decapitation, and its logic was explained by a *Foreign Policy* article, "Victory Is Possible": "The United States should be able to destroy key leadership cadres, their means of communication, and some of the instruments of domestic control. The USSR, with its gross overcentralization of authority, epitomized by its best bureaucracy in Moscow, should be highly vulnerable to such an attack. The Soviet Union might cease to function if its security agency, the KGB, were severely crippled. If the Moscow bureaucracy could be eliminated, damaged, or isolated, the USSR might disintegrate into anarchy." By the end of 1980, a Tomahawk cruise missile in a test struck within sixteen feet of its target, meaning Decapitation was feasible.

Oppenheimer had once compared the US and USSR to scorpions in a bottle, but the Cold War superpowers acted more like a rapid-fire version of evolution between predator and prey. The Soviets responded to Decapitation by engineering an automatic attack, run by computers, which, after their society's head had been cut off, could be triggered by a small band of surviving Russians from an underground bunker—this strategy called Perimetr, or Dead Hand. Minuteman officer Bruce Blair explained that if Dead Hand "senses a nuclear explosion in Russian territory and then receives no communication from Moscow, it will assume the incapacity of human leadership in Moscow or elsewhere and will then grant a single human being deep within Kosvinsky Mountain the authority and capability to launch the entire Soviet nuclear arsenal. . . . Communication rockets, launched automatically by radio command, would relay fire orders to nuclear combat missiles in Russia, Belarus, Kazakhstan, and Ukraine. The doomsday machine provides for a massive salvo of these forces. Weapons commanders in the field may be completely bypassed. Even the mobile missiles on trucks could fire automatically, triggered by command from the communications rockets."

Dead Hand became operational in 1985, just one more ingredient in the superpowers' nuclear operations that might bring *Dr. Strangelove*'s accidental Armageddon to life. Several times in the 1950s, American defense forces' alarms warned of incoming Soviet missiles, which turned out to be flocks of Canada geese. In 1960, nuclear alerts were sounded over meteor showers, and radar reflections. In 1997, Soviet defense forces set off an alarm over incoming US missiles, which turned out to be a Norwegian weather satellite. On at least three occasions in 1980 alone, US forces were set on combat alert when the nation's early-warning computer system failed. These errors were so egregious that the KGB announced they must be CIA plots to create avenues of surprise attack. National security adviser Zbigniew Brzezinski was woken from a dead sleep at 3:00 a.m. on November 9, 1979, to be told that 220 Soviet missiles had been first-strike fired against the United States. Brzezinski said he would wait for a second confirming call before informing President Carter, and that call came immediately. But the confirmation call said the first was mistaken; it was not 220 missiles, but 2,200 incoming missiles—a Soviet Killing a Nation. Brzezinski sat quietly in the dark for some time not waking his wife, assuming that they would both be dead in about thirty minutes anyway. Mere seconds before Brzezinski was going to telephone President Carter, there was a third call. Someone had put training videos into the system without informing the watch crew. It was all a mistake.

In 1961 during the Kennedy administration, NATO installed medium-range Jupiter missiles in Turkey and Italy pointing at Moscow and Leningrad—missiles that needed so much prep work before launch that they could not be used reactively, but only to attack. In a February 1961 meeting of the Senate Foreign Relations Committee, Senator Albert Gore Sr. told Secretary of State Dean Rusk that these were a "provocation. . . . I wonder what our attitude would be" if the Soviets put similar missiles directed at US cities, say, in Cuba?

When in the autumn of that year, the United States told the Soviets that a missile gap had been discovered by American spy satellites—a gap in favor of the United States—Nikita Khrushchev combined this revelation with the recent American-led Bay of Pigs invasion of Cuba to believe that Kennedy was threatening him. His response was exactly as Senator Gore had predicted, installing Soviet missiles pointing at America from Cuba. Nikita Khrushchev: "We had no desire to start a war. On the contrary, our principal aim was only to deter America from starting a war. We were well aware that a war which started over Cuba would quickly expand into a world war."

An American U-2 surveillance plane discovered the Cuban medium-range missile bases on October 14, 1962. A few days later, the U-2 uncovered IRBMs—intermediate-range missiles—with a range of twenty-two hundred miles, meaning they could reach any target in the continental United States save Oregon and Washington State. The discovery was debated at the White House by a group that would be called the executive committee of the National Security Agency—ExComm. During the first ExComm meetings on October 16, Kennedy said the situation was "just as if we suddenly put in a *major* number of MRBMs in Turkey. Now that'd be *goddamned dangerous*, I would think." National Security Advisor McGeorge Bundy replied, "Well, we did it, Mr. President."

Bundy agreed with CIA director John McCone that Khrushchev "knows we don't really live under fear of his nuclear weapons to the extent that he has to live under fear of ours." They were right, for Khrushchev felt that now Americans "would learn just what it feels like to have enemy missiles pointing at you; we'd be doing nothing more than giving them a little of their own medicine." Though publicly Kennedy called the Soviet missiles "an explicit threat to the peace and security of all the Americas," on that first day of the crisis, he said, "It doesn't make any difference if you get blown up by an ICBM flying from the Soviet Union or one that was ninety miles away. Geography doesn't mean that much." When Bundy asked, "How gravely *does* this *change* the strategic balance?" Defense Secretary Robert McNamara said, "Not at all. . . . I don't think there is a military problem here. . . . This is a domestic, political problem." Meaning that the administration couldn't afford to look soft politically about the Kremlin's planting missiles in Cuba pointed at American cities. The group did, however, worry about precedence, as McNamara later remembered: "There was a fear that if we did not force the missiles out, the Soviets would move aggressively elsewhere in the world against Western interests, and it was a very deep-seated fear."

"During its entire history Russia had been within range of hostile weaponry," Nikita's son, Sergei, said. "Russia had to rely on sound judgment on the part of opposing political leaders, on an American president's not sending his squadrons to bomb Moscow without good reason. Father assumed that Americans—not just the president but ordinary people—would think more or less the same way . . . but Americans thought otherwise. They were fortunate. For more than two centuries wide oceans had protected their land from enemies. Unlike Russians, they were used to living in security and were horrified by the possibility, however remote, of any vulnerability. The presence of Soviet ballistic missiles near America's borders evoked shock, and even psychosis. The press further inflamed emotions; the country lost its bearings; and the Cuban Missile Crisis became primarily an American psychological crisis. It seemed to Americans that they could continue to live as before only if the missiles were removed from Cuba, and removed at any price."

After nonessential personnel were evacuated from Guantánamo and American forces arrived en masse to begin combat training in Florida, on October 22 at 7:00 p.m., Kennedy told the world in a televised speech that nuclear missile sites had been detected in Cuba, that he had ordered a naval blockade of sixty American ships to quarantine the country, and that this quarantine would continue until the sites were dismantled and the missiles removed—the blockade was a tactical echo of what the Russians had done in Berlin. The speech said nothing about the missiles pointed at Russia from Turkey, or that the Soviets' nuclear arsenal totaled 36 ICBMs, 138 bombers ferrying 392 warheads, and 72 SLBMs (sub-launched ballistic missiles), while the United States was armed with 203 ICBMs, 1,306 bombers with 3,104 warheads, and 144 SLBMs—nearly 600 percent more.

SAC escalated from DefCon-3 (Defense Condition 3) to DefCon-2, the only time this happened in US history; DefCon-1 means nuclear war. LeMay ordered 136 ICBMs readied to launch and added 54 SAC nuclear bombers to the 12 already on around-the-clock alert monitoring the Pacific, the Mediterranean, the Arctic . . . anywhere he could detect in advance a Soviet-launched strike.

JFK's special counsel Ted Sorensen remembered of one ExComm session that "Curtis LeMay called [the quarantine] 'almost as bad as the appeasement at Munich' and demanded 'direct military intervention right now,' while the Marine Corps' commandant insisted that 'You'll have to invade . . . as quick as possible.' . . . The Joint Chiefs discussed an annihilation bombing run by the 82nd Airborne which would 'mop up Cuba in seventy-two hours with a loss of only ten thousand Americans, more or less.' " LeMay thought Kennedy was a coward, that Cuba was an excellent excuse to teach the Russians their place in the world: "The Russian bear has always been eager to stick his paw in Latin American waters. Now we've got him in a trap, let's take his leg off right up to his testicles. On second thought, let's take off his testicles, too." LeMay suggested the navy and SAC surround Cuba and if necessary "fry it," then invade the island with ninety thousand American troops. "LeMay talked openly about a [nuclear] first strike against the Soviet Union if the Russians ever backed us into a corner," McNamara said, and even after the crisis was resolved, LeMay insisted he was right: "During that very critical time, in my mind there wasn't a chance that we would have gone to war with Russia because we had overwhelming strategic capability and the Russians knew it. . . . We could have gotten not only the missiles out of Cuba, we could have gotten the Communists out of Cuba."

What the hawkish wing of ExComm did not know was that Cuba already had twenty nuclear warheads on its IRBMs ready to launch, as well as nine tactical nuclear missiles to use against invading American troops. If Washington had attacked as LeMay insisted it must, without question these would have been fired on American cities and American soldiers. Additionally, the United States thought there were eight thousand Soviet troops waiting to defend the island with the Cuban military. Instead, there were forty-three thousand. McNamara: "It wasn't until nearly thirty years after [the Cuban missile crisis] that we learned . . . that the nuclear warheads for both tactical and strategic nuclear weapons had already reached Cuba . . . 162 nuclear warheads in all. If the president had gone ahead with the air strike and invasion of Cuba, the invasion forces almost surely would have been met by nuclear fire, requiring a nuclear response from the United States." Presidential adviser Dean Acheson believed that, if the Americans bombed the Cuban missile sites, the Soviets would retaliate by bombing NATO missile sites in Turkey and Italy. The United States would then be forced by treaty to retaliate, bombing missile sites within the Soviet Union. And the real war to end all wars—as it surely would have resulted in the end of the world—would have begun.

The Americans also did not know that the Soviet ships now arriving in the Caribbean were accompanied by nuclear-torpedo-bearing submarines. Just as there was to be a confrontation on October 24, Moscow ordered the Russian ships carrying military cargoes to reverse course. Then, on October 25, a US destroyer was ordered to seize a Soviet sub, which dove to escape. McNamara: "Later we learned that the submarine commander was likely out of communication with Moscow, and under those circumstances he had the authority to launch [his nuclear torpedo] if he believed it necessary. He could have started a world nuclear war. We were that close. That's not wise management that avoided nuclear war, that's luck."

On Friday, October 26, Khrushchev drafted a letter to Kennedy offering to pull the missiles from Cuba if the Americans guaranteed they would never again invade the island and withdrew NATO's missiles from Turkey and Italy. Almost immediately after that letter was sent, Soviet military intelligence told the Kremlin that the Americans would invade Cuba in two days. Khrushchev was so alarmed by this news that he sent another offer, which removed the American missiles from the equation and only included the pledge never to invade. But there was a problem at the Central Moscow Telegraph Office. Sergei Khrushchev: "Everything came to a halt. Technical problems piled up, one after another. The letter that might decide the fate of the world could not reach Washington for at least six hours."

The arrival of two conflicting telegrams worried everyone at ExComm—was this a sign that Khrushchev had been deposed? The president sent his brother Robert to meet Ambassador Dobrynin at the Soviet embassy. The Russian confirmed that the Soviets would remove the Cuban missiles if Washington promised to never attack Cuba and to remove the US missiles from Turkey. Robert Kennedy went to the next room, called his brother, and came back to reply, "The president said we are prepared to examine the question of Turkey. Favorably."

At ExComm on Saturday morning, JFK said of the premier's offer, "To any man at the United Nations, or any other rational man, it will look like a very fair trade," but everyone else in the room—McNamara, Bundy, Rusk, Bobby Kennedy—opposed it. The president let them bicker, then said, "Now let's not kid ourselves. Most people think that if you're allowed an even trade, you ought to take advantage of it. . . . I'm just thinking about what we're going to have to do in a day or so . . . five hundred sorties . . . and possibly an invasion, all because we wouldn't take the missiles out of Turkey. And we all know how quickly everybody's courage goes when the blood starts to flow, and that's what's going to happen in NATO . . . when we start these things and the Soviets grab Berlin, and everybody's going to say, 'Well, this Khrushchev offer was a pretty good proposition.' "

On October 27, another American U-2 approached the coast of Cuba to take surveillance photographs. Now, though, Soviet troops were manning SAMs (surface-to-air missiles), with orders to fire if the island was attacked. The men who had found the American plane on their radar were uncertain. They fired two missiles at the plane and destroyed it.

Robert Kennedy again met with Dobrynin and explained that, after an American plane was attacked and its pilot killed, everyone was telling the president he should invade Cuba. Publicly, if the missiles were removed, the United States would promise not to invade. Privately, it would also remove the missiles from Turkey. Dobrynin reported back to the Kremlin. Sergei Khrushchev: "Father sensed that he was losing control of the situation. Today one general fires an antiaircraft missile; tomorrow another may launch a ballistic missile. As Father said later, it was at that moment that he understood intuitively that the missiles had to be removed, that real disaster was imminent."

On October 28 the Soviet leaders were meeting outside Moscow, as it was a Sunday, where an urgent message arrived from Havana: "Castro thinks that war will begin in the next few hours and that his source is reliable. They don't know exactly when, possibly in twenty-four hours, but in no more than seventy-two hours. In the opinion of the Cuban leadership, the people are ready to repel imperialist aggression and would rather die than surrender. Castro thinks that in the face of an inevitable clash with the United States, the imperialists must not be allowed to deliver a strike. . . . [that Havana should be] allowed to be the first to deliver a nuclear strike."

At the Soviet meeting, Nikita Khrushchev flatly said, "That is insane. We deployed missiles there to prevent an attack on the island, to save Cuba and defend socialism. But now not only is he ready to die himself, he wants to drag us with him. . . . Remove them, and as soon as possible. Before it's too late. Before something terrible happens."

At four that afternoon, Moscow Radio announced a new letter from the premier to the president: "The Soviet government has ordered that these weapons . . . which you have characterized as offensive, be dismantled. We supplied them to prevent an attack on Cuba, to prevent rash actions. I regard with respect and trust the statement you made in your message of October twenty-seventh, 1962, that there will be no attack, no invasion. . . . In that case, the motives which induced us to render assistance of such a kind to Cuba disappear."

Never again would there be a moment when the world came so close to nuclear war. "In my seven years as [defense] secretary, we came within a hair's breadth of war with the Soviet Union on three separate occasions," Robert McNamara summarized. "Cold War? It was a Hot War. . . . [In Cuba] we literally looked down the gun barrel into nuclear war. LeMay was saying, 'Let's go in, let's totally destroy Cuba.' At the end, we lucked out. It was luck that prevented nuclear war. We came that close. Rational individuals . . . came that close to total destruction of their societies. And that danger exists today. The major lesson of the Cuban Missile Crisis is this: the indefinite combination of human fallibility and nuclear weapons will destroy nations. Is it right and proper that today there are seventy-five hundred strategic offensive nuclear warheads, twenty-five hundred are on fifteen-minute alert, to be launched on the decision of one human being? . . . Any military commander who's honest with himself will admit that he's made mistakes in the application of military power. He's killed people . . . unnecessarily . . . through mistakes, through errors of judgment. [But] there is no learning curve with nuclear weapons. You make one mistake, and you're gonna destroy nations."

When the crisis ended, LeMay called it "the greatest defeat in our history." But in another case of a miracle with two faces, Kennedy and Khrushchev's brush with holocaust led to a remarkable improvement in US-Soviet relations. In August 1963, the two superpowers signed the Partial Test Ban Treaty, halting atomic tests in the air, in the oceans, and in outer space. Soon after, they installed a hotline teletypewriter between Moscow and Washington so that it wouldn't take six hours to get a telegraph during a crisis.

But the most dramatic change was the American public's attitude about nuclear war. Before Cuba, it was common US wisdom that another world war was in our future, and that armed conflict with the Soviet Union, likely nuclear, was certain. For thirty years, the *Bulletin of the Atomic Scientists* was published with a doomsday clock set at minutes to midnight; in 1960, C. P. Snow called atomic war a mathematical certainty, and many others, including Albert Einstein, had a similar outlook. In 1959, 64 percent of Americans said "war, especially nuclear war" was their country's biggest problem . . . but by 1965 it was 16 percent. The *Readers' Guide to Periodical Literature* cites north of 400 articles on the subject "nuclear" for each year from 1961 to 1963 . . . but by 1967, there are around 120. Atomic worries revived somewhat in the 1980s, a confluence of NATO's 1979 plan to install nuclear missiles in Europe, Reagan's comments that these could be used without targeting either superpower, Jonathan Schell's bestselling contemplation of nuclear holocaust, *The Fate of the Earth*, and ABC's TV movie on the same topic, *The Day After*. But ever since, the US citizen's fear about nuclear weapons has declined.

In America, the removal of the missiles was portrayed as a remarkable triumph. But it so humiliated the Soviets that they ratcheted up the arms race all over again. Soviet lieutenant general Nikolai Detinov: "The results were very painful and they were taken very painfully by our leadership. Because of the strategic [imbalance] between the United States and the Soviet Union, the Soviet Union had to accept everything that the United States dictated to it, and this had a painful effect on our country and our government . . . [to such extent that] all our economic resources were mobilized to solve this problem. . . . [A]fter the Caribbean crisis all production and other areas started going down thanks to the fact that all factors were mobilized in the name of military technology."

Remarkably enough, US military chiefs had the same reaction. If the Soviets could so easily sneak such a threat in under their noses, the nation needed a far greater atomic arsenal to defend itself. Commentator Louis Menand: "What drove the Cold War . . . was not business or science. It was . . . politics—the opportunities for partisan gain made available by gesturing toward the ubiquitous shadow of an overwhelming emergency. And the manipulation was not all on one side. If the United States assigned the Soviets the role of mechanized Enemy Other, the Soviets did their best to play it. The occasional hyperbole of the [American] Committee on the Present Danger was nothing compared with the bluster of Khrushchev and Gromyko, men who had their own domestic constituencies to worry about. It served both sides in the Cold War to take each other's rhetoric at face value. We have yet to learn how not to do this."

PART FOUR

POWER AND CATACLYSM

13

Too Cheap to Meter

FRANKLIN ROOSEVELT always planned to share the Manhattan Project's final blueprints with Britain and Canada. After all, they had contributed scientists and money to the research. But after FDR's death, the United States instead forbade the sharing of secret atomic-energy information with any foreign country, including Britain and Canada, on pain of death. It didn't actually matter for the allies as they had been involved enough to know the fundamentals, but since the USA had a monopoly on uranium enrichment, the British were forced to engineer reactors that used natural uranium metals, moderated by graphite but cooled by gas. France followed Britain's design in their own burners, and Canada used similar fuels, but moderated with heavy water. In the end, America's attempts at safeguarding her atomic secrets hurt only her own allies. When in 1960, Argonne, Westinghouse, and Oak Ridge proudly displayed the first US pressurized- and boiling-water reactors for civilian utilities, they were six years behind the Soviet Union in nuclear power. The American design was much safer than the Soviet's, however, for the simple reason that it was originally created to propel a submarine.

The system began when an American naval officer spent the years before World War II working in a sub that used diesel engines when surfaced and electrical motors when submerged. The officer was disturbed by how frequently his crew's life was imperiled by the battery's willingness to set itself on fire. When this captain, Hyman Rickover, was then assigned duty with the Manhattan Project, he dreamed that one day there might be a nuclear-powered submarine, which he called *Nautilus* after Jules Verne's *Twenty Thousand Leagues Under the Sea*. Since the state-of-the-art wartime sub was the

Nazi *Unterseeboot*, Rickover followed many German design ideas for his own ship, including a twenty-eight-inch-wide hull. This meant that he needed a reactor core about the size of a garbage can. And to keep his crews safe, he used water as both moderator and coolant—meaning that in any radioactive crisis, the engine would automatically shut itself down—with the coolant in a sealed plumbing loop to minimize the danger of radioactive leaks.

Beginning in 1949, the AEC began testing a host of reactor designs, including Rickover's, at Root Hog, Idaho, adjacent to the Craters of the Moon. The town changed its name to Arco, and the commission would eventually spend $500 million there, more than the estimated worth of the state of Idaho as a whole. So many of Arco's experiments were classified that it was said, "Nuclear engineers and physicians are alike. They both bury their mistakes." But Rickover's was no mistake. Launched January 17, 1955, USS *Nautilus* would achieve her twenty thousand leagues under the sea on February 5, 1957, before becoming the first craft to traverse the underside of the north pole. And Hyman Rickover's PWR (pressurized water reactor) would make its way to America's first nuclear power plant, twenty-five miles outside Pittsburgh.

On December 8, 1953, President Dwight David Eisenhower gave a speech to the United Nations that begat both the UN's International Atomic Energy Agency and what Ike hoped would be a significant part of his historic legacy, Atoms for Peace: "Today, the United States' stockpile of atomic weapons, which, of course, increases daily, exceeds by many times the total equivalent of the total of all bombs and all shells that came from every plane and every gun in every theater of war in all the years of the Second World War. . . . It is not enough to take this weapon out of the hands of the soldiers. It must be put into the hands of those who will know how to strip its military casing and adapt it to the arts of peace. . . . The United States pledges before you, and therefore before the world, its determination to help solve the fearful atomic dilemma—to devote its entire heart and mind to finding the way by which the miraculous inventiveness of man shall not be dedicated to his death, but consecrated to his life."

The following September, Eisenhower was on a Colorado vacation where he was presented with a cabinet of electronics and a neutron wand. When the president's hand waved the wand over the cabinet, a neutron beam was captured by the cabinet's rate meter, which sent a current twelve hundred miles to Shippingport, Pennsylvania, triggering a high-lift power jack to begin excavating the foundation of that first US nuclear plant, outfitted with Rickover's burner. By 1955, Eisenhower decided a nuclear merchant ship should be wrought as an Atoms for Peace global ambassador. Featuring Raytheon's Radarange (the first commercially available microwave oven), NS *Savannah* was christened by first lady Mamie on July 21, 1959, and, when it docked in New York City, inspired a "Nuclear Week" of educational events, which included two episodes of the *Tonight* show. Joining the Atoms for Peace agenda with his Plowshare Program was none other than Edward Teller, who studied the use of fusion bombs to dredge harbors and canals, nuclear explosions for fracking shale oil fields, and firing a nuclear rocket into the moon. This last proposal, Teller said, was to "observe what kind of disturbance it might cause." He told the University of Alaska in 1959, "If your mountain is not in the right place, just drop us a card" and "We're going to work miracles."

Back at the Pentagon, if the navy was going to have nuclear submarines, then goddammit the air force would get nuclear-powered jets. But much of what made *Nautilus* brilliant was unsuitable in the air. Rickover's reactors were encased in lead to protect his seamen, and if a naval reactor needed to shut down, it could be restarted while the sub was at rest—neither an option for a jet. Even so, the USAF spent $30 billion trying to make nuclear air power work, until Kennedy shut it down in 1961 . . . but perhaps it's for the best that the program never took off. While the Soviets have had a classified number of submarine disasters, the American air force has suffered a classified number of incidents with its nuclear-armed bombers. Sandia Labs itemized at least twelve hundred "significant" accidents with nuclear devices from 1950 to 1968. One piece of lasting evidence of these "Broken Arrows" can be seen in a twenty-five-foot crater created on May 22, 1957, by a B-36's accidentally dropping a ten-megaton thermonuclear bomb onto New Mexico.

One of Kennedy's great legacies as president would have its own nuclear history. In 1944, Chicago's Met Lab worked with Los Alamos to design an atomic rocket—its reactor heated hydrogen gas until it exploded from an exhaust nozzle. In the 1950s, Freeman Dyson continued this work for the Pentagon with Helios, an egg-shaped spacecraft with the crew in a small front cabin shielded by lead. A series of atomic bombs would explode one after the next in the main sphere, their plasma shooting through a nozzle in the back and shoving the craft ever forward. With Ted Tyler at General Dynamics, Dyson then designed Orion, which exploded five nuclear bombs every three seconds two hundred feet behind itself to reach a thrust of 3,000 mph. Even though a great deal of thought had gone into Orion's sophisticated shock absorbers, the idea had a flaw in that a liftoff from earth would leave behind a cloud of radioactive exhaust.

On June 29, 1961, the first atomic satellite powered into orbit—the US Navy's *Transit 4A*—using plutonium-238 in a radioisotope thermoelectric generator (RTG) to fuel a battery, the System for Nuclear Auxiliary Power—SNAP. RTGs have since made their way into the satellites that explore the universe for NASA: Pioneer, Viking, Voyager, Galileo, Cassini, New Horizons, Curiosity—as well as in the experimental apparatus left on the surface of the moon by Apollos 12–17. Russia has sent about forty nuclear-powered reconnaissance satellites into low-earth orbit; one crashed into Canada on January 24, 1978, irradiating six hundred miles, while on April 21, 1964, an American satellite collapsed, releasing seventeen thousand curies from its SNAP over the skies of Madagascar.

As fears of being incinerated by thermonuclear war subsided in the twilight of the Cold War, worries of an attack from a very different source intensified. On the banks of the Susquehanna River on March 28, 1979, at 4:00 a.m., pumps moving steam to the electrical generator and returning water to the nuclear reactor in the thee-month-old Unit 2 of Metropolitan Edison Company's Three Mile Island Nuclear Generating Station (TMI), which warped and woofed electricity for the residents of Dauphin County, Pennsylvania, stopped functioning. This triggered an automatic shutdown of the turbine, and a reactor SCRAM—control rods automatically inserted, and all fission halted. The safety procedures worked, and everything was as it should be.

The radiant core, however, remained hot. Auxiliary coolant pumps engaged, but Met Ed workers had violated a federal rule by shutting those pumps' valves for maintenance without in tandem shutting down the reactor. As the steam built and pressure increased, a valve automatically opened to release it, and then shut as it was designed to do, according to the indicator lamp in the control room. But that lamp was broken, the valve was stuck, and through this leak, unknown for several hours by anyone, thirty-two thousand gallons of the core's liquid coolant evaporated away.

In the reactor, steam pockets created both additional heat on the fuel plates and pressure on the instruments, which misled the operators into thinking adequate water was available. In time, that steam combined with fission into an atomic heat that eroded the fuel rods' zirconium cladding, and they began to crack open (uranium dioxide pellets look like black versions of the silica desiccants packaged with electronics to reduce moisture). The radiation monitor flashed—but the operators, not seeing any reason why this should be happening, decided this instrument was the broken one. At 4:15, radioactive coolant began to stream into the containment building, where a sump pump drove it to a building outside the containment dome's walls.

The staff believed that turning off the circulating coolant would then be the right thing to do. It was not. By 6:00 a.m., the top of the core was fully exposed, and the nuclear heat began to melt the fuel rods' sleeves, greatly irradiating the escaping coolant and creating an ever-growing cloud of volatile hydrogen. At the same time, the control room changed shifts. One of the new arrivals noticed the stuck valve, but by then, thirty-two thousand gallons of coolant, three hundred times as radioactive as normal, had leaked out in twenty-five minutes, striking emergency detectors. Radiation alarms rang throughout the complex. Unknown and unsuspected by the estimated twenty to sixty employees now working the control room, twenty tons of rods had already melted and were pooling into a radioactive lava.

To an outsider this cascade of human incompetence might seem Three Stooges laughable. But in fact this series of small, seemingly unrelated acts, combined with a control room festooned with over six hundred alarm lights, meant, in a serious accident, so many alarms were ringing or buzzing or flashing that practically no employee could grasp the underlying problem, or its solution.

Following federal regulations on radiation alarms, Met Edison declared a general emergency, meaning area residents faced a "potential for serious radiological consequences." This required an appearance by Lieutenant Governor William Scranton III, representing the state of Pennsylvania, who passed along the utility's assurances at a morning press conference that "everything is under control." That afternoon, he admitted that the crisis was "more complex than the company first led us to believe." After learning that Met Ed had vented the plant's radioactive gases without informing them, state regulatory officials asked for immediate intervention from the federal Nuclear Regulatory Commission. One of those commissioners later said, "We didn't learn for years—until the reactor vessel was physically opened—that by the time the plant operator called the NRC at about eight a.m., roughly one-half of the uranium fuel had already melted."

On the following day (March 29) an NRC spokesman told the public that the "danger was over." In fact, those at the site faced a terrifying conundrum. Control room instruments showed that the reactor had been pressurized by two atmospheres, which meant that the steam hitting the zirconium cladding had produced hydrogen. The released gas had formed into a bubble, which if it came in contact with oxygen could, like the *Hindenburg*, explode, breaking through the containment shell—that iconic dome of nuclear power in the United States that protects a plant's neighbors from radiant infection in case something goes wrong—and release untold amounts of radioactive gas. Emergency-cooling expert Roger Mattson explained to the NRC, "They can't get rid of the bubble. They have tried cycling and pressurizing and depressurizing; they have tried natural convection a couple of days ago; they have been on forced circulation; they have steamed out the pressurizer; they have liquided out the pressurizer. The bubble stays." They were stymied. Mattson argued that to shut down the reactor, the pressure had to be reduced, but lowered pressure could expand the bubble, which might force all the core's water out and lead to meltdown.

For two days federal officials had accepted Met Ed's assurances and grossly underestimated the danger. Now that they knew what was really happening, they overcompensated. The Food and Drug Administration woke chemical-manufacturer executives in the early-morning hours to immediately requisition a quarter million bottles of potassium iodide, which fills the human thyroid gland so that it won't sup the radioactive iodine, which is the most dangerous human carcinogen in a damaged reactor's by-products. The NRC then advised Pennsylvania governor Thornburgh to evacuate pregnant women and preschoolers from a five-mile radius and announced at a press conference that a full evacuation of between ten and twenty miles might be necessary to prevent harm from radioactive effluvia, especially in the case of young children. The Environmental Protection Agency immediately sampled the area's soil, water, and plants for contamination, including the milk of cows and goats and the tongues of white-tailed deer (which concentrate residue from the leaves they lick—the first sign of polluting fallout).

On day three, an AEC representative arrived, decided that an incorrect formula had been used in assessing the bubble's risk, and directed efforts to burn away the hydrogen out of the containment towers and extinguish the bubble. Met Ed vented 13 million curies of radioactive gases, which included less than seventeen curies of the iodine-13 that can trigger thyroid cancer—meaning no danger to the public unless the skies were filled with rainstorms and everyone stood around faceup with their mouths open. Even so, Governor Richard Thornburgh broadcast an alert that farm animals should be covered and only given stored feed, and that everyone within a ten-mile radius of TMI should stay indoors. Walter Cronkite's lead story for that evening's CBS evening news began, "The world has never known a day quite like today. It faced the considerable uncertainties and dangers uclear power plant accident of the Atomic Age. And the horthat it could get much worse. The potential is there for the of meltdown at Three Mile Island." Hearing this, alongside ports from the utility, from Thornburgh, and from federal offi- Pennsylvanians fled.

y, April 1, while Thornburgh accompanied President Carter to lant (as a lieutenant in the navy's nuclear submarine program, lone graduate work in reactor technology and nuclear physics), nounced that the accident did not elevate radiation enough to one additional death among the area's residents. This reassured many returned to their homes, and five days later on April 6, h announced that the "crisis had passed."

ermath report, the NRC insisted that no one had been harmed: es are that the average dose to about 2 million people in the area nly about 1 millirem. To put this into context, exposure from a chest ay is about 6 millirem." It would take two years before the reactor could e inspected with remote-controlled cameras, which revealed that half the core had melted, and 90 percent of the fuel rod cladding had dissolved. Yet, its containment shell had worked, the reactor vessel's walls keeping its radioactive effluvia from infecting the outside world.

Though it had no effect whatsoever on human health—the evacuation posed more danger to the public than anything leaking out of the plant—Three Mile Island shocked both the American people and Washington's nuclear powers. Before TMI, the Nuclear Regulatory Commission believed safety meant equipment design, maintenance, and doubling-up, with every crucial mechanical component having a backup. Though part of TMI's failure was the stuck valve and its broken indicator light, the real danger lay with poorly trained workers, corporate mismanagement, a warning-lamp and buzzer system that overwhelmed human comprehension, and government ineptitude. The lesson learned by American citizens, meanwhile, was that, when it came to nuclear power, the utility companies didn't know what they were doing, and neither did local or national bureaucrats, and when it came to public safety, the evidence was plain: no one was in charge.

In a remarkable coincidence, a movie had been playing in American theaters for twelve days before the first sign of trouble on the Susquehanna. *The China Syndrome* told the story of TV journalist Jane Fonda and nuclear plant employee Jack Lemmon facing meltdown because of a California reactor's inadequate coolant, which would render an area "the size of Pennsylvania permanently uninhabitable." The film's title came from the fear that uranium melting from an unquenchable atomic fire could burn th a con-tainment building's concrete flooring, then dig its way, lavalike, reality, meltdown could burn through the floor and contaminat a. In water table, rendering local territory unfit for agriculture as will al pen on the other side of the world.

The movie started with Fonda's PowerPoint-like explanation a nuclear reactor produced electricity, which was very likely the firs many understood exactly what happened inside an atomic power Americans had been introduced to nuclear science from Hiroshima Nagasaki, and now they would learn about nuclear power from Three M Island and *The China Syndrome*. The combination of movie and meltdo triggered an outburst of public activism, with sixty-five thousand march ing on Washington in May 1979, followed by a series of Madison Square Garden "NO NUKES" concerts, and two hundred thousand gathering in Central Park in September; three years later would see the largest protest in American history when an estimated 1 million gathered at a Central Park no-nuke demonstration on June 12, 1982. One of those speaking to the 1979 crowds was *China*'s Jane Fonda, whose activism in turn inspired Edward Teller to publicly lobby in favor of nuclear power, until he suffered a heart attack . . . which he blamed on her: "You might say that I was the only one whose health was affected by that reactor near Harrisburg. No, that would be wrong. It was not the reactor. It was Jane Fonda. Reactors are not dangerous."

In the wake of TMI and *The China Syndrome*, the containment dome—the prominent safety feature of American nuclear power plants, which had kept Three Mile Island from contaminating an area the size of Pennsylvania—became ominous, and mythic. Electrical generating stations that everyone drove by without a second thought now seemed to deserve many second thoughts, and nuclear power in America, begun from a science fiction novel, *The World Set Free*, would be strangled by a science fiction movie, *The China Syndrome*. Physicist James Mahaffey: "The nuclear power expansion was already dead years before the TMI disaster, and TMI was merely the last nail in the coffin. . . . Nuclear power went into a coma. Electrical power delivered in the United States by nuclear reactions stopped in 1977 at 20 percent, where it has remained ever since."

A few years after TMI's meltdown, engineer Stanley Watras was working in eastern Pennsylvania on the new Limerick reactor when, one morning, "all the alarms went off," he said. "Sirens went off. Red lights went off. It came out on a digital display that I was highly contaminated throughout my entire body. So, obviously, that kind of set me back." He was scrubbed clean, checked with dosimeters, found to be decontaminated, and allowed to go home for the day. But when he came back to work the next morning, it happened all over again, and it kept happening for two weeks. Finally Limerick nuclear physicists agreed to inspect his house. Watras: "They took air samples, little grab samples. It was the standard norm back in 1984. They took these samples down to the chemistry lab and they found out that it was that the place was highly contaminated with background radon radiation."

Radon is a naturally occurring gas rising from the decay of uranium and thorium beneath your feet. In a horrible irony, it would turn out that the area surrounding Three Mile Island was four times as tainted with radon as the US average. Though many residents are convinced to this day that the disaster contaminated their lands, workers were in fact never irradiated by TMI; instead, they were contaminating Three Mile Island from their radon-tainted basements.

In 2009, the Nuclear Regulatory Commission approved a license extension for Three Mile Island's other reactor, which has never caused any trouble, and TMI-1 will continue providing electricity to Dauphin County until April 19, 2034.

14

There Fell a Great Star from Heaven, Burning as It Were a Lamp

On April 28, 1986, as workers checked in for their morning shift, warning alarms clanged and brayed across Sweden's Forsmark nuclear power plant. The employees' clothes were saturated in radioactive particles, which meant a Forsmark reactor had to be severely damaged. After a morning spent investigating their equipment, however, the Swedes realized their plant couldn't be the reason for the contamination, that this must be a story like Stanley Watras's in Pennsylvania—the workers were getting irradiated from somewhere else. Checking the patterns of the wind for the previous few days revealed only one possible origin—nearly seven hundred miles away.

On April 27, the same alarms had screamed across the Institute for Nuclear Power Engineering, just outside Minsk, Belarus. Institute physicists tramped across their grounds, doing exactly what the Swedes would do, searching for what was wrong with their piles or their plumbing or their contamination shields. And just as in Forsmark, the radiation was everywhere, a stunning amount of contamination—soaking the employees' uniforms and civilian clothes, in their hair and in their shoes, absorbed by air filters tested at one hundred times normal, and even found in the soil and leaves of the trees outside the institute's walls. Lab chief Valentin Borisevich:

> The weather was so wonderful! Spring. I opened the window. The air was fresh and clean, and I was surprised to see that for some reason the bluejays I'd been feeding all winter, hanging pieces of salami out the window for them, weren't around.

> By lunchtime we find out there's a radioactive cloud over all of Minsk. We determined that the activity was iodide in nature. That means the accident was at a reactor. My first reaction was to call my wife, to warn her. But all our telephones at the institute were bugged.
>
> I pick up the phone. "Listen to me carefully."
>
> "What are you talking about?" my wife asked loudly.
>
> "Not so loud. Close the windows, put all the food in plastic. Put on rubber gloves and wipe everything down with a wet cloth. Put the rag in a bag and throw it out. If there's laundry drying on the balcony, put it back in the wash."
>
> "What's going on?"
>
> "Not so loud. Dissolve two drops of iodide in a glass of water. Wash your hair with it."

As the Belarusians immediately knew, iodide could only mean one thing. A reactor, somewhere, had exploded, and a cloud of radioactive fallout was floating across Europe, a toxic brume that would eventually cover the whole of the Continent from Russia to the Pyrenees. Experts would differ in their calculations as to the cloud's radioactive breath, which they described in layman's terms as "Hiroshimas." At first, they said it was ten times the radioactive power of the bomb dropped on Hiroshima. Then they said more like dozens of Hiroshimas. Then, it was two hundred Hiroshimas. And then, four hundred Hiroshimas.

The worried Belarusians called their nearest neighboring power plant, in Lithuania, only to be told that the same mystery was frightening the scientists there. Then, they called the other neighboring plant, in Ukraine. They called again and again, but no one answered the phone.

Eisenhower's Colorado neutron wand did not create the world's first civilian atomic power station. The Soviets did, on June 1, 1954, with the Atom Mirny-1 ("peaceful atom") reactor, 110 kilometers southwest of Moscow, using graphite as a moderator (like Fermi's squash court) but with water for coolant. This combination, Hans Bethe explained, was "fundamentally faulty, having a built-in instability. . . . [A] reactor that loses its coolant can under certain circumstances increase in reactivity and run progressively faster and hotter rather than shut itself down." Additionally, nuclear burners using only water had a moderator and coolant that couldn't burn; those using only graphite had a moderator and coolant that couldn't explode.

The Soviet pile design, used in twelve plants across the USSR including the Ukrainian pile where the phone went unanswered that April 27, had a coolant that could boil into steam and explode, and a moderator that could erupt into nearly unquenchable atomic flames.

Due south of Finland, nestled between Russia to the east and Poland and Romania to the west, the Eurasian Great Plains of Belarus and Ukraine became, under Soviet domination, both a buffer zone against another invasion by Europeans (as an army would have to kill many Ukrainians and Belarusians before reaching Russian souls) and a breadbasket for the empire. Like the topographically similar American Midwest of Iowa, Nebraska, and the Dakotas, this territory was once a great sea, and dog-day temperatures can reach 113 degrees.

In 1970, on the banks of the Pripyat River sixty miles north of Kiev, the Soviets began what was planned as Europe's greatest nuclear energy generator—the V. I. Lenin Atomic Power Station. By 1986, they had four thousand-megawatt reactors burning day and night, with two more under construction and two more being planned. At the same time, Moscow built a new concrete metropolis, Pripyat, home to fifty thousand. The town and its energy complex were all that communism was supposed to achieve: grand, spacious, powerful, yet at the same time gemütlich. Pripyat was notable for having no shortages: plenty of roses, shoes from Czechoslovakia, and enough energy from its plant to provide for itself as well as 2 million residents of Kiev. Surrounded by cherry orchards, the Lenin's nuclear smokestacks were painted in red and white stripes, like peppermint candy canes from Santa. It was all so unlike the little farming town of twelve thousand next door, Chernobyl—Ukrainian for "mugwort" or "wormwood," the leafy source of vermouth and absinthe—which had been a predominantly Jewish enclave for three hundred years, with a dynasty of illustrious wise men and two Hasidic zaddik shrines still in use. Perhaps this history is why many locals insist that the Wormwood mentioned in the Bible's book of Revelation was a Chernobyl prophecy from the hand of God:

And the third angel sounded, and there fell a great star from heaven, burning as it were a lamp,
and it fell upon the third part of the rivers, and upon the fountains of waters;
And the name of the star is called Wormwood:
and the third part of the waters became wormwood;
and many men died of the waters, because they were made bitter.

In case of a power failure, the Lenin Atomic Station had emergency diesel engines to power its coolant pumps, but it would take forty seconds to bring those engines up to speed—time enough for the reactor to overheat. On April 25, 1986, plant electrical engineers, naive to the ways of physics, decided to conduct a safety experiment. They wanted to see if, after powering down Reactor #4 to 2.5 percent, sufficient inertial spin was left in its turbines to generate enough electricity to power the coolant pumps for those forty seconds.

The test was supposed to take place during the day, but given the unusually high electrical demand that afternoon, it was postponed to 11:00 p.m. that night. The engineers decided to drop the power precipitously to make up for lost time, but did not follow the correct procedure, and the reactor's fission stopped entirely. To fix that, they withdrew most of the 211 control rods to restart the chain reaction, but this only brought #4 back to thirty megawatts . . . a level at which the pile was at its most unstable. The correct thing to do was to wait twenty-four hours for the neutron balance to right itself, but the deputy chief working that shift wanted this experiment over and done with and demanded the reactor be powered up. By 1:00 a.m. on the twenty-sixth, they had withdrawn all but six rods—though knowing full well the minimum reserve in place for such a reactor was thirty rods—raising it to 200 megawatts. Additionally, for this "safety" experiment, they had disabled Reactor 4's emergency electrical backup systems and the emergency core-cooling system. One operator was confused by the instructions and called over another employee. "In the program there are instructions of what to do, and then a lot of things crossed out," he pointed out. The other operator insisted they should "follow the crossed-out instructions."

At 1:23 a.m., the turbine was shut down for the experiment, halting power to the coolant pumps. The coolant started to boil. As the 205 control rods were then reinserted, they drove out water, increasing both heat and fission, as Bethe had noted. As alarms began to ring, the engineer in charge of operations, Leonid Toptunov, pushed the emergency SCRAM button. But it was already too late—now the control rods refused to fully enter their slots, perhaps due to heat warping the rods' channels. Instead of shutting down, Reactor 4's temperature rose to a hundred times normal. The core began to melt.

A foreman looked down to see the reactor's shield, made of 770-pound cubes, shaking and rattling "as if seventeen hundred people were tossing their hats into the air." Turbine operator Yuri Korneev: "At the moment when the turbine stopped working, there was a sudden explosion in the area of the tubing corridor. I saw it with my own eyes, heard it with my own ears. I saw pieces of the reinforced-concrete wall begin to crumble, and the reinforced-concrete roof of our Turbine 7 began to fall. In a few seconds the diesel apparatus kicked in, and emergency lights went on. I immediately looked at the roof of the engine room. It was crumbling in layers. Falling pieces of concrete were slowly coming closer to my turbine. . . . It was all so unexpected. It was difficult to figure out what was happening."

Two explosions then struck back-to-back, blowing a hole in the thousand-ton concrete shield, revealing a reactor engulfed in nuclear fire, spewing radioactive gas, graphite, and uranium debris into the sky. The pile continued to explode "like a volcanic eruption," Toptunov said, raining down lead cubes and flaming graphite, which in turn set the asphalt roofs of the complex's other buildings aflame. The explosions also broke the reactor's fuel packs, which fell into the paths of the control rods, keeping them from being inserted beyond a third of the way in, followed by a massive burst of steam rupturing the two-thousand-ton reactor case.

Besides the worrisome combination of graphite and water, the Soviets used concrete covers for their reactors, but not the fortified containment domes that keep other nations' atomic accidents from contaminating their neighbors, "since all Soviet nuclear facilities were designed so they could also produce weapons-grade plutonium," as Richard Rhodes explained. "This was in fact the reason why the facilities did not have the traditional containment shells protecting the public from just this sort of accident; it had a removable lid for ease of fuel change and the production of nuclear weapons. Once the lid blew off, that was the end of containment. So the history of thinking of this as a civilian nuclear disaster is quaint; it was a Cold War–era military nuclear disaster."

The core exploded all over again, either from steam, hydrogen, or out-of-control fission. "Flames, sparks, and chunks of burning material went flying into the air above the #4 unit . . . red-hot pieces of nuclear fuel and graphite," physicist Grigori Medvedev remembered. "About fifty tons of nuclear fuel evaporated and were released by the explosion into the atmosphere . . . about seventy tons were ejected sideways from the periphery of the core, mingling with a pile of structural debris, onto the roof. . . . Some fifty tons of nuclear fuel and eight hundred tons of reactor graphite remained in the reactor vault, where it formed a pit reminiscent of a volcanic crater."

The air, ionized, glowed in a purple haze, radiance visible, evolving to a neon pink so vibrant and so vast that everyone in Chernobyl and Pripyat came out of their homes to watch. "I can still see the bright crimson glow, it was like the reactor was glowing. This wasn't any ordinary fire, it was some kind of emanation," Nadezhda Vygovskaya said. "It was pretty. I've never seen anything like it in the movies. We were on the ninth floor, we had a great view. People came from all around on their cars and their bikes to have a look. We didn't know that death could be so beautiful. Though I wouldn't say that it had no smell—it wasn't the spring or autumn smell, but something else, and it wasn't the smell of earth." Andrei Sakharov: "Do you know how pleasantly the air smells of ozone after a nuclear explosion?" It smells fresh and clean and electric, like the air of great natural hydrodynamics, of Angel, of Niagara, of Victoria.

"In the darkness we made our way through piles of rubble and went up to the landing. Everything was in shambles, steam was coming out in bursts, and we were up to our ankles in water," radiation monitor Nikolai Gorbachenko said. "Suddenly we saw [a man] lying unconscious on his side, with bloody foam coming out of his mouth making bubbling sounds. We picked him up by the armpits and carried him down. At the spot on my back where his right hand rested I received a radiation burn. He died at 6:00 a.m. in the Chernobyl hospital, never having regained consciousness. The two guys who looked for him with me later died in a Moscow hospital."

"There was a loud thud that made the windows rattle," fireman Leonid Shavrej remembered. "I jumped up immediately. The emergency signal kicked in almost at the same moment. We jumped out on the street, ran toward our trucks, and heard the dispatcher yell that there was a fire at the atomic station. We looked up and saw a mushroom cloud; it also looked like the chimney above the Unit 4 reactor was half gone. We were never instructed on how to work in radioactive conditions—despite the fact that the fire station was attached to a nuclear power station."

By 4:00 a.m., 186 firemen in eighty-one engines were fighting a fire that couldn't be extinguished. In such a crisis, eyewitness memories often conflict. "We didn't know it was the reactor because no one told us. We thought it was just a normal fire," one emergency worker insisted, but Anatoli Zakharov said, "I remember joking to the others, 'There must be an incredible amount of radiation here. We'll be lucky if we're all still alive in the morning.' Of course we knew! If we'd followed regulations, we would never have gone near the reactor. But it was a moral obligation—our duty. We were like kamikaze." One described his experience of the radiation as "tasting like metal," and feeling pins and needles all over his face. For ninety minutes they tried to control the flames, and one by one almost every man collapsed, vomited, and passed out. The nuclear fire that produced enough effluvia cloud to cover a continent in toxic fallout would take two weeks to extinguish. Lyudmilla Ignatenko:

> One night I heard a noise. I looked out the window. Vasily saw me. "Close the window and go back to sleep. There's a fire at the reactor. I'll be back soon."
>
> At seven I was told he was in the hospital. I ran there, but the police had already encircled it, and they weren't letting anyone through. Only ambulances. The policemen shouted: the ambulances are radioactive, stay away!
>
> I saw him. He was all swollen and puffed up. You could barely see his eyes. Many of the doctors and nurses in that hospital, and especially the orderlies, would get sick themselves and die. But we didn't know that then.
>
> The doctor came out and said, yes, they were flying to Moscow, but we needed to bring them their clothes. The clothes they'd worn at the station had been burned. The buses had stopped running already and we ran across the city. We came running back with their bags, but the plane was already gone. They tricked us. So that we wouldn't be there yelling and crying.
>
> There's a fragment of some conversation, I'm remembering it. Someone is saying: "You have to understand: this is not your husband anymore, not a beloved person, but a radioactive object with a strong density of poisoning. You're not suicidal. Get ahold of yourself." And I'm like someone who's lost her mind: "But I love him! I love him!"

After fourteen days, Vasily died and, being so radiant, was buried in a series of *matryoshka* exequies, his body wrapped like a nesting doll in a cellophane bag, then settled within a wooden coffin, then covered in another heavy bag, then set inside a zinc coffin, and finally entombed in a concrete slab.

Two hours after the Lenin Atomic Station blew up, Moscow received a coded signal—"1, 2, 3, 4"—meaning the absolute highest state of emergency. Gorbachev called the Politburo into crisis session, and by Saturday noon a fact-finding team of doctors, physicists, and government officials were flying to Kiev and being ZIL-limousined to Pripyat. The group report mentions "a white pillar several hundred meters high" of fire and smoke marked by "individual spots of deep crimson luminescence . . . of burning products constantly flying from the crater of the reactor" and local officials who had no idea what they were supposed to do in such a catastrophe, as "they had no guidelines written earlier and were incapable of making any decisions on the spot."

By May 8, firemen's pumps had drained 5 million gallons of radioactive water out of the reactor's basement, but this exposed more graphite to air, which continuously ignited, and the nuclear fire continued. Finally it was understood that water alone could not extinguish this blaze, so the decision was made to try to smother it. At 11:00 p.m., the town organized 150 of its residents to go to a quarry by the river and fill bags with sand. But they forgot twine to close up the bags, so calico strips from holiday ornaments were used instead—a perfect counterpoint to the plant's peppermint-candy-cane smokestacks. Soviet Mi-8 helicopter teams then hovered 110 meters (360 feet) above the fire, the crews leaning out their side hatches to dump those holiday bags of river sand—in ninety-three flights, they whelmed the atomic volcano with nearly one hundred thousand pounds. Until the end of June, helicopters blanketed the fire with 37 million more pounds of sand, clay, boron, and dolomite. Word passed among the pilots that if they ever wanted to have kids, they needed to shield their testicles with lead. "By May 4 the pilots had buried the reactor core in sand despite conditions that were difficult and dangerous," chemical defense chief Colonel Anatoli Kushnin said. "The dosimetric devices on these helicopters measured radiation levels of up to five hundred roentgens an hour." A lethal dose of radiation is about a hundred roentgens per hour for five hours; some Chernobyl sections measured twenty thousand roentgens per hour. "We started out wearing protective suits in Chernobyl, but it made us move very slowly, because they're so heavy," American oncologist Robert Gale wrote. "So people ended up getting more radiation because they were wearing these heavy clothes. It was better to work very fast, without protection, than very slowly with protection. In the end, we didn't wear any protective clothing."

The graphite tamper, now enflamed at more than 1,200°C, began to burn through the reactor floor, mixing with concrete to meld into corium, a radioactive lava. Underneath the reactor itself were two floors of bubbler pools, reservoirs for the emergency cooling system, which could now at any minute boil away and explode into steam. Three men went into tunnels to open the pools' sluice gates, but their sole lamp failed, and they had to find the valve by touching their way along a pipe, like three blind men. They returned to the control room and announced their success at finding and opening the drain. In time, all three would contract acute radiation sickness, and all three would die.

Though the possibility of a basement steam explosion had been circumvented, the corium lava could still burn its way into the water table below the reactor and contaminate even more territory than had already been poisoned by atmospheric fallout—a real-life *China Syndrome*. As physicist Shan Nair explained, "The water table will start leaching actinides and fission products from the melted glob of fuel into the environment. So you will end up with some radioactive contamination of water supplies and ultimately crops and other products. That's a major problem because radioactive particles are much more dangerous when digested—they cause internal irradiation of organs with resulting increased cancer risks."

To prevent this, a team began to daily inject fifty-five thousand pounds of liquid nitrogen to freeze the earth beneath the reactor to -100°C, stopping the lava flow and stabilizing the collapsing foundation. But eventually this proved unworkable, so the basement rooms were pumped full of concrete, with the liquid nitrogen used to quench the fire from beneath. Fifty-six miles of dams with polyethylene shields were then installed to keep rainwater from surging the contained waste into the water supply.

Six hundred thousand workers, called liquidators, arrived from across the empire to fight the crisis. Thirty-four hundred rushed in wearing protective suits to quiet the fire and excise the poisonous debris—for forty seconds at a time, absorbing a lifetime of dosage. They were called the bio-robots. Others removed miles of radioactive topsoil and then planted hundreds of thousands of trees to hold the earth and reduce the spread of toxic dust (their cars and trucks are still in the plant's parking lot to this day, too radiant for human touch).

"It was a real war, an atomic war," one liquidator said. "In those times the Russian shows how great he is. How unique. We'll never be Dutch or German. And we'll never have proper asphalt and manicured lawns. But there'll always be plenty of heroes." The final tally: 206 days of cleanup, and three engineers getting ten-year prison sentences for criminal mismanagement. Of the 237 workers and firefighters who contracted ARS (acute radiation sickness), 31 died in the first three months.

Contaminated food and equipment was supposed to be buried, but much of it made its way to the black market and was sold, along with supplies sent in for the victims—oranges, coffee, buckwheat. Resident Anna Artyushenko: "There was a Ukrainian woman at the market selling big red apples. 'Come get your apples! Chernobyl apples!' Someone told her not to advertise that, no one will buy them. 'Don't worry!' she says. 'They buy them anyway. Some need them for their mother-in-law, some for their boss.' "

On April 27 at 2:00 p.m., over 336,000 people were told they would need to evacuate their homes for three days, and a ten-mile-long convoy—1,216 yellow school buses and 300 supply trucks—arrived in Pripyat from Kiev. Instead of three days, though, the residents were permanently exiled. The Nazis destroyed 619 Belarusian villages in World War II; Chernobyl emptied another 485 villages, with 70 having to be buried beneath the earth. Liquidator Arkady Filin: "One collective farm chairman would bring a case of vodka to the radiation specialist so they'd cross his village off the list for evacuation; another would bring the same case so they'd put his village on the list—he'd already been promised a three-room apartment in Minsk." Not everyone agreed to evacuate. Resident Zinaida Kovalenko: "The soldiers knocked. 'Ma'am, have you packed up?' And I said: 'Are you going to tie my hands and feet?' Old women were crawling on their knees in front of the houses, begging. The soldiers picked them up under their arms and into the car. But I told them whoever touched me was going to get it." Anna Artyushenko: "The police were yelling. They'd come in cars, and we'd run into the forest. Like we did from the Germans." Zinaida Kovalenko: "Everyone up and left, but they left their dogs and cats. The first few days I went around pouring milk for all the cats, and I'd give the dogs a piece of bread. They were standing in their yards waiting for their masters. They waited for them a long time. The hungry cats ate cucumbers. They ate tomatoes." Arkady Filin: "Gangs of men were sent to kill the household pets to keep epidemics of disease from springing up. It was very easy at first, since the dogs weren't afraid; they ran towards the human voices, thinking they were going to be taken home. Then afterwards, they grew wary and ran into the forests at the sounds of people coming. And the cats learned how to hide." The atomic no-man's-land covered eleven hundred square miles, with a name translated three ways: the Zone of Exclusion. The Zone of Estrangement. The Zone of Alienation.

By May 12, 10,198 people in the region had been hospitalized, and that autumn, when the chestnuts shed their leaves, three hundred thousand tons of them were bagged and buried. By November, five hundred thousand cubic yards of rebar concrete covered the reactor, and though it continued to burn, it could no longer infect anything but itself . . . but the containment turned it into an oven, and the nuclear fire rose to 4,500°F. By December 1986 a $768 million battleship-gray concrete sarcophagus was set in place to keep the melted-down two hundred tons of atomic fuel and corium lava from leaking out. Planned to last twenty years, it began disintegrating almost immediately, with cracks and holes letting in rain and snow. Most of the gaps have now been plugged, and the sarcophagus will supposedly be replaced in October 2015 by the New Safe Confinement (NSC), an $800 million steel arch that will be longer than a football field and taller than the Statue of Liberty and the largest movable structure ever made by human hands.

Still, a number of Belarusians and Ukrainians refuse to leave. Current resident Elena Shagovika: "They come around here and ask us why we never left. Where were we going to go? And what would we do there? When my old neighbors come back to see us, they just stand in the road and weep. We don't belong anywhere else. We belong here." Here means tilling your soil with potassium (to block the cesium-137 from infecting your crops) and lime (stopping the strontium-90). You can grow plenty if your soil is clay based (which soaks up most radionuclides), but only potatoes if it's peat. Shagovika's neighbor Anna Artyushenko: "If we kill a wild boar, we take it to the basement or bury it ourselves. Meat can last for three days underground. The vodka we make ourselves. I have two bags of salt. We'll be all right without the government! Plenty of logs—there's a whole forest around us. The house is warm. The lamp is burning. It's nice! I have a goat, a kid, three pigs, fourteen chickens. Land—as much as I want; grass—as much as I want. There's water in the well. And freedom! We're happy." Zinaida Kovalenko: "Sometimes it's boring, and I cry. The whole village is empty. There's all kinds of birds here. They fly around. And there's elk here, all you want. [Starts crying.] . . . Death is the fairest thing in the world. No one's ever gotten out of it. The earth takes everyone—the kind, the cruel, the sinners. Aside from that, there's no fairness on earth. I worked hard and honestly my whole life. But I didn't get any fairness. God was dividing things up somewhere, and by the time the line came to me there was nothing left."

Chernobyl was merely the fourteenth most lethal nuclear accident in USSR history, with the other thirteen kept classified until the empire fell. A far worse incident, for one example, happened in the south Urals, on September 29, 1957, when cooling equipment for nuclear waste at the Mayak Plutonium Facility malfunctioned, the waste ignited in fire and exploded, irradiating 270,000 people and fourteen thousand square miles of land with remarkable vigor—2 million curies. As the plant's employees had already spent the previous seven years disgorging 2.75 million curies of waste into the Techa River, today, a half century later, the territory remains one of the most radioactive regions in the world.

The United States has had its own trouble spot in Rocky Flats, Colorado, a fusion-bomb trigger plant that suffered a series of plutonium fires in 1957, 1965, and 1969, then was discovered to have been lackadaisical with leaky waste drums. Federal officials were forced to shut the plant down in 1989.

The Soviets have never released official mortality figures for the Lenin Station's two-week atomic fire. Onetime foreign minister and Georgia president Eduard Shevardnadze famously said of the disaster that it "tore the blindfold from our eyes and persuaded us that politics and morals could not diverge." It was a financial catastrophe—hundreds of thousands relocated, billions paid for liquidation, and Belarus and Ukraine still spending around 5 percent of their yearly federal budgets on Chernobyl victims. The cost is so high that the majority of Belarusians opposed the dissolution of the USSR and to this day want to reunite with Russia. Belarus Radiobiology Institute director Yevgeny Konoplya: "We are the great guinea pigs of modern times. We are getting to prove for the world what radiation can do to humans. We have suffered from the policies of a country that no longer even exists. We have suffered from lies. And we have suffered from other people's belief in technology. We once had a beautiful country. What we have now is pain."

From 1992 to 1995, Johan Havenaar, chief of emergency psychiatry at Utrecht University Hospital, oversaw a study comparing fifteen hundred residents of Gomel (previously Belarus's most agriculturally productive region, but now with twenty of its twenty-one districts rendered infertile by Chernobyl), with fifteen hundred from nearby Tver, Russia, where no Ukrainian radioactivity has ever been found. The Belarusians said they were five times as sick as the Russians and argued that almost all of these illnesses were due to "the station"—Chernobyl. Forest administrator Volodya Ronashev, forty-eight years old: "My teeth are falling out, and I can't see too well anymore. I used to be healthy. What else could it be but the station?"

The Dutch then gave the two groups of residents extended medical exams and found that their health was, physically speaking, nearly identical. Psychologically, though, the difference between the two groups was astounding. The Gomelites described themselves as weak and helpless victims with a predetermined, disastrous future. They tried to repair this by either being extremely careful and dramatically exaggerating any health worries, or freely eating the fruits, mushrooms, and animals from the state-warned contaminated zones while screwing, drinking, shooting up, and smoking like it's 1999.

Chernobyl Forum radiologist Fred Mettler found that after twenty years "the population remains largely unsure of what the effects of radiation actually are and retain a sense of foreboding. A number of adolescents and young adults who have been exposed to modest or small amounts of radiation feel that they are somehow fatally flawed and there is no downside to using illicit drugs or having unprotected sex." The pregnant of Gomel are so afraid that their children will be born defected that even today they have three abortions for every live birth—more than twice the rate of the rest of Belarus. Johan Havenaar: "These people are sick. It's just not the type of illness they think. We have to realize that the psychological damage here runs very deep. And we need to treat that every bit as vigorously as we need to treat cancer." Harvard physicist Richard Wilson: "It's not too much to say that Chernobyl helped destroy the Soviet Union and end the Cold War. What it did to Belarus is hard to describe. But the worst disease here is not radiation sickness. Except for children, the physical effects are not easy to measure. The truth is that the fear of Chernobyl [radiophobia] has done much more damage than Chernobyl itself."

The evacuated, meanwhile, seem just as miserable as anyone remaining behind in Gomel. The town of Slavutych was built solely to replace Pripyat; the major form of litter on the roads outside its housing projects are empty vodka buckets; its central square is a black-marble memorial to the station's victims, where surviving relatives regularly gather to lay wreaths and light candles as a boys choir sings *Gospodi, Gospodi, Gospodi*—"my God, my God, my God." Tamara Lusenko was one of those forced to move from her family farm to what she says is a prison: "If I knew it would be this bad, I would have chained myself to the gates back home. Is the danger really so bad there now? Isn't it time we all went home?"

Despite everything, Ukraine continued running the other Lenin Station reactors for the next fourteen years. Eight times, the government decided to shut down the plant, and each time it reversed the decision, as it would destroy five thousand jobs in an economically cratered nation. Most of the remaining V. I. Lenin power plant employees have developed a special form of Slavic bravado. "Radiation is good for you," one said. "I work here so when I come home glowing, my wife will think I'm a god." Another shrugged, saying, "Life itself is dangerous, my friend." On December 15, 2000, before a coal-black statue of Prometheus (the Titan who stole fire from the gods and had an eagle eternally chewing his liver in punishment), Ukrainian president Leonid Kuchma placed a wreath in honor of the dead and announced that the last Chernobyl reactor, Unit 3, would finally be shut down. Workers in jumpsuits protested with black armbands and unfurled banners, but forty-two hundred were laid off, leaving four hundred to maintain the site.

The government is now trying to revive the area economically by turning the Zone of Alienation into a tourist attraction—*visit the end of the world, circa 1986, for a mere $150 a day.* The ghost city Pripyat includes Soviet apartment towers listing in torpor, peregrine falcons nesting in high-rise balconies, schools sprouting stalagmites of mold, the ruins of an amusement park in faded kindergarten colors, black storks perching in the great oaks of the cemeteries, and in the harbor a graveyard of river ships—something like an atomic Detroit. While the town of Chernobyl has replaced its post office's time and temperature display with a dosimeter monitoring radiation levels in different sectors of the Zone, VIPs can visit the control room of Reactor #4, with missing ceiling tiles, exposed wires and cables, and its walls covered in decontaminant, which has dried to the color of human blood. With sidewalks, roads, and building foundations sprouting in flora, every town in the Zone, like a clock of civilization running backward, is reverting to forest, becoming Soviet ruins like Cambodia's Angkor Wat or Guatemala's Tikal. But KEEP OFF THE GRASS has a whole new meaning here; after the massive decontamination effort of the liquidators, "it's safe where we are," Sergei Saversky, deputy chief of zone management, explained to a recent group of tourists. "Just don't walk where you're not supposed to."

With the wilderness, comes the wild creatures. Contrary to expectations of nuclear winter and atomic desert, after the evacuation of ever-hungry people with their eternal agricultural war against predators, the Zone of Alienation's 1,660 square miles became a wildlife sanctuary teeming with cormorants, cranes, herons, and sixty-six different species of mammals—bears, wild boar, wolves, red deer, roe deer, beavers, river otter, foxes, lynx, thousands of elk, and a surfeit of *barsuk*, the badger of central Europe. "Northern Ukraine is the cleanest part of the nation," an Academy of Sciences official explained. "It has only radiation." When the waterways were overrun with thousands of beaver, their woodworking dammed the canals that drained the fields, returning them to marshland, and becoming once again a home for otters, fish, moose, badgers, bear, boar, and waterbirds. Since so much of the Zone is forbidden to human trespass, two endangered species, bison and wild horses, were reintroduced here, and because no one remains to fish and eat them, catfish living in the station's canals now grow to ten feet in length, their giant whiskers twitching in the air for the bread tossed by tourists.

It is a perfect illustration of the world without us.

The Zone's other major business is science, with teams of radioecologists turning the region into an alfresco laboratory, studying the effects of radiation. Yes, the plants and animals are thriving, but what is going on inside their cells, and what about their DNA? The findings are wildly mixed. University of Georgia geneticist Ron Chesser studies chubby, mouselike voles: "Chernobyl represents a huge mystery, and scientists love mystery. . . . The mutation rate in these animals is hundreds and probably thousands of times greater than normal. . . . You wouldn't want to keep one of those voles in your pocket for any length of time." But except for enlarged spleens, the scientists have yet to find anything biologically wrong with their charming voles. Immediately after the disaster, an entire four-square-kilometer forest turned from pine green to deep red. When its birch and pine seedlings were grown elsewhere, they became bushes instead of trees, with giant needles and a feathery mien. A local population of dormice—the exquisite hamster and historic Roman empire delicacy—has been studied for fifteen years, and though 4–6 percent have genetic abnormalities, the population in general is healthy. Moose-bone leftovers from a wolf meal revealed fifty times the normal amount of radiation as late as 2010, yet studies comparing wolf populations in irradiated versus non-station-infected territories found no significant overall differences.

A University of South Carolina team investigating cobalt-headed barn swallows has found brains 5 percent smaller than usual, and vestigial albinism in 13 percent instead of the normal 4 percent. Yet, a colony of white mice seem to be developing a resistance to radiation that is passed on to their descendants, which could be of great benefit to humans. And from inside the still-burning Reactor #4 itself, one robot emerged covered in a black goo—radiographic fungus, growing ecstatically on the unit's very walls.

15

Hitting a Bullet with a Bullet

AFTER his speeches as a presidential candidate repudiated the live-and-let-live détente of Nixon, Ford, and Carter in favor of hawkish Cold Warrior aggression, Ronald Reagan's ascension to the Oval Office in 1981 alarmed the Kremlin. Then in the first days of his administration, CIA director William Casey reprised LeMay's bear-baiting by throwing bombers over the pole into Soviet airspace until Russian radar took notice. Similarly, NATO fighter jets crossed over the empire's Eurasian border every week, performed a variety of erratic maneuvers, and then vanished . . . until starting up all over again.

In May of 1981, after Reagan had been president for five months, KGB chief Yuri Andropov told his superiors he had information from the highest sources that the United States was preparing a nuclear first strike against the Soviet Union. After becoming premier, he ordered Soviet spy agencies KGB and GRU to work overtime for two years to uncover the details of this forthcoming assault. Instructed to look for alarming evidence, Soviet operatives found just that, only further terrifying an already frightened Kremlin. Tensions were so extreme that, when South Korea's Flight 007 touched into Soviet airspace by accident on September 1, 1983, Moscow had it attacked, killing all 269 people on board.

After John Hinckley Jr. tried to assassinate him in the spring of 1981, Reagan said, "[Cheating death] made me feel I should do whatever I could in the years God has given me to reduce the threat of nuclear war." He would regularly discuss eliminating atomic bombs in private, but no one in his administration supported this, and the president did nothing about it in practice, either militarily or diplomatically. When he met with his political comrade-in-arms Margaret Thatcher at Camp David on December 22, 1984, and told her about this goal, she was "horrified." Thatcher was one of many who had come to believe that nuclear arms were what kept the Cold War cold, telling Soviet general secretary Mikhail Gorbachev three years later, "Both our countries know from bitter experience that conventional weapons do not deter war in Europe whereas nuclear weapons have done so over forty years."

On March 1, 1982, President Reagan watched the National Military Command Center rehearse a nuclear attack. A screen displayed a map of the United States, and as the missiles arrived, and the warheads fell, red dots bloomed, over and over, growing together into a bloody cloud—in a mere thirty minutes, America was no more. A book on the bestseller lists that year was Jonathan Schell's *Fate of the Earth*, which foretold, "In the first moment of a 10,000 megaton attack on the United States, flashes of white light would suddenly illuminate large areas of the country as thousands of suns, each one brighter than the sun itself, blossomed over cities, suburbs, and towns. . . . The thermal pulses could subject more than 600,000 square miles, or one-sixth of the total land mass of the nation, to . . . a level of heat that chars human beings. Tens of millions of people would go up in smoke. . . . In the ten seconds or so after each bomb hit, as blast waves swept outward from thousands of ground zeros, the physical plant of the United States would be swept away like leaves in a gust of wind. . . . virtually all the inhabitants, places of work, and other man-made things there—substantially the whole human construct of the United States—would be vaporized, blasted, or otherwise pulverized out of existence. Then, as clouds of dust rose from the earth, and mushroom clouds spread overhead, often linking to form vast canopies, day would turn to night. . . . Shortly after, fires . . . would simply burn down the United States. . . . Then comes radioactive fallout, ultraviolet radiation, destruction of the ozone layer, extinction of species."

From November 2 to 11, 1983, NATO conducted a coordinated multinational nuclear exercise: Able Archer. The Soviets knew NATO's plan of launching a nuclear first strike disguised as an exercise, and when Soviet intelligence reported an unusual amount of civilian leadership involved in Able Archer, Andropov believed nuclear attack was imminent, triggering a crisis nearly as dangerous as Cuba. These "war games" were so realistic that they included the drafting of a speech for Queen Elizabeth to read to the nation as atomic missiles fell on England's mountain green: "I have never forgotten the sorrow and the pride I felt as my sister and I huddled around the nursery wireless set listening to my father's inspiring words on that fateful day in 1939. Not for a single moment did I imagine that this solemn and awful duty would one day fall to me. We all know that the dangers facing us today are greater by far than at any time in our long history. The enemy is not the soldier with his rifle nor even the airman prowling the skies above our cities and towns but the deadly power of abused technology."

In his diaries, Ronald Reagan's sole mention of feeling sad occurs on October 10, 1983, after watching a preview of a TV movie on nuclear horror, *The Day After:* "It's powerfully done, all $7 mil. worth. It's very effective and left me greatly depressed. So far they haven't sold any of the 25 spot ads scheduled & I can see why. . . . My own reaction was one of our having to do all we can to have a deterrent & to see there is never a nuclear war." Nearly 40 million US homes—half the nation—watched *The Day After*, including journalist Alexander Zaitchik: "It finally settled my internal debate about what to do in the thirty minutes between test pattern and first impact. For months, I had debated whether to try and run and hide, or climb the nearest roof. The movie decided it for the roof. It answered Nurse Brower's question, asked in the raw cut Reagan saw, but removed from the final edit, of whether it was the living that envied the dead."

On November 11, 1983, Reagan gave a speech to Japan's Diet that reprised the history of US and USSR leaders' pronouncements: "I believe there can be only one policy for preserving our precious civilization in this modern age. A nuclear war can never be won and must never be fought. The only value in possessing nuclear weapons is to make sure they can't be used, ever. I know I speak for people everywhere when I say our dream is to see the day when nuclear weapons will be banished from the face of the earth" (if only Reagan, and all the presidents and premiers before him who gave speeches saying the exact same thing, were in positions to do something to help solve this terrible problem). At the time of that speech, the Pentagon's SIOP listed five thousand foreign targets for decapitation alongside forty-five thousand military, industrial, and economic targets needing destruction in the case of war. After Reagan was briefed on it he wrote, "In several ways, the sequence of events described in the briefings paralleled those in the ABC movie. Yet there were still some people at the Pentagon who claimed a nuclear war was 'winnable.' I thought they were crazy. Worse, it appeared there were also Soviet generals who thought in terms of winning a nuclear war."

Someone else was speaking similarly. In December 1984, the USSR's parlimentary delegate Mikhail Gorbachev said to Britain's legislature: "Whatever is dividing us, we live on the same planet and Europe is our common home—a home, not a theater of military operations. . . . The Soviet Union is prepared . . . to advance towards the complete prohibition and eventual elimination of nuclear weapons." On January 15, 1986, Mikhail Gorbachev directly proposed to Ronald Reagan "a concrete program, calculated for a precisely determined period of time, for the complete liquidation of nuclear weapons throughout the world . . . within the next fifteen years, before the end of the present century. . . . Over a period of five to eight years the Soviet Union and the United States will halve the nuclear arms which can reach each other's territory." Reagan told his secretary of state, George Shultz, that this was "a hell of a good idea." Then in July, Gorbachev announced that the USSR was unilaterally halting all nuclear tests and asked Reagan to follow suit. After all his public and private declamations on ending nuclear arms, though, the American president wouldn't do it. On September 27, the general secretary suggested both sides cut their long-range stockpile by half and agree to a ban on weapons in space. Reagan wouldn't do that, either. While this may seem one more instance of a superpower leader's being all hat and no cattle, Reagan's intransigence had a secret reason, and his name was Edward Teller. Teller had convinced Reagan he could create a technology that would forever protect the United States of America from nuclear attack—the Strategic Defense Initiative—Star Wars. Inspired by Teller, Ronald Reagan's SDI fantasies would keep the United States and the USSR from reaching the great dream that so many have had since Hiroshima . . . of zero nuclear arms.

Isidor Rabi had sent President Eisenhower an October 28, 1957, memo explaining that, as the arc of a ballistic missile was a mathematical signature, the location of its origin could be identified. If so, incoming rockets could be destroyed by atomic bombs in outer space before ever reaching American soil, creating a force field, a shield. Eisenhower did nothing with this idea, but Edward Teller, having used Fermi's comment to launch a career in fusion, would use Rabi's concept to launch another in missile defense.

As governor of California, Reagan visited the Lawrence Livermore labs in 1967, where the facility's director, Teller, briefed him on the difficulties of keeping Americans safe in their homes under onerous test ban treaties that kept physicists from freely studying nuclear science. As president, Reagan visited NORAD—North American Aerospace Defense Command, fifteen buildings of carbon steel plate on 1,319 thousand-pound shock-absorbing springs accessed by five-ton, blast-proof doors and protected by two thousand feet of granite in Cheyenne Mountain, Colorado, that monitors incoming missile attacks—where he was told that the Soviets had just developed a new missile, the SS-18, against which NORAD was defenseless. Reagan became obsessed that America was nakedly vulnerable to ICBMs, and that as president he must do something about it. Whenever he discussed this fear with anyone else in Washington, however, apparently no one explained that the United States was far from defenseless, that in fact a missile attack on American soil would be met with an epic nuclear retaliation beyond the ken of Hollywood's biggest budgets.

Even as chief of Livermore, Edward Teller was still scorned by many of his professional colleagues, at the same time that antiwar protesters made it increasingly difficult for him to lecture. He was still a piece of work, having, for decades, promoted experiments that revealed flies living longer after being mildly irradiated, claiming this proved radiation was beneficial for living things. Leo Szilard accused him of knowing the truth, that the real reason was that the radiation killed a parasite infecting the flies, but Teller kept publicly making this argument without mentioning the underlying parasites.

When Teller had an Oval Office meeting with Reagan on September 14, 1982, followed by a White House dinner six months later, he described to the commander in chief how he had worked on the first two generations of nuclear science—fission and fusion—but that a third generation was yet to come, one that would use atomic propulsion to create enormous lasers and microwave beams. A giant X-ray satellite, floating in space, could do what Rabi had told Eisenhower decades before—intercept and destroy incoming missiles before they could kill American citizens. Instead of Mutual Assured Destruction, with Teller's help the president could offer the nation Mutual Assured Survival with an atmospheric shield so powerful it would render all nuclear weapons obsolete. It is unclear to this day whether Teller explained to Reagan that all of this was based on enormous nuclear weapons floating continuously overhead in low-earth orbit.

Just as Eisenhower had great hopes with Atoms for Peace, Reagan became enthralled with the Strategic Defense Initiative. But in many quarters, the news was not well received. After hearing the president announce his new program in a televised speech, Gorbachev met with the Kurchatov Institute of Atomic Energy's deputy director, Yevgeny Velikhov, who told him that Russian physicists had tried for decades to create exactly the weapons Teller and Reagan were talking about, including the same space laser cannon, as well as antimissile rockets fired from satellites. They didn't work. No matter what intercept method was used, launching an assault with a cloud of metallic chaff and dummy missiles overran any so-called shield's capacity, and SDI—Star Wars—would never succeed. Even so, Gorbachev knew that the Soviet military would insist on having their own outer-space technology to match Washington's, and that the Soviet economy couldn't afford a whole new avenue of arms race. They already had nuclear weapons in the oceans with submarines, on the land with ICBMs, and in the air with bombers; now they would need billions of dollars and rubles spent on weapons in orbit? He also believed that, if Velikhov was wrong and Teller's SDI was a success, a MAD-free America might launch a nuclear first strike. To Reagan, SDI was the greatest of dreams, and to Moscow, the worst of threats.

As Teller's SDI group proceeded, they followed almost exactly the technological footsteps that the Soviets had trod. First they experimented with a nuclear-fired X-ray laser; by 1986, it was clear this would not work. The Pentagon spent $11 billion testing a vast array of satellites that would fire antimissile rockets—the notorious Brilliant Pebbles—which failed as much for Teller as they had for the Soviets. But Reagan could not let go of his wonderful dream; in his 1985 inaugural address, the president explained all over again that Star Wars "wouldn't kill people. It would destroy weapons. It wouldn't militarize space, it would help demilitarize the arsenals of the earth. It would render nuclear weapons obsolete."

On January 15, 1986, Gorbachev offered to stop all atomic testing for five to eight years; to limit warheads to six thousand apiece; to remove all medium-range missiles from Europe; to enact a ban on space-strike weapons; for China, France, Britain, the United States, and the USSR to ban tactical nuclear weapons and reduce their arsenals over a five-to-seven-year period; and finally, to ban all nuclear weapons over fourteen years. After hearing of this, Reagan wrote in his diary: "We'd be hard put to explain how we could turn it down." Yet, as Moscow halted testing for ninety days to try to shame the United States into following suit, on March 22, the AEC detonated a twenty-nine-kiloton bomb at the Nevada Test Site.

Mikhail Gorbachev and Ronald Reagan—one instantly recognizable from his ocher birthmark peninsula; the other from his shiny black macassar helmet—then met on October 11, 1986, at Reykjavík, Iceland, where the first secretary raised his January offer to now include the two superpowers' eliminating all offensive nuclear arms—the triad of ICBMs, bombers, and sub-launched cruise missiles. Gorbachev: "So let me precisely, firmly, and clearly declare, we are in favor of finding a solution that would lead eventually to a complete liquidation of nuclear arms. Along the way to that goal, at every stage, there should be equality and equal security for the USA and the Soviet Union. Anything less would be incomprehensible, unrealistic, and unacceptable." Reagan agreed to everything, but then, when he was told which aspects would interfere with Star Wars, he refused to accept those elements, and instead he offered to share Teller's space weapons technology with the Kremlin. Gorbachev's voice rose: "Excuse me, Mr. President, but I do not take your idea of sharing SDI seriously. You're not willing to share with us oil well equipment, digitally guided machine tools, or even milking machines. Sharing SDI would provoke a second American Revolution! And revolutions don't occur all that often. Let's be realistic and pragmatic."

Reagan: "Let me ask, do we mean by the end of the two five-year periods all nuclear explosive devices will be eliminated, including bombs, battlefield weapons, cruise missiles, sub-launched, everything? It would be fine with me if we got rid of them all."

Gorbachev: "We can do that. We can eliminate them all."

American secretary of state George Shultz: "Let's do it!"

Reagan: "If we agree that by the end of the ten-year period, all nuclear weapons would be eliminated, we can send that agreement to Geneva. Our team can put together a treaty and you could sign it when you come to Washington."

Gorbachev then offered a compromise of limiting SDI to the laboratory, meaning no nuclear testing. Reagan asked his adviser Richard Perle about it, and Perle, who had fought nuclear arms reduction and control during the whole of his political career, said that agreeing to those limits would destroy the program. So Reagan, still dreaming of Star Wars, turned down Gorbachev.

The historic agreement to end all nuclear weapons, which would have been their immense historic legacies as president and first secretary, was dead.

But the conversation had been started and it, somewhat, continued. On September 27, 1991, Reagan's successor, George H. W. Bush, completed both the START treaty instigated by Gorbachev and Reagan as well as START II, the biggest arms reduction in history, topping the arsenal for each superpower to thirty-five hundred warheads apiece. Bush then unilaterally eliminated both US chemical and tactical (battleground) nuclear weapons—artillery shells, naval torpedoes, ground-missile warheads. On October 5, Gorbachev did the same.

Still, as of 2013, the Pentagon has spent $157.8 billion on the Strategic Defense Initiative and its successors (including the present-day Missile

Defense Agency), even though fifty Nobel laureates signed a 2001 petition to Congress pointing out that, outside of laboratory conditions, the goal of "hitting a bullet with a bullet" was absurd. MDA enthusiasts pointed to the 90 percent success rate of Israel's Iron Kippah (and the American agency is working with Israel on a system known as David's Sling), but the Kippah's targets are artillery only. US tests of a national missile shield to protect the homeland have, meanwhile, achieved a pathetic 53 percent.

The entire point of missile defense for the continental United States is nonsensical for the same reason that was never given to President Reagan: the nation is already well enough secured by everything else in the Department of Defense's arsenal. Any incoming missiles would easily be identified, and their source would then suffer a military retaliation unparalleled since Nagasaki. Lieutenant General Robert Gard directly made this point: "What country is suicidal enough to launch a weapon of mass destruction on a long-range missile when it leaves a trail of where it came from?" We don't need to add billions in research to the tens of billions already sluicing through the Pentagon for a concept that hasn't worked in two decades on a problem that we have already solved. But the bouncing, bouncing, bouncing, goes on, apparently unstoppable. Under Reagan, missile defense was annually budgeted at $2–$4 billion. Twenty years on with nothing to show for it, under Bush and Obama, that figure has more than doubled, to $8–$9 billion.

On December 8, 1991, in a hunting lodge outside Brest, the leaders of Russia, Ukraine, and Belarus declared the end of the Union of Soviet Socialist Republics. The lodge didn't have a Xerox machine, so to make copies they had to fax the agreement back and forth. The Soviets still had 15,000 tactical nuclear warheads, 822 in bombers, 6,623 in ballistic missiles, and 2,760 in torpedoes and cruise missiles, and the United States still had its tremendous nuclear arsenal aimed at a Soviet Union, which no longer existed. The arms race, which could've been strangled in its crib by the right attitudes of Secretary of State James Byrnes and which could've been stopped nearly two decades ago if Ed Teller had been more honest in his glowing forecasts, continues . . . even though one side of the race has folded its tents and slipped away.

Today America continues to spend $55 billion a year on atomic weapons that have never and will never be used. Cutting this arsenal in half would save $80 billion over the next ten years, and even then the Pentagon would have fourteen times as many warheads as the nearest competitor, China. Congressman Barney Frank asked the DoD to pick one of the three methods they had of atomic-striking the Soviet Union—ICBMs, bombers, or submarines—and eliminate it, to save around $10 billion a year. The Department of Defense refused, insisting it must keep the tripartite weapons system, for which there is no enemy. Bounce, bounce, bounce.

Over the past decade, Republican secretaries of state Henry A. Kissinger and George P. Shultz, Democrat secretary of defense William J. Perry, and Democrat senator Sam Nunn have joined physicist Sidney D. Drell to promote an abolition of all nuclear weapons, period. Shultz was at the table in Iceland when Reagan backed away from the Zero Option to safeguard Star Wars, and Shultz has regretted Reagan's choice ever since. In 1991, Nunn worked with the Pentagon and the Department of Energy to secure the nuclear weapons left behind by the collapsed USSR. One problem is that the Soviets were so lax in their waste disposal that the surrounding territory is nearly luminous. Dosimeters carried by American staff looking for sequestered nuclear arms have been set off by woodpiles, deer, and fish.

During that period, a Russian Foreign Intelligence Service defector revealed in his debriefing that as of 1991, at least one Russian oligarch had his own Bomb at an exurban Moscow dacha. When the spy said he thought this sounded suspicious, the "businessman" said, "Do not be so naive. With economic conditions the way they are in Russia today, anyone with enough money can buy a nuclear bomb. It's no big deal really."

Without the iron hand of the Kremlin in place, the West had a new worry—that terrorists might realize easy access to nuclear materials from the collapsed empire and manufacture "dirty" bombs—which was then used as an excuse for a new arms race in order to retaliate. But dirty-bomb-wielding terrorists would require about a hundred pounds of HEU (highly enriched uranium) at minimum 90 percent U-235 (which is traceable and in short supply; nuclear reactors run on 4 percent HEU), a sophisticated workforce, and the ability to keep the two fifty-pound sets of fuel away from each other, or they might spontaneously chain-react before being delivered to their target. As for buying nuclear weapons from a rogue state, analyst Matthew Bunn explained, "A dictator or oligarch bent on maintaining power is highly unlikely to take the immense risk of transferring such a devastating capability to terrorists they cannot control, given the ever-present possibility that the material would be traced back to its origin." The worst terrorist attack in history was engineered with box cutters and a few weeks of flying lessons. Dirty bombs are a myth and an absurdity.

In the end, the biggest victims of the Cold War arms race turned out to be American and Soviet citizens. After Nevada's underground nuclear tests ended in 1992—though the US Congress never ratified the Comprehensive Test Ban Treaty, the agreement was followed in practice—the Department of Energy measured the worst of the region's water at millions of picocuries per liter (the federal peak for human consumption? Twenty picocuries). In 1979, the *New England Journal of Medicine* found a rise in leukemia mortality in local children born between 1951 and 1958, and a 1997 National Cancer Institute report found that Nevada Test Site explosions had left 5.5 exabecquerels of radioactive iodine-131 across nearly the whole of the continental United States from 1952 to 1957, enough to produce somewhere between ten thousand and seventy-five thousand cases of childhood thyroid cancer. Women's breasts and ovaries are unusually susceptible to radiation cancers—the misogyny of the 1950s CDC come to bitter truth. Today, an American's health is more likely to be impaired by the leftover pollution of those tests than any other form of radiation except suntanning, an especially galling bit of news since, as the world has never suffered a nuclear attack since Nagasaki, the vast majority of those tests were essentially pointless.

Native American miners working for the AEC from the 1940s to the 1960s in the Southwest were not informed that the dust they breathed was contaminated with radon gas, as was much of their well water. Led by Stewart Udall, a group of miners tried to sue the federal government, but their case was dismissed. Eventually, however, the long arc of the moral universe bent to justice in 1990 with the Radiation Exposure Compensation Act, giving Nevada Test Site downwinders with medical proof $50,000, uranium miners $100,000, and test participants $75,000. It has been difficult, however, for the widows of Navajo miners to amass the required paperwork and receive their due.

The United States of course is not alone in polluting the world with near-worthless atomic tests. The Soviet Nevada in Kazakhstan was the Polygon, a territory that suffered 456 detonations between 1949 and 1989—twenty-five hundred Hiroshimas. The tests "smelt . . . you know, like hair. Like hair burning. The smell came back from the earth every time it rained," a woman living nearby said.

While Fukushima dropped an estimated 10 million curies, Chernobyl 100 million, and Three Mile Island 50, the over five hundred open-air nuclear bomb detonations of those decades saturated the planet with 70 billion curies. The Centers for Disease Control has reported that every person born after 1951 in the continental United States has been exposed to radioactive fallout from the Nevada Test Site, that "all organs and tissues of the body have received some radiation exposure," and of the nearly six hundred thousand Americans dying of cancer every year, eleven thousand will die from the remnants of bomb tests.

During the last years of his life, Enrico Fermi repeatedly asked the question now known as the Fermi paradox: "Where is everybody?" He was asking why, despite the great size and age of the universe, no extraterrestrial civilizations had been discovered. At times, he believed that the answer might be nuclear annihilation. Journalist Adam Gopnik: "Spengler may have been right about the foreordained blossoming and decay of civilizations, on a far more cosmic scale than he could've imagined: once a society reaches to sun power, and makes nuclear weapons, it destroys itself. That's why we feel ourselves to be alone in the universe. What we see staring at the vast night sky is not a mystery but a morgue, full of suicided civilizations."

However dramatic and evocative these mournful thoughts might be, they are today nostalgic tailings of the fading Atomic Age, as dated as corsets. What used to be a prestigious symbol of national achievement—in 1965 French president Charles de Gaulle insisted, "No country without an atom bomb could properly consider itself independent"—is now taken up only by pariah states such as North Korea and Pakistan. These two, along with Israel and India, are the only countries to acquire nuclear arsenals since Lyndon Johnson's foreign policy breakthrough of July 1, 1968. The Treaty on the Non-Proliferation of Nuclear Weapons now has 183 countries vowing to never produce or acquire atomic weapons while the five original nuclear powers—the United States, the USSR, Britain, France, and China—agreed to reduce and in time eliminate their arsenals. In 1993, South Africa gave up its atomic stockpile left over from apartheid, and in 1996, Ukraine, Belarus, and Kazakhstan gave up theirs left over from the Soviet Union.

Since the NPT, every instance of a state's going nuclear has been a defensive posture. After she was defeated by China in 1962's Sino-Indian War over their borders, India wanted its own warheads; when Pakistan was defeated by India in Kashmir in both 1948 and 1965 and then again in Bangladesh in 1971, it wanted its own arsenal; and the constant mentioning by the neighboring politicians of Israel of how wonderful it would be to wipe that nation from the face of the earth led to Tel Aviv's nuclear cache. After successfully building nuclear weapons for Islamabad, Abdul Qadeer Khan—the Johnny Appleseed of rogue-state plutonium—went on to help start atomic weapons R&D for the threatened governments of Libya, Iran, and North Korea. In July 2002, the Myanmar junta told the UN's International Atomic Energy Agency (IAEA) that it was building a nuclear reactor with Russian advisers, but two Burmese exiles insisted that a parallel complex, this time with North Korean advisers, was being built to engineer bomb-worthy plutonium. In 2011, a Saudi prince/government official said his country would consider creating its own atomic weapons program if it found itself cornered by Israel's and Iran's, while Iran's leaders developed a renewed surge of interest in nuclear weapons after the United States invaded Iraq, the Teheran mullahs reasoning that the Americans would never have entered Baghdad if Saddam Hussein was armed with the Bomb. To confirm that point of view, US news reports in 2012 were filled with Washington insiders discussing whether to invade Iran before it was atomic powered, yet there has never been similar televised debate about attacking nuclear-enabled North Korea. Then on April 9, 2013, after Pyongyang once again rattled its warheads, a prominent member of South Korea's parliament insisted in Washington that it was now time for Seoul to have its own atomic defense.

The costs of going rogue in the eyes of the Non-Proliferation Treaty are severe—investigation by the IAEA, international boycotts, and censure—and some signatories have supplemented those measures with assassinations and cyberattacks. On November 29, 2011, Iran's Atomic Energy Organization director, Fereydoon Abbasi, was on his morning commute when a man on a motorcycle pulled up and attached a bomb to his car. Abbasi and his wife escaped more or less unharmed, but one of his colleagues was killed by a similar attack, as was an Iranian particle physicist in January 2010, an electronics specialist in July 2011, and a manager at the Natanz uranium enrichment plant in January 2012. Teheran blamed Tel Aviv and Washington for the assassinations, as well as for the malware viruses known as Flame and Stuxnet, which were discovered in the spring of 2012 infecting Iran's uranium enrichment computers. Flame is lithe spyware that turns on computer microphones and Skypes the recorded conversations; scans the neighborhood's Bluetooth gadgets for names and phone numbers; and takes pictures of the computer's screen every fifteen to sixty seconds. Stuxnet infected Iran's uranium-enriching centrifuges and sped them up until they committed suicide.

A Russian nuclear executive summed up that after the fall of the USSR, "the great powers were stuck with arsenals they could not use, and nuclear weapons became the weapons of the poor. . . . [The] technology has become a useful tool especially for the weak. It allows them to satisfy their ambitions without much expense. If they want to intimidate others, to be respected by others, this is now the easiest way to do it."

Journalist Fareed Zakaria countered: "Does anyone really think that North Korea or Pakistan are regarded as fearsome adversaries, countries to emulate, countries with great influence in the councils of the world? No. They are regarded as basket cases—failed states that are dangerous largely because they are unstable and are run by irresponsible governments that are willing to do destabilizing things in their region. The result is they are more watched, cordoned off, and contained than ever before."

Dr. Homi Jehangir Bhabha, father of India's nuclear technology, baldly stated the solution back in 1965: "A way must be found so that a nation will gain as much by not going for nuclear weapons as it might by developing them."

16

On the Shores of Fortunate Island

IN the fading-ember days of World War II, Kiwamu Ariga was one of untold dozens of Japanese schoolchildren whose education was postponed. Instead, he and his classmates were sent off into the forests to look for brown or black-spotted rocks. Day after day they dug with small picks and bare hands, their feet bloody from clambering across the jagged hillside. Finally, an army officer explained, "With the stones that you boys are digging up, we can make a bomb the size of a matchbox that will destroy all of New York." This was the state of 1940s Japanese nuclear science—instead of the world's greatest scientific minds gathering at Los Alamos, little boys were digging out bits of uranium with toy shovels—and this childhood mining operation took place in a part of the country that reminds many Americans of Maine, with its undulating pine forests, salt-air towns, and a Miss Peach championship. This is the prefecture of "fortunate island"—Fukushima.

After fallout from the Bikini Mike test infected *Lucky Dragon Number 5*'s Japanese crew and its tuna cargo, Eisenhower's State Department reported, "The Japanese are pathologically sensitive about nuclear weapons. They feel they are the chosen victims." Illinois representative Sidney Yates recommended the United States apologize to the fishermen and build the Japanese their own Atoms for Peace reactor. Many in Japan simultaneously were coming to think that the paucity of their homegrown energy, with the need to constantly import oil and coal, triggered the imperial conquests of the 1930s and 1940s that calamitously thrust their nation into the Axis powers. The answer for them was also nuclear. While the CIA then worked with Japanese baseball's founding father, Matsutaro Shoriki, and his newspaper chain to promote such stories as "Finally, the Sun Has Been Captured," Japan's atomic plants were fashioned to include visitor's centers—"PR buildings," in Japanese—that held IMAX theaters, swimming pools, Disneyfications of Albert Einstein's and Marie Curie's homes, and programs designed to appeal to young mothers, as surveys had shown them particularly wary of nuclear contamination. In one of his cartoons, anime variation Little Pluto Boy told children it was A-OK to drink liquid plutonium since "it's unthinkable that I could cause any effects on the human body!"

Atoms for Peace included a revision of the US Atomic Energy Commission's agenda. Now, the agency would develop nuclear science and technology; oversee the United States' atomic stockpile; promote the expansion of nuclear energy; and regulate the nation's nuclear power utilities. These bedrock conflicts of interest usually ended up meaning regulation took a backseat to promotion and development, a state of affairs mirrored in Japan's own nuclear agency. In the 1970s, the AEC was split into the Department of Energy, which promotes and develops, and the Nuclear Regulatory Commission, whose mission is clear. Japan's government did not make these changes. Additionally, the United States has a history of civilians and journalists questioning and criticizing state policies. Both Robert Oppenheimer and consumer advocate Ralph Nader, among others, warned of the dangers of civilian nuclear power plants at the industry's very birth. "The Atomic Energy Commission was licensing unsafe reactors operating near major metropolitan areas, and they clearly were aware of this lack of safety," Nader said. "The press wasn't critical. The Congress bought into the Atomic Energy Commission party line. There was a huge taxpayer-funded propaganda for how good nuclear power was, going right into the high schools and elementary schools in our country with traveling road shows. The scientific community was part of the industry itself and there was no outside critique. There was no government critique. And there was secrecy above it all." Japanese society has no similar tradition of public dissent. Finally, as private US insurance companies would only cover nuclear plants to $65 million (about one-tenth of what a major accident would cost), in 1957 Washington created a special insurance pool. Such public largesse for private enterprise would apogee in 2012 when the Japanese government was forced to nationalize with taxpayer yen a "too big to fail" utility destroyed by corporate malfeasance and nuclear meltdown.

In 1971 next to a popular surfing spot on the shores of Fukushima, Tokyo Electric Power Company (TEPCO) built a nuclear power-generating plant bigger than the Pentagon and staffed by six thousand—Daiichi (*Dye-ee-chee*). Fortunate Island was the first of many nuclear plants Japan built along its coastline, and hanging from the bridge over the main road's entry into Futaba, Daiichi's nearest town, was an exhortation to the plant's neighbors: NUCLEAR ENERGY: A CORRECT UNDERSTANDING BRINGS A PROSPEROUS LIFESTYLE!

In the late 1980s, paleontologist Koji Minoura discovered an ancient poem that included:

Do you remember our sleeves
wet with mutual tears
in oath never to leave each other
as the famed waves of Sue-No-Matsuyama
never go beyond the cliffs . . .
Now you're gone,
it seems,
leaving me alone in grief.

Minoura decided that *famed waves* meant an enormous tsunami, and looking through civic records, he found it. On July 13, 869, "the tsunami hit Tagajo and killed more than a thousand people," Minoura learned. "But people soon forgot about the tsunami. I visited Sendai, where the tsunami hit, and found geological evidence." Excavating rice paddies two and a half miles from the coastline, he discovered a layer of marine sediment. The paleontologist kept digging, uncovering more pelagic deposits nestled between layers of terrestrial soils, and realized that Japan had suffered a massive tsunami attack every thousand years or so. In the early 1990s, Koji Minoura showed these findings to representatives of Tokyo Electric, insisting that their seaside nuclear plants were imperiled. He was ignored.

On Friday, March 11, 2011, at 2:46 p.m. under the Pacific sixty-two miles due east of Japan's Honshu Island, the heat of radiation from deep within the earth pushed the Eurasian and Pacific tectonic plates against each other—the Pacific sliding beneath the Eurasian—until they reached that point where one could bend no more, and there was a tectonic snap. Fifteen seconds later, the first shake of the earth, seismic P waves traveling 4 miles per second, struck Japan's northeast coastline, and automatic, computer-generated warnings appeared at the bottom of every television screen. In one hundred seconds, the second shake, slower S waves, hit the forty-year-old Daiichi plant, ninety-three miles from the epicenter, swaying the ground for a full five minutes (typical earthquakes in Japan, which strike hundreds of times a year, last a few seconds). Paved walkways rolled in watery undulations, windows exploded, and speakers carried an urgent female voice across the plant: "Please evacuate! Please evacuate!" Nuclear technician Carl Pillitteri was one of forty Americans working at Daiichi that day:

> I still remember it. The first shock of it. It was just one big hammer. We were in a turbine building that is built, for lack of a better term, like Fort Knox. The entire building was shaking.
>
> I heard that this earthquake lasted six minutes. But for me, it felt like a lifetime. I'm still living it ten months later.
>
> You could feel it under your feet. It was this entire enormous building moving at once. A lot of things were falling. We lost almost every light in the room. The structural steel was moving overhead. The lights were crashing everywhere. In one nanosecond, the entire floor went black. Every light went out. You would expect some emergency lighting would come on, but there wasn't a one. And there was this most welcome beam of white light coming from the gap under the door. I made my way over to the door, and the one and only light in the room. It was swinging violently, and then at the same time I opened the door it busted free and shattered on the floor. It was pitch-black again. I remember thinking, "None of you are getting out of here."
>
> One of the Japanese guys had grabbed me around the waist. I put my arm up on his shoulder. With every jolt I squeezed his shoulder. I remember praying aloud for him, for all of us. I thought, "We're going to perish in this turbine building." I can still hear the turbine making its most unwelcome sound. I had many thoughts. But one of them was "Good God. I got up this morning just to go to work. And this is how it's written for me? Dying is a fact of life. We all have to do it sooner or later. But this is how it's written for me? March 11? On a Friday? On a turbine deck? In Fukushima? At work?"

The earthquake broke the reactor's cooling pumps, but emergency sensors automatically shut down the station's six piles, and backup diesel generators powered up. Fission stopped, but a great deal of energy remained as afterheat, what Princeton physicist Robert Socolow described as "the fire that you can't put out, the generation of heat from fission fragments now and weeks from now and months from now, heat that must be removed." Physicist Louis Bloomfield: "They aren't being heated by fission chain reactions, they're being heated by their own intense spontaneous radioactivity. Only time will reduce that self-heating. Until that time, they have to be cooled and contained. Water slows neutrons and absorbs heat, but that water has to be cooled as well so that the radioactive fuel rods don't heat it to boiling."

That the earthquake did no damage to Daiichi is the official story, but that story is in doubt. "After the second shock wave hit, I heard a loud explosion that was almost deafening," one worker remembered. "I looked out the window and I could see white smoke coming from reactor one. I thought to myself, 'This is the end.'" "There's no doubt that the earthquake did a lot of damage inside the plant," a maintenance worker in his twenties said. "I personally saw pipes that came apart, and I assume that there were many more that had been broken throughout the plant. I also saw that part of the wall of the turbine building for Unit 1 had come away. That crack might have affected the reactor." Nuclear critic Katsunobu Onda: "If [Tokyo Electric Power Company] and the government of Japan admit an earthquake can do direct damage to the reactor, this raises suspicions about the safety of every reactor they run."

What Tokyo Electric did not seem to know was centuries-old lore passed down through generations of the area's fishermen: *After the earthquake, the tsunami.* When the Eurasian and Pacific tectonic plates rebounded from their collision out in the Pacific, that force generated a sixty-mile-wide, three-foot-high ripple in the ocean waters that traveled at over 500 mph—jet speed. As it approached land, the coastal shallows slowed the ocean wave's forward bottom while the top rear continued, rising ever higher into the air to crest into a smooth, black swell, a wall of water twenty-six to thirty feet high, rolling forth at 50 mph.

"I saw the tsunami coming. I stood there, and as it came in, I thought, 'You gotta be kidding me,'" Carl Pillitteri said. "This thing was huge. It didn't resemble a wave. It resembled this huge swell of the ocean. This huge hump in the ocean coming your way. It rolled up over everything. It rolled uphill. It did come up over parking lots and took cars away in front of me. The first one receded back and took enough water to expose the seabed. This big, black ominous front came rolling down on us, so much so that it began to snow. I've never heard anyone mention the snow. I mean, it didn't snow and accumulate. But it snowed. And the wind was—it was like a vacuum going by. There's a harbor in front of Fukushima. And it was totally drained."

Fisherman Yasuo Uchida tried to save his boat. "I went straight to the harbor and headed out to sea. We went over three waves that came directly from the east. They were about fifteen meters high. They were like mountains." The biggest was over forty feet high and traveling at 100 mph.

As it moved forward, the tsunami swallowed up into itself boats and cars, homes and stores, all that there is and all that there was, until the water wall was a nearly solid, rolling glacial mass of soil, plants, animals, human bodies, and societal detritus. Finally, as all waves do, it receded back into the sea, leaving behind churning whirlpools, broken gas lines that set debris afire, decimated villages, a series of five hundred aftershocks, and an incomprehensible catastrophe. One-half million were homeless; 1 million had no power; 2 million had no water; and so many were dead, and so many were missing—twenty thousand? It would take a year before Japan knew anything remotely accurate about the numbers of the dead.

The Japanese rural coastline is, like Florida, a home of the retired and the elderly. One resident heard the tsunami alert siren and immediately got his aged mother and dog into a truck and drove inland as fast as he dared for two miles, the glacier wave rising inexorably toward them in the rearview mirror. Afterward, a translator said, "He would like to think there will be some assistance from the local government. But all he could think was, the city-assembly office is gone. The mayor could be dead. The only thing he can turn to is the government. But his local government is gone."

The earthquake had lowered the coastline by as much as three feet, meaning that the first wave, at 3:27, was still held back by Daiichi's thirty-three-foot concrete wall. But eight minutes later, the second wave, four stories tall, overcame the wall as well as sixty thousand concrete barriers and flooded into the plant's basement, where the emergency diesel generators that were cooling the reactor cores were installed. Backup batteries automatically engaged . . . but they only had eight hours of charge. On March 12, those batteries ran out of power. "They sent people out in the parking lots to scavenge batteries from automobiles, and they hooked them together, and they got some critical DC power for valve operation and instrumentation," MIT Nuclear Science professor Neil Todreas explained. "But you shouldn't have to do that."

The core's temperatures rose to 4,800°F; the coolant boiled into steam and evaporated. TEPCO admitted in a press release, "The emergency water circulation system was cooling the steam within the core; it has ceased to function." With forty thousand US citizens in Japan, American nuclear experts tried to monitor the situation from the bare-bones information that the corporation and the government released to the public. Physicist Ken Bergeron said at the time, "We don't know exactly how they're getting water to the core, or if they're getting enough water to the core. We believe, because of the release of cesium, that the core has been exposed above the water level, at least for a portion of time, and has overheated. What we really need to know is how long can they keep that water flowing." Nuclear analyst Michael Allen: "It'll be like somebody dropped a bomb, and there'll be a big cloud of very, very radioactive material above the ground." Winds could carry that cloud 150 miles south, to the heart of Tokyo.

Tokyo Electric tried to bring in fire trucks to water the cores, but the earthquake and tsunami had thrown a storage tank into the middle of a road and no pumping truck could reach the buildings. First thing the next morning, plant manager Masao Yoshida decided that the only solution was to flood the reactors with seawater, stopping their nuclear fires but destroying them forever. But it took all day for TEPCO executives back in Tokyo to agree to this last-chance solution, and only after their plant employees were confronted by a terrifying chain of events.

The melting uranium, breaking through its casing, had combined with the watery vapor to generate plumes of hydrogen, which would, when they contacted the oxygen of the outside world, violently explode. Just as at Three Mile Island, the steam, now highly radioactive, would have to be vented into the outside atmosphere to release a moderate amount of toxic gas since, if there was an explosion, it might release a devastating amount. To do this, TEPCO needed the official permission of Prime Minister Naoto Kan, who agreed, but to open the vent manually, the workers needed the plant's blueprints. The building housing the blueprints had collapsed, and no one could get inside it. And before any venting could take place, the region had to be evacuated. Residents within three kilometers were checked for exposure, and all within ten kilometers (six miles) of the plant were told to leave. Public broadcaster NHK told those remaining nearby to close their doors and their windows; to put a wet towel over their noses and mouths; and to cover up as much of their skin as possible.

When the prime minister heard no news for six hours—still, no one had figured out the procedure for manually opening the vent—Kan decided TEPCO was keeping information from him and helicoptered to Daiichi to see for himself what was happening. Finally a work-around for the vent was engineered, and the plant manager told Kan he would send in a suicide team.

On March 12 at 9:04 a.m., after swallowing potassium iodide tablets, two teams of six workers each donned firefighting suits with oxygen tanks and attached dosimeters to their belts. No clothing could protect their bodies from gamma rays, however, so they would tag-team relay to prevent any single employee from being exposed beyond a lethal seventeen minutes. Those willing to work with such high risks were derisively called gamma sponges, glow boys, jumpers, and dose fodder. One said, "At Chernobyl, you know, the workers received medals. We'll be lucky if we get a commemorative towel or a ballpoint pen. We are taboo."

In a pitch-black tunnel and temperatures of over a hundred degrees with radiation sirens blaring, the first team found the vent crank, but could only get it a quarter of the way open in the eleven minutes they had remaining. One dosimeter showed its crewman exposed to 106 millisieverts, even beyond the 100 dose generously set by TEPCO. At the time, Unit #1's control room showed levels at a thousand times normal, while the plant's maingate readings were eight times normal. The second team entered to find the radiation so severe that they could only spend six minutes inside and weren't able to even reach the crank. The third team reached the valve, opened it in nine minutes, and at two thirty in the afternoon, the venting released white bursts of radioactive steam into the skies.

As everyone signed in relief and passed around congratulations, the earth began, again, to move underfoot. At 3:36 p.m., Reactor #1 exploded so violently its metal roof shot into the sky. Jiro Kimura, who had spent his entire adult life working at TEPCO, said, "I thought this country was finished." Had the core blown itself to bits? In the control room, gauges said if everyone there would live or die, since no one could survive that level of radiance. But then the monitors fell, and everyone knew it was once again a volatile cloud of hydrogen. Ken Bergeron: "The hydrogen is being vented with the steam, and it's entering some area, some building, where there is oxygen, and that's where the explosion took place. They're venting in order to keep the containment vessel from failing. But if a core melts, it will slump to the bottom of the reactor vessel, probably melt through the reactor vessel onto the containment floor. It's likely to spread as a molten pool—like lava—to the edge of the steel shell and melt through. That would result in a containment failure in a matter of less than a day. It's good that it's got a better containment system than Chernobyl, but it's not as strong as most of the reactors in [the United States]. . . . There is a great deal of concern that, if the core does melt, the containment will not be able to survive. And if the containment doesn't survive, we have a worst-case situation."

Explosions shook the plant, one after the next. Later that night, after one paroxysm tore the wall and roof off a reactor building—though the containment walls remained secure—the government decided TEPCO was out of its depth. Following the technique used at Chernobyl, Colonel Shinji Iwakuma was sent in with a special army team to hose down Reactor #3 and at least begin to get control of the crisis. The men had uniforms that protected them against the worst of the radiation, as well as helicopter bellies plated in lead to shield out gamma rays. Iwakuma: "Just as we were to get out of the jeep to connect the hose, [the reactor] exploded. Lumps of concrete came ripping through the roof of the jeep. Radioactive matter was leaking in through the bindings of our masks. Our dosimeter alarms were ringing constantly." The men were at least able to evacuate before anyone received a fatal dose, but the readings were so strong that it was perilous to approach even by air, so they tried again the next day. But then, strong winds kept the water from hitting its targets, and they had to give up. Smoke and debris made clear photographs of the site needed for analysis impossible to take from the copters.

Daiichi stored its spent radioactive fuel rods in water-filled pools on the top floors of each reactor building outside the containment shields, with each pool cooling 548 fuel assemblies—four times the size of the cores themselves. Since after dozens of years and billions of dollars, the United States still has no long-term storage facility for nuclear waste, 104 of its atomic plants similarly store sixty-five thousand metric tons—a football field stuffed twenty feet high—of radioactive leftovers, with at least two dozen utilities housing radiant leftovers in aboveground pools, just like Daiichi's.

On March 13, Fortunate Island's pools began to boil and evaporate. As two of the buildings had lost their roofs, their pools were now fully exposed to the outside world. If the afterheat grew strong enough, the pool could boil away and expose those rods igniting and radiating far more powerfully than even a full core failure. "It's worse than a meltdown," said nuclear engineer David A. Lochbaum. "The reactor is inside thick walls, and the spent fuel of Reactors 1 and 3 is out in the open."

Then the water in the spent-fuel-rod pool of Unit #4 also began to boil. Shigekatsu Oomukai of Japan's NISA (Nuclear and Industrial Safety Agency) insisted that the pool was so deep that this wasn't of imminent concern, but he also admitted that the temperatures were so high that workers couldn't reach the pools to replenish them. The gauges in Units 5 and 6 revealed that their own pools were getting hotter and hotter.

Another hydrogen explosion tore through Daiichi at 3:40 p.m. on the fourteenth, destroying Unit 3's outer building, injuring four workers, and raising into the air a pink cloud. Some onlookers became terrified when winds pushed it into what to their eyes was a mushroom shape. The emergency cooling system for Reactor #3 stopped working, and its core also began to melt. More radioactive vapor had to be vented, and workers began to flood that reactor with seawater and boric acid.

Washington became so disturbed by the lack of information coming from either the Japanese government or TEPCO that on March 15 it flew a Global Hawk surveillance drone overhead. Reconnaissance photos revealed the fuel-rod pools in serious trouble, as well as chunks of fuel rods scattered all over the plant's grounds. Clearly, much of Daiichi was now lethally contaminated.

Though Tokyo Electric insisted it didn't need any help, the United States next sent in a team of nuclear consultants. "Everything in their system is built to build consensus slowly," one of them said. "And everything in this crisis is about moving quickly. It's not working." The Americans waited until the crisis so worsened that they warned of countless emergency workers losing their lives to safeguard Japan from nuclear catastrophe. One adviser, Robert Gale, had worked in Chernobyl and described how of the two hundred workers there who were overdosed, thirteen had to get bone marrow transplants. He recommended TEPCO harvest and store blood cells from its workers beginning immediately. Though the Japanese nuclear industry had developed robots to use in accidents, all but two of them were given to a college and a museum in 2006, and Fukushima had to instead use loaner American ones from iRobot, maker of Roomba.

The US navy then arrived; the USS *Ronald Reagan* Carrier Strike Group ferried out a posse of H-60 helicopters to deliver supplies and rescue the desperate. But after a mere eighteen hours, sailors had been stricken with thirty times the normal dose of radiation, and the ships had to be repositioned away from the prevailing easterly winds.

A broken valve had kept workers from venting Reactor #2 and pumping in seawater that would stop a meltdown. By Tuesday morning, March 15, that valve was fixed, but earthquake-caused leaks kept the fuel rods from being fully covered. The reactor then exploded. Later that day, spokesman Edano announced, "Number Four is currently burning, and we assume radiation is being released. We are trying to put out the fire and cool down the reactor. There were no fuel rods in the reactor, but spent fuel rods are inside. [This] did not pose an imminent threat." After discovering that 70 percent of Unit 1's rods were damaged and 33 percent of Unit 2's, the government announced there had been a partial meltdown in the cores.

In the wake of the explosions, the evacuation zone increased from ten to twenty kilometers (twelve miles). Twenty kilometers means 185,000 refugees; eventually the Japanese evacuated 210,000 people from a twelve-mile radius around the plant (the Americans suggested the zone be fifty miles, which was what British, French, Italian, and US citizens in Japan followed). A full meltdown would mean 150 miles, and that would include Tokyo's 35 million people. At the same time as Daiichi, three other TEPCO nuclear power stations were in trouble, one, Daini, a mere seventy-five miles outside the capital. To protect against radioactive-iodine-induced throat cancer, 230,000 doses of potassium iodide were given to residents near both Daiichi and Daini.

At the Fukushima town of Namie, thousands began to flee. With no instructions from Tokyo, local officials believed that prevailing southerly winter winds meant the townspeople should caravan north. For three nights, as four reactors exploded in hydrogen leaks and sent off radioactive plume after plume into the air, the people of Namie exiled themselves to Tsushima, making rice with water from a mountain brook, their children playing out in the open air.

Two months later, town leaders found out that a government computer program had shown radioactive winds bearing down directly on Tsushima—but no one was informed. Politicians in Tokyo didn't want the public to know how severe the crisis was and they didn't want to expand the evacuation zone beyond the twelve-mile radius so they kept all the data to themselves. Tsushima would have to be evacuated after tests found clouds had rained down radioactive cesium and iodine. Namie's mayor said there was a word for withholding such information: murder. Meanwhile, in the abandoned lands, farm animals and family pets that didn't starve to death were scavenging through homes, fields, and lawns, becoming feral.

The story of Namie was only the start of business and government, of TEPCO and Tokyo, withholding information from the public. Two months after the crisis was resolved, government inspectors admitted that they had detected tellurium-132—a signature by-product of a reactor meltdown—the day after the tsunami. They had publicly raised and lowered the radiation levels considered safe for schools—as seen with Chernobyl, kids and emergency workers are the hardest-hit victims of nuclear power disasters—causing mayhem for parents. In late March, 45 percent of 1,080 children in the surrounding areas tested positive for thyroid exposure to radiation, but the government claimed these readings were so low there was nothing to worry about. Even 150 miles to the south in the heart of Tokyo, however, radiation was measured at twenty times normal.

In the face of TEPCO's and the Japanese government's anodyne public announcements, the *hibakusha*—Hiroshima's and Nagasaki's nuclear-attack survivors—came forward, begging the authorities for "more sense of crisis." Prime Minister Kan then called the devastation the worst event in Japan's history since the end of World War II, and Emperor Akihito felt he needed to give his first televised address, using a courtly language few could understand, offering "heartfelt hope that the people will continue to work hand in hand, treating each other with compassion, in order to overcome these trying times." His appearance was so extraordinary that it was compared to the sole broadcast given by his father, Hirohito, announcing in 1945 that *senso owari*, "the war is over," and begging the Yamato people to "endure the unendurable, bear the unbearable."

Three reactors were now on the cliff of simultaneous meltdown. In Washington, the Nuclear Regulatory Commission believed Daiichi was so out of control that it ran a computer simulation on "the popcorn scenario." If the Japanese were so overwhelmed by radiation exposure they couldn't get water to the cores, the entire complex could explode in a series of toxic plumes and fuel meltdowns. "One will pop and then another one and then another one and then another one," one NRC member said. If that happened, Tokyo could be contaminated with clouds of radioactive fallout.

Another hydrogen explosion ripped open a roof, leaving the nuclear waste ponds exposed to the air. Reactor #4's fire then ignited and exploded, blowing out a twenty-six-foot-wide hole in the wall. Then an eruption broke the inner containment walls of Reactor #2, releasing eight hundred times the highest recommended levels of radiation.

During one conference call with a government liaison and TEPCO's president, Masataka Shimizu, site manager Masao Yoshida was told *not* to cool the reactors by spraying them with seawater. The Tokyo executives heard Yoshida repeat this order to his subordinates, but they did not hear him whisper to one employee at the same time that everyone should ignore it. This corporate insubordination, unheard of in hierarchy-obsessed Japan, may have saved Daiichi from becoming a global disaster. Then Shimizu told Prime Minister Kan's chief cabinet secretary on March 15, "We cannot hold on to the site!" In a moment of sociopathic corporate malfeasance, Tokyo Electric decided to evacuate everyone and desert the fiasco, leaving the region and perhaps the nation to a *que sera sera* fate. "If they withdrew, six reactors and seven fuel pools would be abandoned," the prime minister said. "Everything would melt down. Radiation tens of times worse than Chernobyl would be scattered. . . . I could not let [an evacuation] happen. It just wasn't an option."

Another government official called site manager Yoshida, to see if he agreed with the boss. Yoshida did not: "We can still hold on, but we need weapons, like a high-pressure water pump." Kan told the TEPCO president that his corporation had a moral obligation to try everything it could to end the crisis, to fight this battle with every weapon and every strength. The PM insisted that he was ready to give up his own life if need be.

The decision was made that over 750 plant employees would evacuate, with 50 remaining. Kan announced to the people of Japan, "They are ready to die." These became the legendary Fukushima 50, treated as nuclear samurai, restoring honor to their nation through their selfless, even suicidal, acts, men who to this day deserve the world's thanks for keeping a horrible tragedy from turning into a global disaster. Even in Japan, though, they are now wholly forgotten.

The fifty were actually part of a group that eventually numbered four hundred who cycled in and out of the plant to avoid being overexposed. A fair percentage were day jobbers from the Tokyo and Osaka slums, and a number were Japanese gangsters—yakuza. In December 2011, journalist Tomohiko Suzuki revealed the extensive relationship between Japan's "nuclear mafia" of utilities and government agencies with the real Japanese mafia. The last refuge for members of the "dark empire" is the nuclear industry, as one explained: "When a man has to survive doing something, it's the nuclear industry; for a woman, it's the sex industry." One midlevel yakuza executive was quick to insist, "The accident isn't our fault. It's TEPCO's fault. We've always been a necessary evil in the work process. In fact, if some of our men hadn't stayed to fight the meltdown, the situation would have been much worse. TEPCO employees and the Nuclear and Industrial Safety Agency inspectors mostly fled; we stood our ground."

In the wake of 3.11, corporate and government leaders proclaimed, in that time-honored fashion, that they were battling "acts of God." But both the Japanese government and Tokyo Electric knew perfectly well that putting emergency generators in basements was a poor strategy on an island with a history of major earthquakes.

In December 1999, France's Blayais nuclear power plant was flooded and its power supply destroyed. This triggered a reassessment of plant design in Europe, and a reworking of emergency electrics. TEPCO and NISA knew this history, but did not apply that lesson to themselves. In 2004, University of Tokyo seismologist Kunihiko Shimazaki warned that Japan's AEC—the Nuclear and Industrial Safety Agency—and Tokyo Electric had grossly underestimated the risk of tsunamis at Fukushima. Regulators on the panel immediately dismissed these comments as speculative and categorized them as "pending further research." In 2008, TEPCO ran a computer simulation predicting a fifteen-meter tsunami wave could breach Daiichi's sea-storm defenses and render havoc. TEPCO executives decided their own computer's predictions were fanciful and ignored them. On June 24, 2009, another pair of seismologists appeared before a government committee to explain that Daiichi could be destroyed by tsunami. Even an employee remembered thinking, "I always wondered why you would build a nuclear site this size in an earthquake zone right on the ocean."

This close association between government and industry in Japan is no coincidence. Called a "culture of complicity" by outsiders, *amakudari*—"descent from heaven"—means retired government workers are routinely given senior positions at the companies they once regulated, and vice versa. After winning a seat in parliament, one Tokyo Electric vice president demanded the nation's textbooks delete mentions of the antinuclear protest movements in Europe and the United States.

Even after America's AEC was split into the Nuclear Regulatory Commission and the Department of Energy to repair just this kind of inbreeding, the relationship between federal agencies and private industry in the United States is nearly as cozy as it is in Japan. "In the year 2002, the [NRC's] reactor oversight process gave the Davis-Besse plant [directly south of Detroit] the highest marks possible, basically straight A's, even though it was then discovered to have come closest to an accident since the Three Mile Island accident in 1979," nuclear expert David Lochbaum said. "So, anytime a system can't distinguish the best from the worst, there's still some work left to be done on it." A 2011 report from the Union of Concerned Scientists detailed fourteen "near-miss" safety or security incidents from the prior two years, noting a number of these where both the US utility and the American government knew about safety violations and did nothing.

Like Marie Curie using her husband Pierre's quivering gold leaf inside a vacuum bulb to sense the presence of radiance, a plant disaster means that no one can see, directly, what is going on. The trouble can only be deduced remotely from various instruments, controllers, alarms, and dosimeters. Since the quake and the tsunami had destroyed nearly all of Fukushima's monitoring equipment, in many ways its employees were now deaf and blind. Since all of the government nuclear agency's on-site inspectors had fled, no one was left who could evaluate the toxicity and advise the workers on protective gear or time limitations in different areas of the site. So when it came to exposure, they were essentially naked.

Reactor #2's control room was, by March 15, so infectious—a thousand times normal, with 170 people exposed and 22 with symptoms of radiation sickness—that no one could work in it for any length of time. Under the current Japanese regulations covering nuclear employees' radiation-exposure limits, Daiichi was about to run out of staff. Luckily, the country's Ministry of Health amended the limit from 100 millisieverts, over five years, to 250. In America, that limit is 50.

Each of the Fukushima 50 knew that if their dosimeters registered more than 50 millisieverts over a year, they would be banned from working in the nuclear industry, so they took off their monitors to avoid being made unemployable. Others took off their masks in the radiation-drenched buildings to light up smokes. Otherwise, they worked in blue protective suits and respirators at grueling schedules. Their breakfast was emergency crackers and vegetable juice; dinner, instant rice and a can of mackerel; no lunch. In two days they ran out of protective bootees, and had to make do with plastic bags and masking tape. Since there was no plumbing, there was no way to wash; only hands could be cleaned with an alcohol spray. They couldn't contact their families since phone lines and cell towers were down. Besides covering the plant in wreckage, the tsunami dropped hundreds of fish that died all over the grounds, and hungry seabirds were swooping in to feed. The workers slept wherever they could find room in a "clean" building with their assigned blanket. TEPCO employee Emiko Ueno: "My town is gone. My parents are still missing. I still cannot get in the area because of the evacuation order. I still have to work in such a mental state. This is my limit."

The continuing explosions; the fires erupting everywhere; the tsunami-scattered debris; the odor of dead sea life, of burning fuel rods, of overworked and fearful and unwashed men. In all the Cold War fantasies about how the world will end in nuclear holocaust, no one imagined the details of a power plant run amok, an industrial utility collapsing in atomic chaos. But to those in the middle of Daiichi in its worst hours, it appeared that these were the Last Days. "In the control room, people were saying we were finished," one employee remembered. "They were saying it quietly, but they were saying it. We felt we had to flee. This was the end."

As TEPCO employees work to this day under strict nondisclosure contracts, most of the public information about the Fukushima 50 comes from outsiders caught up in the crisis. Koichi Nakagawa was a subcontractor doing routine maintenance and inspections at the plant on March 11 and could easily have been evacuated at the very beginning; when Japanese TV broadcast the first explosion, Nakagawa's boss called him and demanded, "What the hell are you doing? Get out of the place!" But Nakagawa stayed, thinking he might lose his job in a cutback: "I couldn't say no to [TEPCO] because they'd give my company work in the future." While almost every Tokyo Electric employee walked the grounds in protective gear, Nakagawa was still in his standard-issue cotton uniform. After the explosion in Reactor #2, everyone started talking about how all six reactors might blow up, and that all of them would die. Nakagawa began wondering if he could escape without being noticed: "My mind went blank. I thought I was toast. I couldn't take it anymore."

Nakagawa slipped out of the plant and drove home. But the town where he lived was deserted, and he didn't have enough gas to drive to where everyone had been evacuated. A little food was left in the house, and there was electricity, at least. Then his boss called, saying he'd pay ten times Nakagawa's regular salary if he returned to Daiichi. He called his TEPCO friends, and they agreed to come and pick him up: "I thought about changing jobs, but it isn't easy. We have no other industry but the power plant, and I have a family to feed. I had no other choice."

On the night of March 15, firefighters, soldiers, and TEPCO employees met at J-Village, a sports training facility twelve miles south of Daiichi. Since the power was out there as well, they arranged a dozen or so trucks into a circle and turned on the headlights. The meeting did not go well. Everyone had a different and conflicting plan on how to solve the disaster. Phone service had still not been restored, so TEPCO executives in Tokyo could not advise on or consent to any decision anyway. One TEPCO manager admitted, "There were so many ideas, the meeting turned into a panic. There were serious arguments between the various sections about whether to go, how to use electrical lines, which facilities to use, and so on."

Nuclear Regulatory Commission chair Gregory Jaczko warned the US Congress on March 16 that little or no water remained in #4's waste-suppression pool. TEPCO spokesman Hajime Motojuku insisted, "We can't get inside to check, but we've been carefully watching the building's environs, and there has not been any particular problem." In fact, if a worker was sent in to repair the damaged plumbing and spent more than sixteen seconds standing next to the pool, he would be dosed enough to die. The Japanese did admit that the containment shells of Reactors 2 and 3 had cracked, releasing radioactive vapors. TEPCO then made an especially unnerving announcement that it had found radiation at 10 million times normal, then retracted that, saying it was actually a mere one hundred thousand times normal. Global security scientist Edwin Lyman: "If you can't make accurate measurements, if you ignore alarms . . . it's a sign of chaos."

Later that day, the NRC's computer program RASCAL finished diagramming the Japanese plumes of radiation. RASCAL said that the danger zone had a fifty-mile radius—about the size of Chicago—terrible news, but if there were no more explosions and if the core temperatures could be controlled, at least Tokyo would not have to be evacuated. The US Energy Department's team of atomic forensics deduced that 70 percent of one core and 33 percent of a second had melted.

Workers trying to repair Unit 3 were found to have elevated radiation levels on March 25, which implied that its containment walls were leaking. As Unit 3 was an advanced design using MOX fuel—a reprocessed combination of uranium and plutonium—an escaping plume would be even more toxic than that from a common core meltdown. The Tokyo police were finally able to bring in a water cannon—normally used against rioters—and sprayed Unit 3 down with thirty tons of seawater.

That day, Japan recommended widening the "voluntary evacuation" zone from twelve to nineteen miles. Yet, twenty-five miles away from the plant, the UN's International Atomic Energy Agency found cesium-137 in amounts up to twice as high as the Soviet Union's Chernobyl cutoff point for judging land unfit for human habitation. Spinach and milk had been contaminated, as was, disturbingly, the tap water of Tokyo, testing positive for radioactive iodine. Seawater collected three hundred yards away from Daiichi had iodine-131 at 3,355 and 4,385 times the legal limit. From its history with tests in Bikini, however, the Americans knew that the ocean has a remarkable ability to absorb and disperse radioactive contaminants. Radioecologist F. Ward Whicker: "The most likely effects would be reductions in reproductive potential of local fishes."

Inspectors finally uncovered the source of Unit 2's leaks on April 1—a twenty-centimeter crack in its maintenance pit, which had ten thousand times the legal concentration of iodine-131. They tried to plug it with sawdust and polymeric gel (the secret ingredient of disposable diapers). This effort failed.

That same day, the Tachikawa squad, a specially trained unit of the Tokyo fire department, finally reached the plant after driving their equipment across seventy miles of earthquake- and tsunami-destroyed roadways. The two requirements for being one of the thirty-two members of this team were to be over the age of forty and not hoping to sire any children.

The squad needed to set up a pumping truck next to the ocean. These pumps were designed to quench severe jet-fuel-ignited aviation fires. Squad deputy commander Kenichi Kunisawa: "When I heard that the operation was to spray water to the reactors, I felt that we were the right people to do it. We are good at spraying water. . . . We firefighters, even if we feel fear, we never show it to the others. We have high morale and were resolved to accomplish the duty. We only think how to accomplish our duty—even when we are afraid to do it. It would be a lie if we say we did not feel the fear. I admit we had concerns about [Reactor #3], but we'd been trained to do this. That's how we became members of a special unit. We're proud of being special guys, like the Green Berets." The team had to lay eight hundred yards of hose, connect the system, and begin to spray down the containment pools with seawater. To keep from being dangerously contaminated, they had a mere hour to accomplish all of this. And they did it.

Kunisawa: "We had to decide so many things at the site. Things were changing all the time. There was a lot of commotion and people were yelling at each other, to be heard. But the older members over forty took charge and things went okay. We were basically cool. Our minds were on the job and we didn't really feel fear. We didn't have time for that."

Fixing the pools dropped the radiation enough so that TEPCO workers could return. Miles of pipe were laid connecting the sea to each out-of-control reactor. After twenty-two days of battle, the worst was over.

On April 4, TEPCO started dumping eleven thousand tons of water one hundred times as radioactive as the legal limit into the Pacific. Edano: "Unfortunately, the water contains a certain amount of radiation. This is an unavoidable measure to prevent even higher amounts of radiation from reaching the sea." Daiichi detected new levels of hydrogen accumulating in Unit 1 on April 5 and decided to inject nitrogen into its containment shell to keep any combustible oxygen from entering. It also plugged the maintenance pit leak and tested the ocean three hundred yards from the plant, which had previously registered radiation levels 4,000 times the legal limit. Those levels had fallen to 280.

Accused of mismanaging the crisis, Prime Minister Kan was forced to resign.

By August 25, the Ministry of Education, Culture, Sports, Science, and Technology announced that the cesium-137 contaminating the country from Daiichi was equal to 168 Hiroshima bombs and admitted that thirty-four of the plant's neighboring districts were more radioactive than the threshold set by the Soviets for Chernobyl, making them uninhabitable for humans. That same month, TEPCO started building a sixty-foot underground wall to keep more radioactive water from seeping out of the ground and into the Pacific.

By December 6, traces of cesium appeared in Japanese milk, beef, vegetables, fish, rice, spinach, tea leaves, and baby formula. Manufacturer Meiji insisted that babies could drink its "formula every day without any effect on their health." The government found contaminated rice over the safety limit grown thirty-five miles away, and beef raised more than forty miles away from Daiichi, and kept both from being sold. But when used cars were tested as too radioactive to be exported to Russia, Australia, and North and South America, some Japanese dealers reregistered them with new plates and sold them, illegally, to their countrymen.

A group of concerned Tokyo citizens paid for soil sample testing of their own neighborhoods and uncovered, in the capital of Japan 150 miles away from the plant, Chernobyl-level readings of radioactive cesium. Of 132 areas sampled, 22 were over the Soviet limit for contamination. Japan's press ignored this story, instead widely broadcasting another about an adorable rabbit baby born near Fukushima that had no ears. Was it a genetic mutation caused by the accident? Would the prefecture be run amok with earless rabbits everywhere, like radiant mascots? It was so cute!

Now Japan faced a choice. She could follow the Soviet lead, cleaning up as much as feasible, then keeping humans from living in her own Zone of Alienation for decades . . . or she could rehabilitate the New Jersey–sized prefecture and return its citizens to their homes. Japan's Diet chose this latter course, at a cost of a trillion yen, but after a year of effort, the results are not promising. A day laborer said, "We are all amateurs. Nobody really knows how to clean up radiation," and decontamination worker Takeshi Nomura insisted, "They tell us decontamination would take three years. It's utter bullshit. It's going to take longer than that before people are going to feel safe enough to come back. . . . That day may never come."

The rehabilitation began by hosing down every building, removing all the leaves from the trees, and digging up the top two inches of earth. Fukushima now has a billion cubic feet of contaminated detritus sitting around in garbage bags like a tidy regional dump—enough to fill thirty-three domed football stadiums. The refuse needs to be landfilled, but no community will accept it, including Fukushima itself.

When it comes to convincing Fortunate Island's residents to move back home, imagine the amount of dust your house reveals every time you clean it, then imagine living in a community where that dust might, over time, give your kids cancer. The chances are remote, but still. After her eight-year-old son, Yuma, was discovered to have been contaminated with cesium, Mitsue Ikeda said she would never return to Fukushima: "It's too dangerous. How are we supposed to live, by wearing face masks all the time?" With Japan's cratered economy, though, Kunikazu Takahashi thought he had no choice but to continue his job as a nuclear technician at Daini, a mere six miles from Daiichi: "They called several days ago, asking for me. I have to go back." When he was asked about contamination, he said, "I try not to think about it." Those trying to start a new life elsewhere have found themselves stigmatized. Being from Fukushima turns out to be as much of a social disgrace as being from Hiroshima or Nagasaki used to be.

On May 9, 2012, the Japanese government announced it was ready to spend 2.4 trillion yen ($30.1 billion) to pay compensation to Fukushima victims and 1 trillion yen ($12.5 billion) on a bailout to temporarily nationalize TEPCO—supplier of all Tokyo's electricity and one-third of the nation's as a whole—a "too big to fail" move of public largesse. All of the company's current directors would resign. Regarding the dark history of plant workers' ties to criminal gangs, a Japanese senator explained, "Nuclear energy shouldn't be in the hands of the yakuza. They're gamblers, and an intelligent person doesn't want them to have atomic dice to play with." When TEPCO has regained enough credibility to sell corporate bonds, the nationalization will end.

On May 28, former prime minister Naoto Kan appeared at a Diet inquiry and testified that the country should forgo nuclear power since, if Daiichi had not been brought under control, it would have meant evacuating the whole of Tokyo, paralyzing the government—"a collapse of the nation's ability to function." Kan accused the nuclear establishment of "showing no remorse" post-3.11 and summed up, "Gorbachev said in his memoirs that the Chernobyl accident exposed the sicknesses of the Soviet system. The Fukushima accident did the same for Japan." Novelist Haruki Murakami: "This is a historic experience for us Japanese: our second massive nuclear disaster. But this time no one dropped a bomb on us. We set the stage, we committed the crime with our own hands, we are destroying our own lands, and we are destroying our own lives."

On July 19, 2011, TEPCO announced it had stabilized the reactors' temperatures, and on December 15, Kan's successor as prime minister, Yoshihiko Noda, declared the crisis over, with Daiichi in "a state of cold shutdown." This was not true at the time, or even as this book was being written. One engineer working at the plant a year later admitted, "The coolant water is keeping the reactor temperatures at a certain level, but that's not even near the goal [of a cold shutdown]. The fact is, we still don't know what's going on inside the reactors."

In the end, for all the world's obsession with Japan's nuclear disaster, Fukushima was almost wholly insignificant compared to the 3.11 tragedy for the nation as a whole. The country's National Police Agency official tally as of December 12, 2012, itemized the horror:

Dead	15,878
Injured	6,126
Missing	2,713
Damages	over $300 billion

As of July 2013, thirty-one countries burn 432 nuclear power reactors generating 13.5 percent of the world's electricity. In the wake of Fukushima, Germany and Switzerland began phasing out nuclear power within their borders, joining Australia, Austria, Denmark, Greece, Ireland, Israel, Italy, Latvia, Liechtenstein, Luxembourg, Malaysia, Malta, New Zealand, Norway, and Portugal in foreswearing nuclear power. In the opposing camp stand Belgium, Bulgaria, the Czech Republic, Hungary, Slovakia, Slovenia, South Korea, Sweden, Switzerland, and Ukraine, each of which gets more than a third of its energy from reactors.

Many Germans are profoundly antinuclear and supportive of the country's turn to wind and solar, but unfortunately, we are currently mired in a fossil-fuel world, and turning to green energy is arduous and expensive. Coal, gas, and oil are so much cheaper than every other source that they have remained the most used energy forms for a century, even though two pounds of coal creates three kilowatt-hours of electricity, while the same amount of oil generates four, and two pounds of uranium can create 7 million kilowatt-hours, with none of fossil fuel's air pollutants or greenhouse gases. Greenpeace cofounder Patrick Moore, Gaia theorist James Lovelock, NASA climate scientist James Hansen, and Earth Institute director Jeffrey Sachs, among many others in the environmental movement, are convinced that we need nuclear power to reduce carbon dioxide emissions, a greenhouse gas.

Like most of the world, "Americans have never met a hydrocarbon they didn't like," journalist Elizabeth Kolbert said. "Oil, natural gas, liquefied natural gas, tar-sands oil, coal-bed methane, and coal, which is, mostly, carbon—the country loves them all, not wisely, but too well. To the extent that the United States has an energy policy, it is perhaps best summed up as: if you've got it, burn it." For of all the ways we have right now of producing electricity, nuclear is in many ways the least of our worries, so much so that perhaps antinuclear activists should refocus on the far greater menace of coal. Thanks to coal, the skies of China are annually filled with 3.2 billion tons of carbon dioxide, and 26 million tons of sulfur dioxide. Political commentator William Saletan: "The sole fatal nuclear power accident of the last forty years, Chernobyl, directly killed thirty-one people. By comparison, Switzerland's Paul Scherrer Institute calculates that from 1969 to 2000, more than twenty thousand people died in severe accidents in the oil supply chain. More than fifteen thousand people died in severe accidents in the coal supply chain—eleven thousand in China alone. The rate of direct fatalities per unit of energy production is eighteen times worse for oil than it is for nuclear power. Even if you count all the deaths plausibly related to Chernobyl—nine thousand to thirty-three thousand over a seventy-year period—that number is dwarfed by the death rate from burning fossil fuels. The Organisation for Economic Co-operation and Development's 2008 *Environmental Outlook* calculates that fine-particle outdoor air pollution caused nearly 1 million premature deaths in the year 2000, and 30 percent of this was energy-related. You'd need five hundred Chernobyls to match that level of annual carnage."

So while France's fifty-eight reactors generate over 80 percent of its electrical power, India plans to build at least five more burners, Vietnam wants at least eight, and China will build fifty. By 2025, Southeast Asia is expected to go from zero to twenty-nine atomic plants. Of the sixty nuclear power plants being built around the world now, fifteen are engineered and constructed by Russia's state-owned nuclear company, the most of any nation.

Would these governments—notably top-down operations such as China and Vietnam—be so enthusiastic about nuclear power if they knew Mikhail Gorbachev's opinion of Chernobyl? That it was, in his words, "perhaps the real cause of the collapse of the Soviet Union . . . [a] turning point [that] opened the possibility of much greater freedom of expression, to the point that the system as we knew it could no longer continue."

Journalist Mark Joseph Stern: "By 1987, the year following Chernobyl, glasnost had taken hold of Soviet society, with sudden openness dominating the press and the public forum. Outrage over the catastrophe began to spread among even loyal citizens who had never questioned the infallibility of their government. This opened the door to comparison with the West, a toxic line of thought in this famously closed society. Soviets had been told for decades they were the best in the world—at everything. Through the mid-1980s, they still believed they were a major superpower, facing only the United States as serious competition. When information about Chernobyl and the public health crisis leaked, though, Soviet citizens realized that their government and industries were startlingly incompetent. Before the explosion, most Soviets were not discontented dissidents; they believed in the Soviet system, forgave its flaws, and hoped for a better future within its confines. But after Chernobyl, the system seemed potentially unredeemable—and actively dangerous. In the early days of glasnost, stories of Stalin's mass murders decades earlier slowly bubbled to the fore, but those generally receded, so far removed were they from everyday life. After Chernobyl, though, every citizen's safety was at stake."

There is one solution: engineers creating technological breakthroughs to end the ever-present threat of atomic and political meltdown. China is testing a "pebble bed" reactor design, where reactor fuel comes not in big rods but as four hundred thousand billiard ball "pebbles," coated in graphite and cooled by helium. If a pebble bed reactor is SCRAMed, the graphite shell quiets the fission without threat of infinite afterheat and meltdown, while the helium is inert, so if it has to be vented, it won't be radioactive.

Another breakthrough may come with the thorium breeder reactor, which through an ingenious weaving of half-lives creates its own fuel. The pile's uranium-233 fissions throw neutrons into the surrounding layer of thorium, which becomes thorium-233. The 233 decays, becoming protactinium-233. The protactinium decays, becoming uranium-233, and the cycle repeats. Unfortunately, the history of breeder reactors is not good; all four of the AEC's test breeders of the 1960s were failures. But a number of physicists and engineers insist that the thorium design will solve all of those problems, with less maintenance, and less waste. Microsoft billionaires Bill Gates and Nathan Myhrvold are, meanwhile, investing in a "traveling wave reactor" process, a type of breeder reactor fueled by ordinary uranium instead of enriched, which, if it works, won't require massive Oak Ridge–like industrial plants isolating near-weapons-grade isotopes.

Breeder reactors are so interesting that Eagle Scout David Hahn decided that, for his 1994 Atomic Energy merit badge, he should build one in his parents' suburban Detroit potting shed. "His dream in life was to collect a sample of every element on the periodic table," Hahn's high school physics teacher remembered. "I don't know about you, but my dream at that age was to buy a car." On August 31 at 2:40 a.m., Clinton Township police were looking for a boy seen stealing tires from cars when they came across David in his Pontiac, acting suspicious. In his trunk they found acids, fireworks, antique clockfaces, rocks, lantern mantles, a box of dismantled smoke detectors, assorted chemicals, fifty cubes of white powder wrapped in foil, and a toolbox sealed with duct tape and a padlock. To build his test reactor, David had amassed americium-241 from smoke detectors, radium-226 from glow-in-the-dark clocks, thorium-232 from the mantles of kerosene lamps, and uraniums-238 and -235 from pitchblende ore. The cops called in the State Police Bomb Squad and radiologists from the Department of Public Health, who found so much radioactivity coming from the Pontiac's trunk that they had to invoke the Federal Radiological Emergency Response Plan, bringing a coterie of agents from the DOE, EPA, FBI, and NRC to Golf Manor.

On June 26, 1995, the Hahn backyard was declared so toxic it required an EPA Superfund cleanup squad. David's story made him a legend and led to a book, *The Radioactive Boy Scout*. One of that book's biggest fans was eleven-year-old Taylor Wilson, who read the entire thing to himself out loud. "Know what?" Taylor told his parents. "The things that kid was trying to do, I'm pretty sure I can actually do them." Taylor then spent much of his allowance that year on a radioactive collectible—a Fiesta dinnerware set too hot to eat from—and began experimenting. The family worried they might be facing a suburban Chernobyl, like the Hahns. "The explosions in the backyard were getting to be a bit much," Taylor's half sister Ashlee remembered.

Taylor then decided to try for the ultimate nuclear dream: Starlight on Earth. If Eisenhower's Atoms for Peace helped transform the Hiroshima fission bomb into nuclear power, what would happen if you tried doing the same with the Teller-Ulam fusion of Mike?

A sun lies inside the sun, the core 10 percent where all solar power originates. Every second, that core, with a force equal to 96 million thermonuclear bombs, transmutes 4 million tons of matter into 385 million million million million watts—the light and the heat that sustains life. In 1967, Hans Bethe won his Nobel for working this out mathematically with "Energy Production in Stars," and for over fifty years, scientists have been trying to create starlight on earth, to engineer a controlled fusion nuclear reactor.

The teenaged Taylor Wilson was struck by the byzantine problem of radioactive oncology. Medical isotopes used for both diagnosing and treating various cancers have to be short-lived, to kill the bad cells without inflicting too much damage on the good, but this makes their distribution time-sensitive and tremendously costly. Instead of shipping isotopes to patients by private jet, Taylor thought, what if a reactor could be made small enough and safe enough to produce isotopes right in the hospital?

With the help of others legally old enough to drink beer, Taylor at the age of fourteen built a reactor bombarding atoms into each other in a shimmering 500-million-degree plasma (not solid, not liquid, and not gas, but a gaslike state that can be magnetically charged into filaments and beams, best known as the inner glow of neon bulbs). The commercial versions cost less than $100,000 and can be rolled right into the patient's room.

But this is one of the few happy endings in a science that has been promising results since 1955 and achieving as much as Ed Teller's Strategic Defense Initiative. Since 1993, Lawrence Livermore has spent over $5 billion on the National Ignition Facility, a stadium-size laser designed to generate power from fusion. As this book was going to press, the NIF announced a breakthrough: it had finally created a fusion reaction that generated more power than it took to initiate. But it is still a long road from that step to the reaction creating enough energy to sustain itself into a source of fusion power. Scientists in this business like to joke that "fusion is always twenty years away," since that's what they've been saying for the industry's entire lifetime. Before trying to ignite fusion with lasers, physicists all over the world tried goliath magnets. At their 1985 Geneva summit, Reagan and Gorbachev agreed to merge their fusion energy R&D programs with France's and Japan's to create the International Thermonuclear Experimental Reactor (ITER), which, thirty years later, is expected to take another $30 billion and twenty years to work. If it ever does.

When we hear that a nuclear plant has collapsed in catastrophic meltdown, we can't help but imagine a *China Syndrome*, with a whole population infested with tumors, a vast territory rendered into nuclear desert, and offspring afflicted with never-before-seen birth defects. Beyond the heroic martyrdoms of power plant and emergency response workers, though, what actually happens after a nuclear plant disaster is so minor compared to our mythic fantasies that it is almost impossible to understand.

Created after the bombing of Nagasaki to study radiation's long-term biological effects, the United Nations Scientific Committee on the Effects of Atomic Radiation (UNSCEAR) worked for twenty-five years with the Chernobyl Forum (a joint effort of such UN agencies as the International Atomic Energy Agency, the World Bank, the World Health Organization, and the governments of Ukraine, Belarus, and Russia) to catalog the Lenin Station's aftereffects. They concluded that fifty-seven people died during the accident itself, including twenty-eight emergency workers, and that from

1986 to 2002, about 6,848 neighboring children were diagnosed with thyroid cancer from drinking the radioactive-iodine milk of tainted cows. This latter tragedy could easily have been prevented with a working public health service in a normal functioning state—Chernobyl fallout has been detected in dairy products as far away as Oak Ridge, Tennessee—but when it was discovered, the USSR was collapsing. Eighteen of those children have died.

Beyond that, there is no clear medical evidence of Chernobyl's adverse impact on human beings, either in cancer rates or mortality rates or nonmalignant disorders . . . and Chernobyl was an accident far worse than nearly anything that could happen anywhere else in the world, as the Soviets did not use a significant containment dome, and their atomic fire, raging for two weeks, covered almost the whole of Europe in a radioactive cloud equivalent to four hundred Hiroshimas. The worst nuclear disaster in human history, then, turned out to be far less catastrophic than such other industrial horrors as the August 8, 1975, Banqiao Dam failure in China, which killed 171,000, or the December 2, 1984, Union Carbide pesticide plant leak outside Bhopal, India, which killed 3,787 and injured 558,125. UNSCEAR's report concluded, "There has been no persuasive evidence of any other health effect in the general population that can be attributed to radiation exposure," and anyone living there now "need not live in fear of serious health consequences from the Chernobyl accident."

A number of people who have closely studied this tragedy are convinced that UNSCEAR and the Chernobyl Forum have grossly underestimated the effects of the disaster, but they have nowhere near the underlying research data to prove it, and considering the social chaos brought by the fall of the Soviet Union, it may be impossible to ever get it. In one example, physicist Bernard Cohen estimated, "The sum of exposures [from Chernobyl] to people all over the world will eventually, after about fifty years, reach 60 billion millirems, enough to cause about sixteen thousand deaths." Even if this were true, every year in the United States, around sixteen thousand people die just from the air pollution of coal-burning power plants.

So what will become of the Japanese people dosed by Fukushima's radioactive clouds? Beyond the UN's Chernobyl efforts, concrete scientific data on the health of human beings exposed to radiation is thin. The most solid and extensive studies arose from the *hibakusha*, the two hundred thousand survivors of Hiroshima and Nagasaki, whose health has been followed for over sixty years by a joint US-Japan effort, the Radiation Effects Research Foundation.

The expectations at the start of that study were the expectations you might have—that the *hibakusha* would be grossly overrun with tumors, that their genes would mutate, that their descendants might be deformed. Instead, the results were startling. The foundation's Evan Douple said that they had predicted cancer would be widespread but, in fact, "the risk of cancer is quite low, lower than what the public might expect." Ninety-eight of 120,000 in one study group had died of leukemia attributable to radiation, while 850 of 100,000 in another had died from solid tumors—and these were not people who had to evacuate from a nuclear plant exhausting meltdown waste; these were people attacked with an atomic bomb.

Radiologist John Moulder analyzed the data from another control group: "Of those fifty thousand people, about five thousand of them developed cancer. Based on what we know of the rest of the Japanese population, you would have expected about forty-five hundred of them. So we have five thousand cancers over fifty years where we would expect forty-five hundred. So we assume that those extra five hundred cancers were induced by the radiation." Five hundred cancers out of a fifty thousand population means a rate of 1 percent.

Additionally, despite the remarkable findings with fruit flies in the 1950s, there was no increase in inherited mutations. *Hibakusha* children, grandchildren, and great-grandchildren have all turned out just fine. As of 2011—sixty-six years later—40 percent of the Hiroshima and Nagasaki survivors were still alive.

For Fukushima, then, the consensus is a 1 percent increase in cancer for TEPCO employees who worked at the site, and an undetectable increase for Daiichi's citizen neighbors. Statistically, the rates for death and injury for real estate agents and stockbrokers remain higher than those for atomic plant workers. Even Daiichi's horrific ocean flushing of toxic runoff, as sorrowful as it was, was minor compared to the Cold War radioactive waste dumped in thousands of now-rusting drums between 1946 and 1994. The Soviet Union alone is responsible for at least 2.5 million curies of ocean dumping, while America since 1946 has thrown 47,500 fifty-five-gallon drums into the waters of the Farallon Islands off San Francisco. In one of the few studied incidents, some of these have burst open, infecting local sponges with plutonium.

If this is the truth of nuclear plant meltdowns, then why, for everyone outside Japan, did the drama of Fukushima Daiichi so completely overshadow the incomprehensible tragedy of 3.11?

Back in 1945, the world was introduced to nuclear power through images and news reports from Hiroshima and Nagasaki, and no matter how distant the technology inside your local utility's containment dome is from Los Alamos and Edward Teller's blackboard, those resonant images live on in our memories. Physicist James Mahaffey: "As they say in nuclear engineering circles, if the first use of gasoline had been to make napalm, we'd all be driving electric cars now." Nuclear's invisible powers, mythic history, and scientific mysteries add up to inspire in the general public a belief in magic—black magic. Reactors are exotic and strange, the stuff of fantasy and science fiction, mainstays of popular culture—which radiologist Fred Mettler thinks has bequeathed the public with "radiation biology lessons." When it comes to the science of radiance, Mettler said, "Children in the United States are inundated with all kinds of nonsense on television from the time they are six months old." Today, especially in the United States, *nuclear* is synonymous with evil. From Meryl Streep, scrubbed raw and naked and then murdered in *Silkwood*, to supernaturally incompetent Homer working for the villainous owner of a nuclear power plant on *The Simpsons*, atomic power is ominous and ever threatening.

It's hard to know which would horrify the nuclear science laureates profiled in this book more: their degraded reputations, or that their hard-earned science had been transformed into a collection of myths.

17

Under the Thrall of a Two-Faced God

THE history of the nuclear dot on the letter *i* in a newspaper story read by a passenger on the atomic bus seems to endlessly repeat the two-faced miracles of X-rays and radium. An epoch that began in a poor man's version of Frankenstein's lab trying to explain what voltage does in vacuum tubes is ending with the real-world civil and military Frankensteins of utility meltdowns and nuclear arsenals. As the age of radiance draws to its close, its history is one of paradox, ambiguity, absurdity, and blessings with menace on a global scale.

When I went online to find analogies for "two-faced miracle," all I got was a mutant kitten. Then I remembered Janus, the lord of beginnings and transitions, gates and doors, Pollyannas and Cassandras, the past and the future, of harvests, marriages, births, change, and of time, a god so important Rome's eleventh month was named in his honor: January. Janus in his two-faced glory reveals the beauty and reality of all our mixed feelings about our up-and-down atomic legacy. To use one analogy unknown by Google, when great writers engineer complicated, engaging characters, their men and women in various ways are evocative matrices of weak and strong, brave and scared, sacred and profane, admirable and despicable—just as we all are in real life. Perhaps the same two-sided beautiful Janus model needs to inform our thinking when the subject is nuclear, similar to Niels Bohr's nostrum about big truths: "You can recognize a small truth because its opposite is a falsehood. The opposite of a great truth is another truth."

At the dawn of the Atomic Age, physicists were elevated in the public mind to the role of secular priests, their study of subatomic particles appearing to lead them, simultaneously, to spiritual and moral truths. Notably,

Albert Einstein and Marie Curie became heroes to millions around the world, role models for a new era. Along with many civilians, E. O. Lawrence, Enrico Fermi, and both generations of Curies believed that their scientific discoveries would inexorably lead to benefits for all humankind. Instead, that sweet hope, along with their current reputations, has been battered by a history of thermonuclear-winter terrors and run-amok power plants. Following the dropping of the atomic bombs on Hiroshima and Nagasaki, the great pacifist Albert Einstein, who had nothing to do with fission beyond his letters to Roosevelt, was depicted on the cover of *Time* magazine against a mushroom cloud and his most beloved equation, $E = mc^2$, while Mme. Curie is today known for her breakthroughs as a woman, not as a scientist. The public's distaste has grown so pronounced that European physicists created a PR organization, Public Awareness of Nuclear Science, to fix their image problem. Today's prejudice against all things atomic is as naive as was the 1920s radium euphoria and the 1950s techno-utopians predicting nuclear-derived electricity as "too cheap to meter."

In the case of nuclear power, most of the world has come to a decision. What Three Mile Island, Chernobyl, and Fukushima all have in common is that the public relations disaster was far worse than the pollution's health effects. In each incident and each nation postdisaster, one common casualty was the truth, at the hands of the nuclear industry; the local and national governments; and even the antinuclear activists. A founder of Physicians for Social Responsibility, Helen Caldicott, wrote in the *New York Times* that in the aftermath of Chernobyl "almost one million people have already perished from cancer and other diseases. The high doses of radiation caused so many miscarriages that we will never know the number of genetically damaged fetuses that did not come to term. (And both Belarus and Ukraine have group homes full of deformed children.)" Based on questionable and roundly criticized papers coming out of Eastern Europe, her assertions seem as threat inflating as anything Teller told Reagan.

Global Fission author Jim Falk: "People have come not only to distrust the safety of the technology but also the authority of those who have assured them so confidently that nuclear power is safe. In this sense people distrust the entire nuclear enterprise—not only its technology, but the public and private organizations, the political parties, and those often prestigious scientists who advocate and assist in the development of nuclear power." With nuclear power calamities, we have learned through harrowing experience that you can't trust the government, you can't trust the industry, and you can't trust the critics . . . or even your own fears. One psychiatrist, Robert

DuPont, has spent years studying radiophobia: "On all four counts, nuclear power generates fear. It's a cataclysmic accident that people are concerned about. It's controlled by 'them,' the utilities or the government or the scientists or whoever it is that is perceived as the bad guys. It's unfamiliar to most people, and most people feel they don't really need nuclear power, that they can get their power from coal or oil or windmills or some other basis."

Physicist James Mahaffey: "Just the naked word *radiation* is enough to make us uncomfortable. . . . You can just be standing there, feeling nothing unusual, while being killed by it, never mind being actively hit with the meltdown or bomb. A major component of the paradox of nuclear power is that far more people die each year of radiation-induced disease from standing out in the sun than have ever died from the application of nuclear power." Probably only one other word is as loaded with freight as *nuclear*: *cancer*.

The atomic utility industry and its governmental affiliates have done such a poor job of both educating the general public and managing their crises that they will be driven out of business. Atomic utilities now require state-funded corporate welfare to build their plants, to insure them, and to nationalize them when disaster strikes. How many politicians can afford to bankroll reactors at every stage? How many parents want a burning radioactive pile anywhere near their young children? Unless some dramatic technological breakthroughs completely rework public opinion, not in your lifetime and probably not in your children's lifetime but eventually, nuclear power will become so insignificant that it will be essentially meaningless. Certain countries, such as France and parts of the developing world, will continue with nuclear power, and we need some reactors to make radioisotopes. But otherwise, as politics and as business, nuclear power has stopped making sense.

The same fate awaits nuclear arms, which like power plants won't vanish entirely but will fade into insignificance. Consider the facts: The United States and the USSR spent $5.5 trillion on 125,000 nuclear weapons for the Cold War, with America alone throwing away an average of $35 billion a year. Yet having the Bomb did nothing to help Russia with its troubles in Afghanistan, or America with its nightmares in Korea and Vietnam . . . or for that matter, Britain with Suez, Israel with the Arabs, or France with Algeria. A Soviet general said after the fall of the USSR, "Hundreds of billions were spent to counterbalance the mutual fear of a sudden nuclear strike when—as we now know—neither side ever conceived of such a strategy because it knew what horrors it would visit on both." Nearly every politician in charge of the codes—every president and premier since Eisenhower and Khrushchev—has admitted after leaving office that only a sociopath or a madman would launch atomic arms, even in retaliation for a nuclear first strike. The men who had their fingers on the nuclear button for all those decades never even came close to using what their $5.5 trillion had bought.

The great myth of nuclear arms is that they are different from conventional weapons in some magical way beyond their radioactive poisons. This was proved to be a fantasy at the very dawn of the Atomic Age, with Oppenheimer's lunchtime comment at Stimson's Interim Committee—he said the sole difference between atomic bombs and conventional explosives was that all living creatures within two-thirds of a mile would be irradiated, and that the look of the explosion, with its ten suns of light, shock wave, purple ionized air, boiling flames, and mushroom cloud was unforgettable. The fantasy that nuclear weapons are categorically different was also disproved through Curtis LeMay's firebombing campaign. On March 9, 1945, 334 Superfortress flights dropped two thousand tons of incendiary bombs, destroying over sixteen square miles of Tokyo . . . more than Hiroshima and Nagasaki combined . . . killing almost eighty-four thousand men, women, and children—more deaths than either atomic bomb dealt. In the aftermath, Hiroshima's atomic deaths were certainly no worse than Tokyo's fire. Besides the iconic multicolored mushroom cloud and the dissemination of fallout, there is no significant difference except for size. LeMay could've inflicted the same damage to Hiroshima as Little Boy with 210 conventional firebomb strikes, and to Nagasaki as Fat Man with 120. Even the physics is telling, as the energy force of radioactive bombs is 40 to 50 percent blast, 30 to 50 percent heat, and 10 percent radiation . . . meaning 90 percent the same as conventional weapons.

Nuclear policy adviser Linton Brooks: "Fifty years of telling ourselves that these things are different has sort of made them different. That is the mystique of nuclear weapons." That, and nothing more. The only distinction nuclear arms have as weapons rises from our fantasies and ignorance, from our self-created myths.

The costs of developing and maintaining nuclear warheads are ridiculously onerous, especially as they are purely symbolic and psychological. Little Boy and Fat Man may have been successful as proofs of concept, but turning their technology into usable weapons has proved impossible. How, beyond the original demonstration of their power, could atomic bombs be used as weapons in war? Oppenheimer again made this point: "It is a job that calls for a great deal of imagination to think what is the atom good for in war."

Seven decades later and we still don't have an answer, with Oppie's question only raised in magnitude by Teller's Super bombs. For if fission bombs can't be used on the battlefield, of what use is thermonuclear, besides threatening your neighbors or destroying the islands of the Pacific and the deserts of Kazakhstan and Nevada? Mike was eight times as powerful and its fireball thirty times the size of Hiroshima's, with a resulting mushroom cloud one hundred miles in width. Mike's successors were even more immense . . . and what could any military commander ever accomplish with such weapons? All of the theorizing from RAND, the Kremlin, and the Pentagon that went into Killing a Nation, Sunday Punch, Massive Retaliation, Flexible Response, Madman theory, the various SIOPs, the Triad Doctrine, Decapitation, Dead Hand . . . no one could answer this basic question. The only real strategy was Robert McNamara's Mutual Assured Destruction—which many believe absolutely true—and Bernard Brodie's point that "thus far the chief purpose of our military establishment has been to win wars. From now on its chief purpose must be to avert them," which both the Pentagon and the Kremlin successfully ignored.

The history of atomic weapons will in time fall in line with the history of another weapon judged equally immoral. In 1911, Fritz Haber was made the head of Kaiser Wilhelm Institute's chemistry division because of the epochal discovery that would bring him the Nobel Prize in chemistry—man-made ammonia, which as fertilizer became the foundation of modern agriculture. In 1914, the German high command begged for Haber's help in synthesizing sodium nitrate, a critical component in munitions that an English naval blockade was keeping from its German foe. Haber found a way to make gun cotton without that ingredient, and this was his turning point as a scientist, for he was no longer bettering mankind with his research. In 1915, the army asked Haber for a method of driving enemy troops out of their trenches. Fritz suggested chlorine vapors, but that could be defeated by breathing through a wet handkerchief. So, he turned to poison gas, and by 1916 he'd developed phosgene, which killed in seconds, and mustard, which when not fatal, rendered men blind and lung-scarred.

Even though across the whole of World War I less than 1 percent of battlefield deaths came from these gases, and it took over a ton to make each kill, many reacted to Haber's weapon with portents of doomsday . . . all exactly as would happen with nuclear arms. Winston Churchill announced that humanity "has got into its hands for the first time the tools by which it can unfailingly accomplish its own extermination," and it was repeatedly explained to the public that in a mere hour Paris could be annihilated by a hundred planes each carrying a ton of gas, while London could be wholly destroyed with forty planes and tons. Just as antinuclear activists would one day scream epithets at the scientists of Los Alamos, so Haber's colleague Albert Einstein called him a pathological criminal. Haber's own wife agreed with Einstein, and after he refused to stop working on weapons, she killed herself. Today, everyone knows that employing poison gas in warfare is completely ineffective since, depending on the wind, it can kill as many of you as it does of them . . . exactly as would happen with atomic fallout. Eventually this kind of common sense will reach the owners of nuclear arms.

But what of the argument that the United States and various other nations need their thermonuclear fists to counter an atomic maniac? The fear of deranged tyrants with nuclear bombs has been with us since Joseph Stalin and First Lightning, but there has already been an acid test. On October 16, 1964, a leader with unlimited powers, who supported global terrorism, threatened the whole of Southeast Asia, battled the United States and India, and insisted that his 700 million citizens did not need to fear nuclear war with America, acquired atomic weapons. At the time, Mao Zedong made comments similar to those made by Fidel Castro to Khrushchev, that if the Chinese had to be martyred by the millions in a nuclear holocaust, it would be worth it to advance global socialism. Yet even Mao at his most ruthless and sociopathic never unleashed the Bomb, and neither did his historic comrade in mass murder Joseph Stalin.

The West was terrified when Russia got the Bomb, when China got the Bomb, when Cuba got atomic missiles, when India and Pakistan went nuclear, when the Kim dynasty of North Korea got theirs, and most recently when the rumblings came from Tehran. Yet, in nearly seven decades, nothing has happened. The only country that has ever attacked another with a nuclear device is the United States of America.

That the Atomic Age is now in its twilight can be seen with two examples: the history of a mountain, and of a mouse pad. Many passengers flying across the continental United States are amazed to discover how much of the country is naked and empty, and one of those big empty places is the Great Basin, home to the Nevada Test Site and within it, eighty miles from Las Vegas, the ridge known as Yucca Mountain.

Since the dawn of nuclear power, a fee has been added to American utility bills, totaling around $750 million a year, to finance a long-term storage facility for high-level nuclear waste. As of 1985, Congress has been redirecting the moneys from that nuclear tariff into Washington's general budget instead of into the waste fund. France, Japan, and a number of other countries recycle their waste into usable fuel, but the United States, suffering from the comic-book fear that either Communists or terrorists might develop a sinister scheme and use our radiant trash against us, never has. From 1978 to 2008, the Department of Energy spent $9 billion studying Nevada's Yucca Mountain as the potential repository for high-level nuclear waste, originally with plans to open for business on January 31, 1998.

Because of delays, the United States now has more nuclear waste than Yucca Mountain could ever hold. When the site was legislated by Congress thirty-five years ago, Nevada had no senior congressional members to oppose it. Now, it has, and everyone in Nevada opposes it. With no atomic facilities of their own, as well as their history of being a target for federal nuclear explosions, Nevadans did not want to be a burial ground for the radioactive garbage of thirty-nine other states.

Back in 1945, six weeks after Little Boy fell on Hiroshima, the city was again attacked, by the Makurazaki Typhoon, killing five thousand people. But in another example of this history's blessed curses, the storm's deluge washed away so much of the American radioactivity that Hiroshima recovered, and by 1963 over a million people were living there. After President Kennedy signed the Nuclear Test Ban Treaty on October 7, 1963, halting atmospheric tests, America's Great Basin was not lucky enough to suffer a typhoon, so today, due to a nuclear history inflicted by Washington, Nevada is in worse shape than Hiroshima. And her residents are supposed to trust the federal government and private contractors with storing nuclear waste for ten thousand years?

Federal negotiator Richard Stallings met with Nevadan residents to discuss their worries: "It was a very hard sell. When people are terrified, they're not concerned about facts. The perception was anything nuclear just scared people to death. Their thought is that it's some kind of green, oozy stuff that's spewing poison, that you get near it and you'll die within minutes or hours. . . . You're dealing with a waste product, and a waste product that has a tremendous half-life. I mean, we're talking about a product that's not going to be just around for a few hundred years, but thousands of years. . . . You can't find any engineer that's going to sign on to a document that this hole in the ground is going to be safe for ten thousand years or safe for even two hundred years. I mean, that's impossible to do."

In 2012, Nevada finally triumphed, and the Yucca facility was canceled. Now the US Department of Energy will pay $16.2 billion over the next seven years for abrogating its storage agreement to utilities. While the navy alone has amassed 100 million gallons of highly radioactive waste—the less potent by-products are being stored in a New Mexican facility known as WIPP—121 of the nation's nuclear utilities are as of this writing stuck with holding 150 million pounds of waste in local pools—such as the 1,432 tons submerged at Indian Point, just outside New York City—exactly the same kinds of tanks that were such a threat during the crisis in Japan. In Washington, the issue is so politically mired beyond even the capital's legendary gridlock that there is no foreseeable answer. After thinking about how trustworthy industry and government have historically been when the subject is nuclear, who wants a plutonium graveyard in his or her state?

Then there's the mouse pad. In September 1953, the Gibertson Company began *Picture Parade*, a comic book to be purchased by grade schools for their students to quickly learn such topics as the United Nations, the first Thanksgiving, and how kids around the world celebrate Christmas. *Picture Parade*'s premier issue, "Andy's Atomic Adventures," was the story of a boy hunting ten butterflies when his dog, Spot, ran off, into the Nevada Test Site. The army refused to let Andy look for his runaway terrier, but when the boy then revealed his fear that an atomic bomb would kill his dog, his dad took him to the hospital where his mom was getting "atomic treatments" and where the lovely Miss Raymond explained the details of nuclear medicine. A week later, the phone rang with good news—Spot had been found! But sadly, the pup couldn't go home for a week, as he was soaked in radioactive dust. The Test Site's Dr. Clark told Andy all about atomic weaponry, and at his father's office at the electric company, the chief engineer explained nuclear power. Finally, Spot and Andy were reunited, and the Wilsons bought their son a damn leash.

When in the years to come *Picture Parade* #1 disappeared from circulation, the cover, of Andy and Spot reunited as, in the background, a test bomb rises in crimson mushroom glory, was reproduced in various forms. Stripped of its cheery educational content, the image became a chilling symbol of nuclear horror. Now, you can buy it on a mouse pad, so what was once an educational endeavor and then a disturbing image of the end of the world is a nostalgic joke.

At some point, this same devolution will hold true for the whole of nuclear history. In 1986 there were sixty-five thousand nuclear weapons; in 2007, twenty-seven thousand. Still too many, but a promising trend. Instead of the nuclear proliferation that was a global fear across the 1950s and 1960s, atomic weapons have become more and more symbolic. In a generation or two, the mushroom cloud will exist as nothing beyond a resonant piece of *Picture Parade* nostalgia, and a family trapped inside its backyard fallout shelter after a misunderstanding that the end is nigh will be the basis for a rip-roaring comedy. After the myth of nuclear has completely faded, when we hear that a country is pursuing an atomic arsenal, instead of worry, we'll feel embarrassed for them. Would you like some trebuchets and blunderbusses to go with those warheads, pal?

After all the trouble nuclear has given us over the past seven decades, it's about time for the cheek to be turned and the benefits to be clearly manifest. When in his 1961 inaugural address John Kennedy echoed Wohlstetter and referred to "that uncertain balance of terror that stays the hand of mankind's final war" (and Canadian prime minister Mike Pearson agreed with, "the balance of terror has replaced the balance of power"), their thinking was prophetic. Soviet and American fighters had a couple of dogfights during the three years of the Korean War, and this is the full extent that the two nations would directly battle each other over their entire history as mortal superpower adversaries. Mutual Assured Destruction kept the Cold War cold, and no conflict in the sixty years since Nagasaki came close to rivaling the carnages of World Wars I and II—with World War II alone slaughtering 55 million people. Nuclear arms made Alfred Nobel's dream came true: "I wish I could produce a substance or a machine of such frightful efficacy for wholesale devastation that wars should thereby become altogether impossible." Appointed to chair the US Institute of Peace by Ronald Reagan, sociologist Elspeth Rostow insisted that the Bomb was such a deterrent to war that it should be awarded the Nobel Peace Prize. As Luis Alvarez wrote his son: "What regrets I have about being a party to killing and maiming thousands of Japanese civilians this morning are tempered with the hope that this terrible weapon we have created may bring the countries of the world together and prevent further wars. Alfred Nobel thought that his invention of high explosives would have that effect, by making wars too terrible, but unfortunately it had just the opposite reaction. Our new destructive force is so many thousands of times worse that it may realize Nobel's dreams." Imagine our atomic arsenals, glowing and unused, not poised for Armageddon, but instead the warm, nucleic source of global peace.

Nuclear as a concept may have an aura as sinister as cancer, but it includes a number of other inarguably benevolent qualities for humankind—beginning with the radiant isotope. Following in the medical legacies of Marie and Irène Curie, Frédéric Joliot, and tracer creator George de Hevesy, Oak Ridge sent its first radioisotopes to St. Louis's Barnard Free Skin and

Cancer Hospital in 1946. Today we have an enormous array of artificially created elements used for diagnosis and treatment, from yttrium-90 for prostate cancer, to calcium-47 for bones, carbon-11 for positron-emission tomography (PET) scans, actinium-225 for leukemia, iron-59 for spleens, rhenium-188 for arteries, erbium-169 for arthritis, and various radioactive iodines, which can both image problems in the thyroid gland and kill its tumors. In the United States, around 17 million people annually are diagnosed or treated with nuclear medicine. Ninety percent of the field is imaging, such as the CT or CAT scan, a form of X-ray; the PET scan, which uses tracers of radioactive sugar; myocardial perfusion, which uses two different radioisotopes for test and stress to analyze artery disease; and specialized scans for bones, livers, lungs, brains, gallbladders, thyroids, and pulmonary embolisms. While X-rays, CTs, and PETs reveal various aspects of human anatomy, nuclear tracer imaging unveils the actual physiological workings of the body. When radioactive iodine concentrates in the thyroid, or phosphorus in the bones, or cobalt in the liver, they reveal processes. Thirty-one radiopharmaceuticals are based on just one isotope engineered by Glenn Seaborg and Emilio Segrè, technetium-99, used against tumors and for imaging the brain, lungs, liver, skeleton, and blood. The technetium variations are especially appealing as it is a product of nuclear power plant waste and has a half-life of six hours, briskly turning wholly inert and harmless. Over 20 million people a year are treated with it in some way.

Tracers can follow the metabolism of everything from fats in people to fertilizers in plants and are even used in the petroleum industry to know which crude belongs to which company in a shared pipe. Americium-241 detects the early stages of a fire and is the essential ingredient in modern smoke detectors, while the gamma rays of cobalt-60 are used to kill bacterial and fungal pests (such as trichina and salmonella) in pork, chicken, dried vegetables, herbs, and spices. When an organism dies, it stops consuming carbon, either through digestion or photosynthesis. Some of that carbon, known as carbon-14, is radioactive, with a half-life that can be measured, the famous carbon dating of archaeologists and crime scene investigators. Nickel-63 is so long lasting it made its way into pacemaker batteries, while for over fifty years NASA has used the hot decay of plutonium-238 to generate electricity for its satellites.

Unfortunately our two-faced god is manifest here—patient overdose from X-rays, CT scans, and everything else in the radiant medical arsenal poses a far greater health peril than any atomic plant meltdown. The research of David Brenner, the director of Columbia University's Center for Radiological Research, shows that radioactive procedures, their technicians, and their equipment are not as monitored as the public believes, and that children getting CT scans have increased cancer rates. Brenner told Congress that Homeland Security's overuse of whole-body X-ray security scanners will result in a hundred more Americans getting cancer every year; his and other protests forced the federal government to replace its imaging machines powered by X-rays with ones served by electromagnetics.

Conversely, even in the wake of the rabid Cold War bomb tests that poisoned the bodies of everyone born between 1951 and 1958, there's some Janus good news. For decades, doctors thought that the cells of the adult heart were different from all other cells in that they lasted a lifetime and never died and regenerated. Swedish scientist Jonas Frisén then realized that a significant percentage of the world's population, contaminated by the Cold War's nuclear tests, had been stamped by carbon-14, and this stamp could reveal the age of heart cells. From his research Frisén determined that in each of us "the heart muscle cells will be a mosaic: some that have been with that person from birth, and there will be new cells that have replaced others that have been lost"—which sounds to me like a resonant fable about the emotional toil of a human heart. The same nuclear-test stamp is starting to be used to save elephants, for it can separate legal African ivory from poached.

Finally, natural radioactivity is what keeps us warm. Around half of the earth's inner heat of forty-four terawatts comes from the decay of uranium, thorium, and their ilk, with the rest either having been trapped in the planet's molten lava core since its birth, or caused by something yet unknown . . . something to be discovered by future geologists. Since it takes millions of years for these elements to half-live their way to nonradiant stability, we'll keep having a warm planet for a bit—the earth cools at a rate of 100°F every half a billion years. From that subterranean nuclear heat a future engineer might devise a source for electrical power more promising than fusion. Our planet's all-natural nuclear force, after all, is so massive that it powers the migration of the continents, and the quakes of the crust.

The Age of Radiance's beginnings showcased an idea of Marie Curie's, "Now is the time to understand more, so that we fear less." Clearly the world has moved past the Cold War's duck-and-cover apocalyptic terrors, but we still fall into bouts of hysteria when an atomic utility fails, while devoutly maintaining a plethora of myths about nuclear science. The future of radiance must be understanding more to fear less.

Can we ever learn to believe that atomic bombs are just another kind of weapon, and that nuclear plants are just another form of utility, instead of imbuing each with mythic powers? Can we learn to accept as common sense that the same radiation that kills diseased cancer cells also kills healthy cells, producing the side effects of nausea and baldness? Or that a meltdown every decade or so is the price paid to save three hundred thousand lives every year from infection by petrochemical effluvia? Or that the same process that produces this clean energy can also be turned by a rogue state into weaponry?

Enrico Fermi late in life said that the "history of science and technology has consistently taught us that scientific advances in basic understanding have sooner or later led to technical and industrial applications that have revolutionized our way of life. It seems to me improbable that this effort to get at the structure of matter should be an exception to this rule. What is less certain, and what we all fervently hope, is that man will soon grow sufficiently adult to make good use of the powers that he acquires over nature."

After seven decades, it is humankind's responsibility to use the two-faced miracle discovered by Curie, Meitner, Fermi, Szilard, Teller, and Ulam correctly, not turn away from it in fear, superstition, and ignorance. It is time to enter a post–Atomic Age, an era where the fearful products of nuclear science are minimized, and its beneficence maximized.

It is time to learn to live with blessed curses.

HEARTFELT THANKS

I WORK WITH the greatest guys and gals in town, baby, and don't you forget it. My agent, Stuart Krichevsky, is so fantastic he gets his own page in this book . . . so if there's anything I've written that you don't like, it's entirely his fault. My editor, Colin Harrison, is like a child's fantasy of a publishing executive: hardworking, caring, thorough, and kind. Who could ask for anything more? Marketing titans Johanna Ramos-Boyer and the Great Kate Lloyd, future Viking-Random-Harper editorial directors Katrina Diaz and Kelsey Smith, Oppenheimer heiress Susan Moldow, the glorious executive titans of Scribner—Nan Graham, Roz Lippel, Brian Belfiglio, Paul O'Halloran, Daniel Cuddy—and the mythic warriors of SKA—Shana Cohen, Ross Harris, Kathryne Wick, and Elizabeth Kellermeyer. My magnificent jacket is by Tal Goretsky and its resonant innards are courtesy of Ellen Sasahara. To all of you, I bow my head in gracious servitude.

Librarians, archivists, and docents—every time I slip on those paper gloves to wear while browsing, I fall in love with you all over again. I'd like to thank for their remarkable professional courtesies and hard work both in person and in absentia the staffs of the American Museum of Science and Energy, Niels Bohr Library and Emilio Segrè Visual Archives of the American Center for Physics, Bradbury Science Museum, Bureau of Atomic Tourism, Churchill College Archives Centre, Library of Congress, Los Alamos Historical Society and Museum, Mandeville Department of Special Collections at the University of California at San Diego, Harvey Mudd College Oral History Project on the Atomic Age at Claremont, National Archives and Records Administration, National Atomic Museum, National Security Archive at George Washington University, Nuclear Weapons Archive, Society of Nuclear Medicine, Special Collections Research Center at the University of Chicago, US Department of Energy, Titan Missile Museum, Harry S. Truman Library & Museum, and Woodrow Wilson Center.

Isaac Newton thanked giants who supported him with their shoulders, and though I'm no Newton, there are indeed giants, starting with the magisterial Richard Rhodes, the fundamental overviews of Amir Aczel, Jim Baggott, David Lindley, James Mahaffey, Marjorie Malley, John Mueller, Jay Orear, Jon Palfreman, N. J. Slabbert, P. D. Smith, Tom Zoellner, the American Institute of Physics, and the US Department of Energy; Ève Curie and Susan Quinn on Marie Curie; Patricia Rife and Ruth Lewin Sime on Lise Meitner; Laura Fermi and Emilio Segrè on Enrico Fermi; William Lanouette and Bela Silard on Leo Szilard; István Hargittai on the other Martians of Budapest; William Laurence on the Manhattan Project; John Hersey and the Radiation Effects Research Foundation on Hiroshima; Svetlana Alexievich and the United Nations Scientific Committee on the Effects of Atomic Radiation on Chernobyl; and the *New York Times* reports on Fukushima Daiichi.

NOTES

1. Radiation: What's in It for Me?

4 *"What spreads the sea floors and moves":* Preuss.
6 *"afterheat, the fire that you can't put out":* Socolow.

2. The Astonished Owner of a New and Mysterious Power

9 *"a tall, slender, and loose-limbed man":* Dam.
10 *"A yellowish-green light spread all over":* Ibid.
10 *"When at first I made the startling discovery":* Nitske.
10 *"I have seen my death":* P. D. Smith.
11 *"Exactly what kind of a force Professor Röntgen":* Dam.
11 *"A more remarkable picture is one taken":* Moffett.
12 *"The Roentgen Rays, the Roentgen Rays":* Nitske.
12 *"Civilized man found himself the astonished owner":* P. D. Smith.
13 *"Röntgenmania":* Nitske.
14 *"I will never be satisfied with explanations":* P. D. Smith.
14 *"One wraps a Lumière photographic plate":* Becquerel, January 24, 1896, lecture to the French Academy of Sciences, cited in Slowiczek and Peters.
15 *"there is an emission of rays without apparent cause":* P. D. Smith.
16 *"My existence has been that of a prisoner":* Quinn.
16 *"Stas is very funny":* Ève Curie.
17 *"If [men] don't want to marry impecunious young girls":* American Institute of Physics.
17 *"I have fallen into black melancholy":* Redniss.
18 *"For the children, the dreadful nature of Czarist occupation":* Ève Curie.
18 *"Constantly held in suspicion and spied upon":* American Institute of Physics.
19 *"While his wife was being treated":* Ève Curie.
19 *"Weak as I am, in order not to let":* Ibid.
19 *"One of the boarders infected Bronya":* Ibid.
19 *"For many years we all felt weighing":* American Institute of Physics.
20 *"We sleep sometimes at night":* Ève Curie.
21 *"My situation was not exceptional":* Curie, *Pierre Curie.*
22 *"Pierre's intellectual capacities were not those":* Ibid.
22 *"I did not regret my nights":* Redniss.
22 *"As I entered the room, Pierre Curie"; "noticed the grave and gentle expression"; "had a touching desire to know"; "It would, nevertheless, be a beautiful thing":* Curie, *Pierre Curie.*
24 *"It is a sorrow to me to have":* Quinn.
24 *"I have the best husband":* Ève Curie.
24 *"I think of you who fill my life":* Curie, *Pierre Curie.*

24 *"practical and dark, so that":* Ève Curie.
24 *"My husband and I were so closely":* American Institute of Physics.
25 *"my dear little child whom I love":* Ève Curie.
25 *"the subject seemed to us very attractive":* Slowiczek and Peters.
26 *"In truth, the red glaze emitted":* Homer Laughlin Company, response to *Good Morning America*'s report on domestic radiation, March 16, 2011.
27 *"from this point of view, the atom":* Mme. Skłodowska Curie.
27 *"Neither of us could foresee that":* American Institute of Physics.
28 *"The life of a great scientist in his laboratory":* Curie, *Pierre Curie.*
28 *"We lived in our single preoccupation":* Ève Curie.
29 *"You can't imagine what a hole"; "Life is not easy for any":* Ibid.
30 *"Radium has the power of communicating":* Mme. Skłodowska Curie.
31 *"Sometimes we returned in the evening":* American Institute of Physics.
32 *"Viewed through a magnifying glass":* Marie Curie, "Recherches sur les Substances Radioactives," *Annales de Chemie et de Physiques* (Paris), 1903.
33 *"read a dozen square yards of newspaper daily":* Ève Curie.
33 *"A phenomenon of such extended malignancy":* Tuchman.
34 *"not only as mass, but also":* Mahaffey.
34 *"would possess a weapon by which":* P. D. Smith.
36 *"anemia, arteriosclerosis, arthritis, asthenia, diabetes"; "Such a light as this should shine":* Redniss.
37 *"I don't see the utility of it":* Ève Curie.
37 *"I have wanted to write to you":* To E. Gouy, January 22, 1904, in Curie, *Pierre Curie.*
38 *"It's pretty hard, this life":* Ève Curie.
38 *"The Radium Water Worked Fine":* Kean, "Radium."
38 *"One girl fainted at the sight":* Harvie.
38 *"The luminosity was brilliant":* P. D. Smith.
39 *"neither very well, nor very ill"; "had grown so accustomed to the idea":* Letter to Bronya, August 25, 1903, American Institute of Physics.
39 *"I have been frequently questioned":* Curie, *Pierre Curie.*
39 *"Did I eat a beefsteak?":* Ève Curie.
39 *"a little heartache":* Quinn.
40 *"His body passed between the feet":* Ève Curie.
41 *"My son is dead":* Ibid.
41 *"I enter the room":* Quinn.
41 *"They brought you in and placed":* Curie, *Cher Pierre.*
43 *"Everybody said Marie Curie is dead":* Borel.
43 *"In science we must be interested":* Ève Curie.
43 *"There is no connection between my scientific work":* Redniss.
43 *"Madame Curie is very intelligent":* Kline.
44 *"The hour when we knew":* Curie, *Pierre Curie.*
44 *"In his scientific thinking Langevin":* "Tribute to Langevin," *La Pensées*, 1946, *Correspondance françaises.*
44 *"Often, during meals, M. Langevin":* Perrin.
45 *"I spent last evening and night"; "I am trembling with impatience": L'Oeuvre,* November 23, 1911.
45 *"You are going to see quite a scandal"; "great friendship angered Mme. Langevin"; "was astounded to see Mme. Curie run"; "shouted threats for everyone to hear":* Perrin.
46 *"Variegated, virulent, turbulent, literary, inventive":* Tuchman.
46 *"I spent yesterday evening and night": L'Oeuvre*, November 23, 1911.
47 *"very difficult to judge the works of Mme. Curie": Le Petit Parisien*, January 5, 1911.
47 *"under the austere scientist":* Borel.
48 *"consorting with a concubine in the marital dwelling":* Perrin.

51 *"I rediscovered in [Pierre Curie's] daughter":* American Institute of Physics.
51 *"well-respected but they are industrialists":* Quinn.
51 *"I miss Irène a lot"; "That young man is a ball of fire"; "My mother and my husband often debated":* Ibid.
52 *"What fools":* Segrè, *From X-Rays to Quarks.*
52 *"An infinitely tiny particle projected":* "Les grandes découvertes de la radioactivité," *La Pensée*, 1957.
53 *"began to run and jump around"; "With the neutron we were too late"; "I will never forget the expression": Republica* (Lisbon), January 10, 1955.
53 *"Then began the harrowing struggle":* Ève Curie.

3. Rome: November 10, 1938

55 *"completely self-confident, but wholly without conceit":* Allardice and Trapnell.
55 *"a steamroller that moved slowly":* Segrè, *Enrico Fermi.*
55 *"There is a Fermi Sea":* Orear.
56 *"a very very common man":* Holton et al.
57 *"They were allowed to use the research"; "Fermi, after much reading of the pertinent literature":* Fermi to Persico, November 29, 1920, *Fermi Papers.*
59 *"Mechanical proficiency and practical gadgets":* Segrè, *Enrico Fermi.*
60 *"The location of the building":* Ibid.
60 *"Fermi worked in the Institute":* Orear.
61 *"I want to mention the 'Fermi Questions'":* Ibid.
61 *"if an atom were the size of a bus":* Aczel.
61 *"Fermi organized a group to do this":* Orear.
62 *"Radon plus beryllium sources were":* Segrè, *Enrico Fermi.*
62 *"The experimenters had to run":* Orear.
63 *"One day, as I came to the laboratory":* Segrè, *Enrico Fermi.*
63 *"Neutron research led to many surprises":* Orear.
63 *"Physics was comprehensible, as long as":* Ibid.
65 *"For the most part, my father":* Ibid.
66 *"Fermi was completely devoted to physics":* Ibid.
66 *"Fermi had always said he wanted":* Holton et al.
67 *"The next winter was the coldest":* Orear.
67 *"Now they are sending away the Jews"; "Re: Anti-Semitic Campaign":* Laura Fermi.
68 *"I'm going to cut the throat":* Kumar.
68 *"not yet hanged":* P. D. Smith.
68 *"For reasons that you can easily":* Fermi to Pegram, September 4 and October 22, 1938, *Fermi Papers.*
70 *"I think that my mother":* Orear.
71 *"She would not take notice":* Ibid.
71 *"having transformed the Physics Institute":* Segrè, *Enrico Fermi.*

4. The Mysteries of Budapest

73 *"Don't you think God already":* Orear.
74 *"It is beneath my dignity":* Lanouette with Silard.
74 *"leaving his country, perhaps for good":* Ibid.
75 *"our Madame Curie":* P. D. Smith.
75 *"I only want to know":* Hargittai, *Martians.*
76 *"a vivid man about five feet":* P. D. Smith.
76 *"They should have arrived here":* Crick.
76 *"Johnny [von Neumann] used to say":* Rhodes, *Dark Sun.*

77 *"an intellectual bumblebee"; "You didn't know what he"; "I never met anybody more imaginative":* P. D. Smith.
78 *"When I worked in the patent office"; "Spinoza's God who reveals himself"; "work in which I myself"; "He tends to overestimate":* Lanouette with Silard.
80 *"Hitler and his Nazis are":* Segrè, *From X-Rays to Quarks.*
80 *"practically everybody who came":* P. D. Smith.
80 *"I'm spending much money":* Hargittai, *Martians.*
81 *"as I was waiting for the light":* P. D. Smith.
82 *"I don't remember him ever":* Lanouette with Silard.
83 *"We are all agreed that":* Dael Lee Wolfle, *Symposium on Basic Research* (New York: National Academy of Sciences, 1959).
85 *"something deeply hidden had to be":* Isaacson.
85 *"how a puddle with a bit of oil":* Rife.
85 *"the most beautiful and stimulating":* Sime.
85 *"bordering on fear of people":* Ibid.
86 *"For many years I never had":* Hahn.
86 *"My strongest and dearest remembrances":* Rife.
87 *"a nose for discovering new elements":* Ibid.
87 *"But I thought you were a* man*":* Hahn.
87 *"it was as if the ground":* Kumar.
88 *"As soon as we were in":* American Institute of Physics.
88 *"It was the passport to scientific":* Meitner, "Status."
89 *"Late":* Sime.
89 *"like an axe in the hands":* Segrè, *From X-Rays to Quarks.*
90 *"she had committed a gross error":* Rife.
90 *"Fermi was a very careful experimenter":* Teller.
91 *"mystics, magicians and religious fanatics":* Rittenmeyer and Skundrick.
91 *"I have nothing against the Jews":* Max Planck, "Mein Besuch bei Adolf Hitler": *Physikalische Blätter* 3 (1947).
92 *"I built it from its very":* Letter to Gerta von Ubisch, July 1, 1947, Meitner Collection.
93 *"After these three laws were read":* Rittenmeyer and Skundrick.
93 *"The Jewess endangers the institute"; "nothing more could be done"; "The great misfortune has happened":* Sime.
93 *"It is considered undesirable":* Rife.
94 *"is like a sensitive child":* Letter to Ehrenfest, May 4, 1920, in Albrecht Fölsing, *Albert Einstein: A Biography* (London: Viking, 1997).
94 *"appeared to me like a miracle":* Paul A. Schilpp, ed., *Albert Einstein: Philosopher-Scientist* (New York: MJF Books, 1969).
94 *"Bohr loved paradoxes":* Teller.
95 *"I wanted her to be provided for"; "The danger consisted in the SS's":* Hahn.
95 *"You have made yourself as famous":* Rife.
96 *"No, but I was told":* Segrè, *From X-Rays to Quarks.*
96 *"Perhaps you cannot fully appreciate":* Meitner to Hahn, September 25, October 15, and December 5, 1938, in Krafft.
96 *"Help us. The Nazis are killing":* Rittenmeyer and Skundrick.
97 *"One always thinks life in this world":* Sime.
97 *"One day, while strolling in the":* Orear.
98 *"Monday evening, in the lab"; "How beautiful and exciting it":* Sime.
98 *"Miss Meitner—Professor Meitner":* "Discovery of Fission."
99 *"Was it a mistake?"; "Perhaps a drop could divide"; "The charge of a uranium nucleus":* Frisch.
100 *"These results, I realized":* Sparberg.
100 *"What is the possibility that uranium 239":* Rife.

100 *"Dear Otto! I am now almost certain":* Sime.
100 *"When I came back to Copenhagen":* "Discovery of Fission."
100 *"Dear Tante, I was able to speak":* Sime.
101 *"When we met on the boat"; I rigged up a pulse amplifier":* "Discovery of Fission."
101 *"We had bad weather through":* Stefan Rozenthal.
102 *"Your results present a wonderful"; "Unfortunately I did everything wrong"; "The uranium work is for me":* Sime.
103 *"We've been such dumb assholes!":* Aczel.
103 *"Fermi wasn't in his office":* Herbert Anderson.
104 *"'a scientist not discovering fission'":* Allison.
104 *"If Fermi had published":* Orear.
104 *"I went up to the thirteenth floor":* "Discovery of Fission."
105 *"God!" he said. "This looks like":* Kinkead.
105 *"All the things which H. G. Wells":* Weart and Szilard.
105 *"It takes one neutron to split":* Laura Fermi.
106 *"We turned the switch"; "That night, there was very little doubt":* P. D. Smith.

5. The Birth of Radiance

109 *"his willingness to accept facts":* Hargittai, *Martians.*
109 *"predict what would happen":* "Discovery of Fission."
110 *"work and dirty my hands":* Rhodes, *Atomic Bomb.*
110 *"with the precision of a prosecuting attorney":* Hargittai, *Martians.*
110 *"You are reinventing the field":* P. D. Smith.
110 *"We went over to Fermi's office":* Weart and Szilard.
110 *"Szilard was not willing to do": Enrico Fermi Papers,* 2:11.
111 *"During the summer of 1939":* Teller.
111 *"On matters scientific or technical":* Weart and Szilard.
112 *"It is like shooting birds":* Feld and Szilard.
113 *"Contrary to perhaps what is":* Fermi, "Physics at Columbia University."
114 *"There's a wop outside":* Rhodes, *Atomic Bomb.*
114 *"Princeton is a wonderful little spot":* Mehra.
114 *"Let's give it up"; "How would it be if we"; "Daran habe ich gar nicht gedacht":* Feld and Szilard.
115 *"The one thing most scientists":* Rhodes, *Atomic Bomb.*
115 *"In the United States of those days":* Orear.
116 *"We did not know just how":* Feld and Szilard.
116 *"he really only acted as a mailbox":* Jungk.
118 *"Alex, what you are after":* Mahaffey.
119 *"Szilard came to the conclusion"; "The question of money arose":* Herbert Anderson.
121 *"We have now reached the conclusion":* Frisch and Peierls.
122 *"For the Lord's sake, don't":* Buderi.
122 *"consider a future device for":* Bush and Conant.
123 *"undoubtedly a Fascist":* Lt. Col. S. V. Constant report, August 13, 1940, *Leo Szilard Papers.*
123 *"oscillated between theoretical and experimental":* Orear.
124 *"He frequently said, he was amazed":* Rhodes, *Atomic Bomb.*
124 *"Cartons of carefully wrapped graphite":* Herbert Anderson.
124 *"Physicists on the seventh floor":* Orear.
125 *"within a few years the use":* Bush and Conant.
126 *"My supervisor at the Met Lab":* Cowan.
126 *"a style of machine that dropped":* Kelly.
127 *"These lines are primarily addressed":* Staff, "Oppenheimer Years, 1943–1945."

130 *"We found out how coal miners feel":* Allardice and Trapnell.
131 *"For the construction of the pile":* Herbert Anderson.
131 *"One aspect of Fermi that wasn't":* Orear.
132 *"The pile, according to plan":* Segrè, *Enrico Fermi.*
132 *"Graphite was an awful material":* Orear.
133 *"It was a great temptation":* Herbert Anderson.
133 *"just in case an important experiment":* Lanouette with Silard.
133 *"Marched into Russia. Murdered the Jews":* Rittenmeyer and Skundrick.
134 *"The whole atmosphere there was":* "To Fermi—with Love."
135 *"Well, what do you do":* "Fermi Facts, Fables."
135 *"The pile is not performing now":* Laura Fermi.
135 *"Pull it to thirteen feet"; "This is not it"; "I'm hungry. Let's go to lunch"; "This is going to do it":* Allardice and Trapnell.
135 *"Fermi had, the night before":* "To Fermi—with Love."
136 *"At first you could hear the sounds":* Rhodes, *Atomic Bomb.*
137 *"I couldn't see the instruments"; "The reaction is self-sustaining":* Allardice and Trapnell.
137 *"Nothing very spectacular had happened":* Rhodes, *Atomic Bomb.*
138 *"Probably for Fermi, however, the real victory":* Segrè, *Enrico Fermi.*
138 *"The Italian navigator has landed":* Bush and Conant.
138 *"One of the things that I shall"; "There was a crowd there":* "Discovery of Fission."

6. The Secret of All Secrets

140 *"proton merry-go-round"; "Oh, sugar":* Herken, *Brotherhood.*
142 *"He was tall, nervous and intent":* Rhodes, "I Am Become Death."
142 *"He wanted everything and everyone":* Conant, *109 East Palace.*
142 *"Oppenheimer was Jewish, but wished":* Rhodes, "I Am Become Death."
142 *"Robert could make people feel":* Rhodes, *Atomic Bomb.*
142 *"Oppenheimer's prestige and ascendancy were great":* Segrè, *Enrico Fermi.*
143 *"J. Robert Oppenheimer, 30, associate professor":* Staff, "Prof Takes Girl for a Ride."
144 *"Hot dog"; "My two great loves are":* Bird and Sherman.
144 *"I remember exactly how I"; "When Alvarez told me the news":* Kelly.
145 *"That's impossible":* Bird and Sherman.
145 *"prospects for useful nuclear energy":* Herken, *Brotherhood.*
146 *"Time is very much of the essence":* Bush and Conant.
147 *"The world would be better":* Conant, *109 East Palace.*
147 *"The Secretary of War has selected":* Baggott.
148 *"When Groves saw that the usual"; "He's a genius":* Conant, *109 East Palace.*
149 *"The investigation of Szilard should":* Groves to Calvert, June 12, 1943, Manhattan Engineer District Records, National Archives.
150 *"The prospect of coming to Los Alamos":* United States Atomic Energy Commission, *In the Matter.*
150 *"People I knew well began":* Kelly.
150 *"I went by train to a place":* Bernstein.
151 *"I drove right through the square":* Conant, *109 East Palace.*
151 *"Here at Los Alamos":* Rhodes, "I Am Become Death."
151 *"In spite of the difficulties"; "It was an unforgettable experience":* Smith and Weiner.
151 *"At great expense, we have gathered":* Laura Fermi.
151 *"Oppenheimer was very patient":* Feynman.
152 *"The water supply had been built":* Bernstein.
152 *"I would like that very much":* Frisch.
152 *"I will have nothing to do":* Sime.
153 *"'Welcome to Los Alamos'":* Zoellner.

153 *"the culmination of three centuries of physics":* Bird and Sherman.
153 *"Behind us lay the Sangre de Cristo":* Kelly.
153 *"The Mesa was indented by":* Segrè, *Enrico Fermi.*
154 *"They had this notion that we"; "Soon Johnny will know so much"; "It is my fault":* Lang.
155 *"Kitty was a very strange woman"; "The General's in a stew"; "You know, usually on a military post":* Conant, *109 East Palace.*
156 *"Though I had an excellent time":* Bernstein.
157 *"The Army routinely leveled every"; "Fermi once told me with hardly"; "wrote an official letter to":* Teller.
158 *"Teller had worked on the bomb project":* Bernstein.
158 *"Throughout the ten years, Oppie":* Rhodes, *Atomic Bomb.*
158 *"God protect us from the enemy":* Conant, *109 East Palace.*
158 *"Fermi was simply unable to":* Mahaffey.
159 *"Below her, she had Rudolf":* Orear.
159 *"he told the horse in a firm voice":* Teller.
159 *"I required a special small laboratory":* Segrè, *Enrico Fermi.*
161 *"someday will be filled with great":* "Atoms in Appalachia."
161 *"Mr. President, I agree that":* Kelly.
162 *"trained like soldiers not to reason why":* Nichols.
162 *"If you walked along the wooden"; "It was so darned bleak"; "Fermi was very discreet":* Kelly.
165 *"Enrico, when I was in":* Segrè, *Enrico Fermi.*
166 *"All science stopped during the war"; "was like tickling the tail":* Feynman.
167 *"The Critical Assemblies Group decided":* Orear.
168 *"Just as a beam of light":* Teller.
168 *"Only he was fully awake"; "Pray, what is the candidate's tailor"; "an almost primitive lack of ability":* George Dyson.
169 *"At the start I had regarded Teller":* Bethe, "Comments."
170 *"The first Sunday I was there":* Laura Fermi.
170 *"While Klaus was a mere"; "somewhat plump, greenish pale":* Rhodes, *Dark Sun.*
174 *"The Mosquito flew at a great height":* Bird and Sherman.
174 *"Both the father and the son":* Kelly.
175 *"You see, I told you":* Teller.
175 *"When Niels and Aage Bohr arrived":* Segrè, *Enrico Fermi.*
176 *"the Nazis' work was held":* Lang.
176 *"main theme was that it":* Staff, "Jewish Spirit."
176 *"Even if my brother steals":* Teller.
176 *"People must learn to prevent":* Baggott.
178 *"With the beginning of the war":* Heisenberg.
180 *"Heisenberg stated that he was"; "stressed how important it was"; "Dear Heisenberg, I have seen":* Niels Bohr Library & Archives.
182 *"He said that his main aim":* Bernstein.
184 *"The big problem was":* Conant, *109 East Palace.*
184 *"Recent reports both through the newspapers": Oppenheimer Papers,* H. A. Bethe folder.
185 *"the only way we could lose":* Bird and Sherman.
185 *"deny the enemy his brain"; "We wouldn't have had the moral"; "I think it was dreadful":* Baggott.
186 *"since there was not even the remotest":* Rhodes, *Dark Sun.*
188 *"émigré's passionate hatred of Hitler was":* Max von Laue to Theodor von Laue, August 19, 1945, Laue Papers, Deutsches Museum Archives, Munich.
189 *"As long as Prof Meitner was":* Meitner Collection.
189 *"They are obliged to try"; "Ah, so you're the little lady"; "was Germany's anointed postwar scientific icon":* Rife.

190 *"nonsense from the first word"; "After the last 15 years which"; "perhaps one cannot be such"; "It is a difficult problem":* Sime.
190 *"That is indeed the misfortune":* Office of Strategic Services: Correspondence; January 5–July 31, 1945; Moe Berg Papers, box 6; Manuscripts Division, Department of Rare Books and Special Collections, Princeton University Library.
191 *"a bitter, disappointed woman":* Hahn.
192 *"Bohr at Los Alamos was marvelous":* Rhodes, "I Am Become Death."
192 *"It is already evident that":* Bird and Sherman.
192 *"Atomic weapons have similar complementary":* Baggott.
192 *"The implication was that Roosevelt":* Conant, *109 East Palace.*
193 *"The war taught us much":* V. I. Lenin, *Polnoe sobrainie sochinenii,* vol. 26, 5th ed. (Moscow, 1958–65).
193 *"worked harder than anyone else"; "It seemed to us that if"; "that it was necessary to promote"; "Where that influence came from":* Rhodes, *Dark Sun.*
194 *"Many of us looked with deep":* Conant, *109 East Palace.*

7. The First Cry of a Newborn World

196 *"We spent several days finding":* Bird and Sherman.
196 *"In May, one hundred tons":* Kelly.
196 *"Comparison with TNT": Oppenheimer Papers,* H. A. Bethe folder.
198 *"Oppenheimer was really terribly worried"; "All the senior scientists who":* "Dan Hornig."
199 *"I started dreaming Kistiakowsky had"; "I was told that [Oppie] came":* Calloway.
199 *"on whether or not the bomb"; "Now we're all sons of bitches":* Conant, *109 East Palace.*
200 *"I think I'm the first"; "The theoretical people had calculated":* Calloway.
200 *"At the instant of the explosion"; "At about thirty seconds, the general"; "After a few seconds the rising"; "The column looked rather like":* "Trinity."
202 *"The enormity of the light"; "Well, there must be something":* Conant, *109 East Palace.*
202 *"The whole country was lighted by":* "Primary Resources."
203 *"I could see that crack"; "Then an amazing thing":* Calloway.
203 *"We've done it":* Bernstein.
203 *"It was like being at the":* Jones.
203 *"When one first looked up":* Bird and Sherman.
203 *"Apparently no one had told":* Kelly.
204 *"A tremendous cloud of smoke":* "Trinity."
204 *"Practically everybody at the Trinity":* Kelly.
204 *"The big boom came about":* Alice Smith, "A Peril and a Hope," *New York Times,* September 26, 1945.
204 *"My grandmother shoved me"; "We saw this huge, huge light"; "I went out there":* Calloway.
205 *"About one hour after the explosion":* Rhodes, *Atomic Bomb.*
205 *"Feynman got his bongo drums":* Conant, *109 East Palace.*
205 *"One man I remember, Bob Wilson":* Feynman.

8. My God, What Have We Done?

206 *"flabbergasted by the assumption":* Richard G. Hewlett and Oscar E. Anderson, *The New World, 1939/1946* (Washington: US Atomic Energy Commission, 1972).
206 *"general demeanor and his desire":* James Byrnes, *All in Our Lifetime* (New York: Harper, 1958).
206 *"If we were to offer to"; "to push ahead as fast as"; "The most desirable target":* Bird and Sherman.

207 *"steps should be taken to sever":* Baggott.
208 *"I was one of those":* Dwight D. Eisenhower, *The White House Years, Mandate for Change, 1953–1956* (New York: Doubleday, 1963).
208 *"Even if the Japs are savages":* Truman's diary, Harry S. Truman Library, http://www.trumanlibrary.org/whistlestop/study_collections/bomb/large/documents/fulltext.php?fulltextid=15.
209 *"You've got to kill people":* Larrabee.
209 *"I heard the huzzle-huzzle":* George Martin, "Black Snow and Leaping Tigers," *Harper's*, February 1946.
210 *"scorched, boiled, and baked to death":* Larrabee.
211 *"I was a little fearful that":* Calloway.
211 *"LeMay said if we'd lost":* Morris.
211 *"The people from Trinity had"; "As we came in from"; "The bomb blast hit us":* Loader, Rafferty, and Rafferty.
213 *"There is a moment of calm":* Langewiesche.
213 *"My God, what have we done?"; "We heard the strange noise":* Hersey.
214 *"I found that there was"; "A woman who looked like":* Matsumoto.
214 *"Hundreds upon hundreds of the"; "I found people who, when":* Burchett.
215 *"What regrets I have about":* Rhodes, *Dark Sun.*
215 *"This is the greatest thing":* Rhodes, *Atomic Bomb.*
215 *"The force from which the sun":* "Henry L. Stimson prepared statement for the public regarding dropping the Atomic Bomb forwarded to President Truman": July 31, 1945, Harry S. Truman Library, http://www.trumanlibrary.org/whistlestop/study_collections/bomb/large/documents/fulltext.php?fulltextid=20.
215 *"A surprising number of the":* Hersey.
216 *"The people of Hiroshima, aroused":* Ibid.
217 *"As I peered through the dark":* William L. Laurence, "Eyewitness Account of Atomic Bomb over Nagasaki FOR RELEASE SUNDAY, SEPTEMBER 9, 1945," Bureau of Public Relations, US War Department.
219 *"At night, the town, the":* Dor-Ner.
219 *"The day after Trinity":* Orear.
220 *"known sin"; "Sometimes, someone confesses a sin":* Hargittai, *Martians.*
220 *"Oppenheimer seemed to lose his":* Conant, *109 East Palace.*
220 *"I have to explain about Oppie":* Orear.
220 *"You see now that the":* Sayen.
221 *"Had I known that the Germans":* Staff, "Atom."
221 *"What would Harry Truman have"; "The firebombing of Tokyo was":* Hargittai, *Martians.*
222 *"completely unnecessary":* Eisenhower.
222 *"Japan would have surrendered":* United States Strategic Bombing Survey.
222 *"because of the vast sums":* Mee.
222 *"In March 1944, I experienced":* Kelly.

9. How Do You Keep a Cold War Cold?

225 *"And Fermi said, thoughtfully":* Rhodes, *Dark Sun.*
225 *"It was a place where":* Hargittai, *Judging.*
225 *"Just to show you what"; "At the yearly Christmas parties"; "One day my father brought":* Orear.
227 *"My general conclusion would be that":* Rhodes, *Dark Sun.*
228 *"I neither can nor will":* Teller and Brown.
228 *"their appetites for weapons work":* Rhodes, *Atomic Bomb.*

228 *"General Groves told me very briefly"; "The time needed [for another country]; "just like poison gases after"; "The years after Los Alamos":* Rhodes, *Dark Sun.*
228 *"We should prefer defeat in war":* Manhattan Engineer District Records, Harrison-Bundy file.
229 *"And God bless General Groves!":* Lanouette with Silard.
230 *"it is not too much to expect":* Lewis Strauss, "Speech to the National Association of Science Writers," September 16, 1954.
231 *"Russia was traditionally the enemy":* Hargittai, *Martians.*
231 *"was rather diffident in his approach"; "You don't know Southerners"; "Stalin summoned [atomic bomb chief]"; "sparkling bluish black"; "I never forgave Truman"; "I soon realized . . . that he was":* Rhodes, *Dark Sun.*
232 *"casually mentioned to Stalin that":* Truman, *Memoirs.*
232 *"I was sure that [Stalin] had no":* Mee.
235 *"invite the nations of the world":* Bush and Conant.
235 *"If the Russians have the weapon"; "In that case, we have no choice":* Bird and Sherman.
236 *"Weisskopf vividly described to me":* Bernstein.
236 *"You have to recall that in 1948":* Rhodes, *Arsenals of Folly.*
237 *"necessary to have within the arsenal":* Memo to GAC, January 13, 1950, nuclearfiles .org.
237 *"that we had to do it":* Eben Ayers Papers, Truman Library.
237 *"but the third time, nobody":* Grimberg.
238 *"that someone (designated by the code)"; "Then, in mid-September [1949]"; "that someone in the British embassy"; "sort of a British Columbo character"; "Were you not in touch"; "reason to believe that someone"; "most logical suspect for [another]":* Lamphere and Shachtman.
239 *"came out of World War II":* Barnet.
240 *"like children lost in the woods":* Weiner.
240 *"After about one minute"; "hot . . . something is happening"; "I'll never be able to read"; "Julius had money"; "The day after my wife"; "I told . . . the FBI right":* Greenglass FBI files, http://vault.fbi.gov/search?SearchableText=greenglass.
241 *"make people realize that this"; "I believe your conduct"; "And what of our children":* Rosenberg trial testimony, http://law2.umkc.edu/faculty/projects/FTrials/Rosenberg/RosenbergTrial.pdf.
242 *"took all the left-wing stuff":* Yourgrau.
242 *"The June evening of the executions":* Hiss.
243 *"the attack will be swift"; "According to information coming":* Gaddis, *George F. Kennan.*
243 *"I would cut them off"; "I would have dropped thirty"; "Our mission called for me":* Cumings.
244 *"Anybody who says twenty thousand":* Rhodes, *Dark Sun.*
244 *"If you say why not bomb":* Hargittai, *Martians.*
245 *"America had a powerful air"; "When America elected General Dwight":* Nikita Khrushchev.
246 *"The war of the future would":* President's Farewell Address to the American People, January 15, 1953, Truman Library.

10. A Totally Different Scheme, and It Will Change the Course of History

247 *"every time he reported, we"; "Both Gamow and I showed"; "What Ulam did was not"; "that the President would not"; "You would swear that the"; "The United States is said to"; "deliberately precipitate war with the USSR"; "There was a look of hatred":* Rhodes, *Dark Sun.*

247 *"The computers of [Polish mathematician]":* Orear.
248 *"What we are creating now":* George Dyson.
249 *"originally came with my friend"; "[Ulam's] suggestion was far from original"; "Our first computer at the"; I went down into the":* Teller.
249 *"adding that perhaps it was"; "Engraved on my memory is"; "From then on, Teller pushed Stan"; "Teller accused the leadership"; "Lawrence believed Edward"; "Of course, we worried about"; "An absolutely insane task"; "When you see the burned birds"; "Even if you strike first":* Grimberg.
249 *"I was well placed to watch"; "Nobody will blame Teller"; "Before the end of the summer"; "Once Teller left Los Alamos":* Bethe, "Comments."
252 *"I felt strongly that that":* Bush and Conant.
253 *"We called the place the Rock":* Michael Harris.
256 *"That's so I don't forget":* Gaddis, *George F. Kennan.*
256 *"We threatened with missiles we":* Sergei Khrushchev.
256 *"On no one did there ever":* Bird and Sherman.
257 *"was a man of great talent":* Hargittai, *Martians.*
258 *"delayed or hindered development of H-bomb"; "in view of the fact that"; "felt that if this case is lost":* Oppenheimer FBI files, http://vault.fbi.gov/rosenberg-case/robert-j.-oppenheimer.
258 *"Teller feels deeply that [Oppenheimer's]"; "had been instrumental in bringing to"; "Lewis, let us be certain"; "The truth is that no matter":* Rhodes, *Dark Sun.*
259 *"a dangerous engineering undertaking":* Herken, *Brotherhood.*
260 *"I have thought most earnestly":* Strauss.
261 *"The day the Oppenheimer case":* Orear.
261 *"had no obligation to subject"; "The trouble with Oppenheimer is":* Pais.
261 *"I felt sick":* Stern and Green.
262 *"There was an approach made":* United States Atomic Energy Commission.
262 *"Teller thought Oppenheimer was somehow":* Brian Kaufman.
263 *"Q. Is it your intention"; "I'm sorry":* United States Atomic Energy Commission.
263 *"The whole damn thing":* Dyson, "Oppenheimer."
264 *"I've never seen [Teller]"; "I really do feel it":* Seife, *Sun.*
264 *"If a person leaves his country":* Hargittai, *Judging.*
264 *"a great debt of gratitude":* United States Atomic Energy Commission.
264 *"I think it broke his spirit"; "I was indignant":* Rhodes, *Dark Sun.*
265 *"The real tragedy of Oppenheimer's":* Dyson, "Oppenheimer."
265 *"One of the last times I saw him":* Segrè, *Enrico Fermi.*
266 *"I personally have discussed":* Orear.
266 *"We've considered every astronomical source":* Mosher.
267 *"ghost towns"; "thought scientists, like other people": Foreign Relations of the United States: 1958–60,* 3:572.
267 *"A lot of the talk of Livermore":* Teller.
268 *"If worse comes to worse"; "The hospital was even more":* Hargittai, *Martians.*
268 *"I'm not strong in English"; "Let's go for a little walk":* Rhodes, *Dark Sun.*

11. The Origins of Modern Swimwear

271 *"keep from contaminating the general":* Cockburn.
271 *"We Marines were brought":* Dor-Ner.
272 *"The bomb will not start"; "The explosion was going to be"; "A mixture of radioactive materials":* Michael Harris.
273 *"I was on a ship that":* Grimberg.
273 *"small and devastating":* Mahaffey.

273 *"An Air Force photographer was":* Cowan.
275 *"Immediately, it felt as if":* Grimberg.

12. The Delicate Balance of Terror

276 *"If you go on with this":* quoted in *Reader's Digest*, December 1954.
276 *"I remember President Kennedy once stated":* Nikita Khrushchev.
277 *"The cost of one modern":* Dwight D. Eisenhower, "The Chance for Peace," Washington, DC, April 16, 1953, http://www.edchange.org/multicultural/speeches/ike_chance_for_peace.html.
277 *"I will give you a simple":* Jerome Wiesner, "We Need More Piefs," SSI Conference Proceedings, SLAC, Stanford University, 1984.
278 *"The atomic bomb makes surprise"; "Thus far the chief purpose"; "If two thousand bombs in"; "We no longer need to argue":* Kaplan, *Wizards of Armageddon.*
278 *"If there is another war"; "a realistic combat mission"; "There was a time in the 1950s":* LeMay, *Mission.*
280 *"During the early to mid-1970s":* Rosenbaum, *How the End.*
281 *"We have to keep the scientists":* Kaplan, *Wizards of Armageddon.*
281 *"The reason for having RAND":* Rhodes, *Dark Sun.*
283 *"Under this concept, the United States":* Secretary of Defense Caspar Weinberger on US Policy of Nuclear Deterrence in a Testimony before the US Senate Foreign Relations Committee, December 14, 1982.
284 *"Our missiles were still imperfect":* Nikita Khrushchev.
284 *"the USSR had a total":* Sergei Khrushchev.
285 *"Of course we tried to derive":* Nikita Khrushchev.
286 *"It would take really very few"; "A fallout shelter for everybody":* Cited in Rose.
286 *"One thing that was rarely":* Roy.
287 *"They called them mannequin families":* Grimberg.
288 *"Well, yeah, we could do that":* Kaplan, *Wizards of Armageddon.*
288 *"And we call ourselves the human race":* Hoffman, citing MaGeorge Bundy, *Danger and Survival* (New York: Random House, 1988).
288 *"learned all the facts of nuclear":* Nikita Khrushchev.
289 *"The cornerstone of our strategic policy":* Secretary of Defense Robert McNamara, "Mutual Deterrence" speech, San Francisco, September 18, 1967.
289 *"the principle of the threat":* "Cold War Strategy," nuclearfiles.org.
289 *"I could go into the next room"; "the need of keeping a berserk president":* Murray Marder, "Two Recall Nixon's A-Remark," *Los Angeles Times,* February 9, 1976.
290 *"All of us, more or less":* "Colonel X's Warning: Our Mistakes plus Your Hysteria," *Détente*, October 1984.
290 *"Let's put it this way":* Steven Kull, *Minds at War* (New York: Basic Books, 1988).
290 *"For thousands of years peoples":* Sergei Khrushchev.
290 *"like a prize-winning pear"; "a thermonuclear Zero Mostel"; "a moral tract on mass murder"; "It doesn't work that way":* Menand, "Fat Man."
291 *"A Navy captain was saying":* Miller.
292 *"The present nuclear situation is":* Southern.
292 *"senses a nuclear explosion in":* Rosenbaum, *How the End.*
293 *"provocation . . . I wonder what our attitude"; "just as if we suddenly put"; "knows we don't really live"; "an explicit threat to the peace"; "It doesn't make any difference"; "How gravely* does *this* change*"; "I don't think there is a military":* Sheldon Stern.
294 *"We had no desire to start":* Nikita Khrushchev.
294 *"There was a fear that if":* Gunther Klein.
294 *"During its entire history Russia":* Sergei Khrushchev.

295 *"Curtis LeMay called [the quarantine]":* Ted Sorensen.
295 *"The Russian bear has always"; "During that very critical time":* LeMay, *Mission.*
296 *"LeMay talked openly about"; "It wasn't until nearly thirty":* Errol Morris.
296 *"Later we learned that the submarine":* Klein, Brauburger, and Knopp.
297 *"Everything came to a halt"; "The president said we are prepared"; "Father sensed that he was losing"; "Castro thinks that war will begin"; "That is insane"; "The Soviet government has ordered":* Sergei Khrushchev.
297 *"To any man at the United Nations":* Kaplan, *Wizards of Armageddon.*
298 *"In my seven years as [defense] secretary":* Errol Morris.
299 *"the greatest defeat in our history":* LeMay, *Mission.*
299 *"The results were very painful":* Nina Tannenwald, ed., *Understanding the End of the Cold War, 1980–87: An Oral History Conference* (Providence, RI: Watson Institute for International Studies, 1999).

13. Too Cheap to Meter

305 *"observe what kind of disturbance":* Seife, *Sun.*
305 *"If your mountain is not":* Hargittai, *Martians.*
307 *"potential for serious radiological consequences"; "everything is under control"; "more complex than the company"; "danger was over"; "crisis had passed":* Peterson.
307 *"We didn't learn for years":* Gilinsky, "Behind."
308 *"They can't get rid of":* "Three Mile Island Emergency."
308 *"The world has never known":* Drain.
310 *"You might say that I":* Teller.
310 *"all the alarms went off":* Palfreman.

14. There Fell a Great Star from Heaven, Burning as It Were a Lamp

312 *"The weather was so wonderful":* Alexievich.
313 *"fundamentally faulty, having a built-in":* Rhodes, "Matter."
315 *"In the program there are instructions":* Hoffman.
315 *"At the moment when the turbine"; "like a volcanic eruption":* Staff, "Word for Word."
316 *"since all Soviet nuclear facilities were designed"; "a white pillar several hundred":* Rhodes, *Arsenals of Folly.*
316 *"Flames, sparks, and chunks of":* Rhodes, "Matter."
316 *"I can still see the bright"; "Do you know how pleasantly":* Alexievich.
317 *"Suddenly we saw [a man]"; "There was a loud thud":* Staff, "Word for Word."
317 *"We didn't know it was":* Sahota.
318 *"One night I heard a noise":* Alexievich.
319 *"By May 4 the pilots":* Staff, "Word for Word."
320 *"The water table will start":* Justin Elliott.
320 *"It was a real war"; "There was a Ukrainian woman"; "One collective farm chairman"; "The soldiers knocked"; "The police were yelling"; "They come around here"; "If we kill a wild boar":* Alexievich.
323 *"We are the great guinea pigs"; "My teeth are falling out"; "These people are sick"; "It's not too much to say"; "If I knew it would be":* Specter.
323 *"the population remains largely unsure":* Mettler.
324 *"Radiation is good for you"; "Life itself is dangerous"; "it's safe where we are"; "Northern Ukraine is the cleanest":* Shukman.
326 *"Chernobyl represents a huge mystery":* Grady, "Countering Radiation Fears with Just the Facts."

15. Hitting a Bullet with a Bullet

327 *"[Cheating death] made me feel"; "horrified":* Paul Lettow, *Ronald Reagan and His Quest to Abolish Nuclear Weapons* (New York: Random House, 2005).
328 *"Both our countries know from":* Margaret Thatcher, *The Downing Street Years* (New York: HarperCollins, 1993).
328 *"I have never forgotten the sorrow":* Sykes.
329 *"It's powerfully done, all $7 mil.":* Reagan diary, October 10, 1983.
329 *"It finally settled my internal debate":* Zaitchik.
329 *"In several ways, the sequence":* Reagan diary, November 18, 1983.
330 *"a concrete program, calculated for": Foreign Broadcast Information Service Daily Report* 3, no. 11 (January 16, 1986).
330 *"a hell of a good idea":* Rhodes, *Arsenals of Folly.*
332 *"We'd be hard put to":* Reagan diary, January 15, 1986.
332 *"So let me precisely, firmly"; "Excuse me, Mr. President, but":* Archives, Gorbachev Foundation.
333 *"Let me ask, do we mean"; "We can do that"; "Let's do it"; "If we agree that by":* Rhodes, *Arsenals of Folly.*
334 *"hitting a bullet with a bullet"; "What country is suicidal enough":* Smith and Ratnam.
335 *"Do not be so naive":* "Nuclear Proliferation."
335 *"A dictator or oligarch bent":* Mueller.
336 *"smelt . . . you know, like hair":* Miklós.
337 *"all organs and tissues of":* Glanz.
337 *"Spengler may have been right":* Gopnik.
337 *"No country without an atom bomb":* Mueller.
339 *"technology has become a useful tool":* Langewiesche.
339 *"Does anyone really think that":* Zakaria.

16. On the Shores of Fortunate Island

340 *"With the stones that you boys":* Fackler, "Fukushima's Long Link to a Dark Nuclear Past."
340 *"The Japanese are pathologically sensitive"; "it's unthinkable that I could":* Osnos, "Aftershocks."
341 *"The Atomic Energy Commission was":* Palfreman.
342 *"Do you remember our sleeves"; "the tsunami hit Tagajo":* O'Brien and Palfreman.
343 *"I still remember it"; "I saw the tsunami coming":* Pillitteri.
343 *"the fire that you can't":* George Johnson.
344 *"They aren't being heated by":* Chris Wilson.
344 *"After the second shock wave"; "If [Tokyo Electric Power Company] and":* Adelstein and McNeill.
344 *"I went straight to the harbor":* Edge.
345 *"He would like to think":* Osnos, "Aftershocks."
345 *"They sent people out in":* O'Brien and Palfreman.
345 *"We don't know exactly how"; "It'll be like somebody dropped":* Mirsky.
347 *"At Chernobyl, you know, the workers":* Osnos, "Fallout."
347 *"I thought this country was finished":* Yokota and Yamada.
347 *"The hydrogen is being vented"; "It's worse than a meltdown":* Mirsky.
348 *"Just as we were to get":* Edge.
349 *"Everything in their system is":* Sanger, Wald, and Tabuchi.
349 *"Number Four is currently burning":* Tabuchi, "A Confused Nuclear Cleanup."
351 *"more sense of crisis"; "heartfelt hope that the people":* Osnos, "Aftershocks."
351 *"the popcorn scenario"; "One will pop and then another":* Osnos, "Fallout."

351 *"We cannot hold on to"; "If they withdrew"; "We can still hold on"; "I always wondered why you":* Powell and Takayama.
352 *"They are ready to die":* Iyer.
352 *"When a man has to":* Adelstein, *Yakuza.*
353 *"In the year 2002, the":* Ariel Schwartz.
354 *"My town is gone"; "There were so many ideas":* Belson, "Panic."
354 *"In the control room, people":* Edge.
354 *"What the hell are you"; "I thought about changing jobs":* Yokota and Yamada.
355 *"We can't get inside to check":* Sanger, Wald, and Tabruchi.
355 *"If you can't make accurate":* Hall and Makinen.
356 *"The most likely effects would":* Staff, "Radiation in Japan Seas."
356 *"When I heard that the operation":* Samuels.
357 *"Unfortunately, the water contains a":* Tabuchi, "Japan Releases."
358 *"formula every day without any":* Tabuchi, "Japanese Tests."
358 *"We are all amateurs"; "They tell us decontamination would":* Tabuchi, "Confused Nuclear."
358 *"It's too dangerous"; "They called several days ago":* Fackler, "Japan Split."
359 *"Nuclear energy shouldn't be in":* Adelstein, *Yakuza.*
359 *"a collapse of the nation's ability":* Fackler, "Japan's Former Leader."
359 *"This is a historic experience":* Tabuchi, "Citizens' Testing."
359 *"a state of cold shutdown":* Fackler, *Devastation.*
359 *"The coolant water is keeping":* Powell and Takayama.
360 *"Americans have never met a hydrocarbon":* Kolbert, "Burning Love."
361 *"The sole fatal nuclear power":* Saletan, "Nuclear Overreactors."
361 *"perhaps the real cause"; "By 1987, the year following Chernobyl":* Mark Joseph Stern.
362 *"His dream in life was":* Silverstein.
363 *"Know what?"; "The explosions in the backyard":* Clynes.
365 *"The sum of exposures":* Rhodes, "Matter."
366 *"the risk of cancer is quite low"; "Of those fifty thousand people":* Grady, "Radiation Is Everywhere."
367 *"radiation biology lessons":* Broad, "Drumbeat."

17. Under the Thrall of a Two-Faced God

369 *"almost one million people have already":* Caldicott, "Unsafe."
369 *"People have come not only":* Falk.
370 *"far more people die each year":* Mahaffey.
370 *"Hundreds of billions were spent":* Rhodes, *Arsenals of Folly.*
371 *"Fifty years of telling ourselves":* Gareth Cook.
372 *"has got into its hands":* Winston S. Churchill, *Amid These Storms: Thoughts and Adventures* (New York: Scribner's, 1932).
379 *"history of science and technology has":* Cronin.

SOURCES

Aczel, Amir D. *Uranium Wars: The Scientific Rivalry That Created the Nuclear Age.* New York: Palgrave, 2009.

Adelstein, Jake. "The Yakuza and the Nuclear Mafia: Nationalization Looms for TEPCO." *Atlantic Wire*, December 30, 2011.

Adelstein, Jake, and David McNeill. "Meltdown: What Really Happened at Fukushima?" *Atlantic*, July 2, 2011.

Adult Health Study and *Life Span Study.* Radiation Effects Research Foundation Publications, 1975–2012. (Successor to the Atomic Bomb Casualty Commission, 1947–75.) http://www.rerf.or.jp/library/index_e.html.

Alexievich, Svetlana. *Voices of Chernobyl.* New York: Picador, 2006.

Allardice, Corbin, and Edward R. Trapnell. *The First Pile.* Los Alamos, NM: Manhattan Engineer District, Autumn, 1946.

Allison, S. K. *Enrico Fermi, 1901–1954.* Washington: National Academy of Sciences, 1957.

Alsos Digital Library for Nuclear Issues. http://alsos.wlu.edu/.

Alvarez, Lizette. "Long-Secret Fallout Shelter Was a Cold War Camelot." *New York Times*, October 1, 2011.

American Institute of Physics. "Marie Curie." http://www.aip.org/history/curie/.

Anderson, Don L. "Energetics of the Earth and the Missing Heat Source Mystery." Seismological Laboratory, California Institute of Technology, MantlePlumes.org. http://www.mantleplumes.org/Energetics.html.

Anderson, Herbert L. "The Legacy of Fermi and Szilard." *Bulletin of the Atomic Scientists*, September 1974. And "Fermi, Szilard, and Trinity." October 1974.

Atomic Energy Commission Records. National Archives. http://www.archives.gov/research/guide-fed-records/groups/326.html.

Atomic Heritage Foundation Oral History Project. http://www.atomicheritage.org/index.php/historical-resources.html.

"Atoms in Appalachia." *Oak Ridge National Laboratory Review.* http://www.ornl.gov/info/ornlreview/rev25-34/chapter1.shtml.

Azimi, Nassrine. "Fukushima in America." *New York Times*, May 10, 2011.

Baggott, Jim. *The First War of Physics.* New York: Pegasus, 2010.

Barnet, Richard J. "The Ideology of the National Security State." *Massachusetts Review* 26 (Winter 1985).

Barringer, Felicity. "A Haunted Chernobyl." *New York Times*, June 24, 1987.

———. "A Rare Isotope Helps Track an Ancient Water Source." *New York Times*, November 21, 2011.

Belson, Ken. "From Safe Distance, U.S.-Japanese Team Draws Up Plan to Demolish Reactors." *New York Times*, April 7, 2011.

———. "Japanese Find Radioactivity on Their Own." *New York Times*, July 31, 2011.

———. "Panic and Heroism Greeted Crisis at Japan Nuclear Plant." *New York Times*, March 30, 2011.

———. "To Japan or Not? Travelers Weigh Risks with Bargains." *New York Times*, July 29, 2011.
Belson, Ken, and Hiroko Tabuchi. "Bodies of 2 Workers Found at Japanese Nuclear Plant." *New York Times*, April 3, 2011.
Bergman, Ronen. "Will Israel Attack Iran?" *New York Times*, January 29, 2012.
Berman, Bob. *The Sun's Heartbeat.* New York: Little, Brown, 2011.
Bernstein, Jeremy. "Master of the Trade." *New Yorker*, December 3, 10, 17, 1979.
Bethe, Hans A. "Comments on the History of the H-Bomb." *Los Alamos Science*, Fall 1982.
———. (Hans Albrecht). Hans Bethe papers, (ca. 1931)–1992. Division of Rare and Manuscript Collections, Cornell University.
———. *The Road from Los Alamos*. New York: Simon & Schuster, 1991.
Biello, David. "Nuclear Fission Confirmed as Source of More than Half of Earth's Heat." *Scientific American*, July 18, 2011.
Bird, Kai, and Martin J. Sherman. *American Prometheus.* New York: Alfred A. Knopf, 2005.
Bissell, Richard M., Jr. *Reflections of a Cold Warrior: From Yalta to the Bay of Pigs.* New Haven, CT: Yale University Press, 1996.
Blackwell, Andrew. *Visit Sunny Chernobyl.* New York: Rodale, 2012.
Bogdanich, Walt. "As Technology Surges, Radiation Safeguards Lag." *New York Times*, January 26, 2010.
———. "Radiation Offers New Cures, and Ways to Do Harm." *New York Times*, January 23, 2010.
Borel, Marguerite, writing as Camille Marbo. *À travers deux siècles*. Paris: Grasset, 1967.
Bosch, Torie. "What Will Turn Us On in 2030?—Forecasting where our energy will come from in 20 years." *Slate*, October 20, 2011.
Boyer, Paul. "From Activism to Apathy: The American People and Nuclear Weapons, 1963–1980." *Journal of American History* 70, no. 4 (March 1984).
Bracken, Paul. *The Second Nuclear Age: Strategy, Danger, and the New Power Politics.* New York: Times Books/Henry Holt & Company, 2012.
Bradshaw, Gideon, director. "Can We Make a Star on Earth?" *Horizon,* 2009.
Bradsher, Keith. "A Radical Kind of Reactor." *New York Times*, March 24, 2011.
Bradsher, Keith, Hiroko Tabuchi, and David E. Sanger. "Workers Strain to Retake Control After Blast and Fire at Japan Plant." *New York Times*, March 15, 2011.
Brams, Steven J. "Game Theory and the Cuban Missile Crisis." *Plus,* January 2001.
Briggle, Adam. "It's Time to Frack the Innovation System—What the history of fracking tells us about our shortsighted R&D system." *Slate*, April 11, 2012.
Broad, William J. "The Bomb Chroniclers." *New York Times*, September 13, 2010.
———. "Drumbeat of Nuclear Fallout Fear Doesn't Resound with Experts." *New York Times*, May 2, 2011.
———. "From Afar, a Vivid Picture of Japan Crisis." *New York Times*, April 2, 2011.
———. "Laser Advances in Nuclear Fuel Stir Terror Fear." *New York Times*, August 20, 2011.
———. "So Far Unfruitful, Fusion Project Faces a Frugal Congress." *New York Times*, September 29, 2012.
———. *Teller's War: The Top-Secret Story Behind the Star Wars Deception.* New York: Simon & Schuster, 1992.
———. "Why They Called It the Manhattan Project." *New York Times*, October 30, 2007.
Broad, William J., and Hiroko Tabuchi. "In Fuel-Cooling Pools, a Danger for the Longer Term." *New York Times*, March 15, 2011.
Broder, John M., Matthew L. Wald, and Tom Zeller Jr. "At U.S. Nuclear Sites, Preparing for the Unlikely." *New York Times*, March 28, 2011.
Brooks, David. "The Alpha Geeks." *New York Times*, May 23, 2008.
Brumfield, Geoffrey. "America's Nuclear Dumpsters." *Slate*, January 30, 2013.
Bryce, Robert. "The Coal Hard Facts—Environmentalists fervently wish for the end of coal. Here's why it can't be replaced anytime soon." *Slate*, May 29, 2012.

Brzezinski, Matthew. *Red Moon Rising: Sputnik and the Hidden Rivalries That Ignited the Space Age.* New York: Times Books, 2007.
Buderi, Robert. *The Invention That Changed the World.* New York: Simon & Schuster, 1996.
Burchett, Wilfred. "The Atomic Plague." *London Daily Express*, September 5, 1945.
Burgan, Michael. *Perspectives on Hiroshima: Birth of the Nuclear Age.* New York: Marshall Cavendish Benchmark, 2010.
Burkeman, Oliver. "Zero Protection from Nuclear Code." *Guardian*, June 16, 2004.
Burke-Ward, Richard, and Robert Strange, writers and producers. "Japan's Killer Quake." *NOVA*, March 30, 2011.
Burr, William, ed. "U.S. Had Plans for 'Full Nuclear Response' in Event President Killed or Disappeared during an Attack on the United States." *National Security Archive Electronic Briefing Book No. 406*, posted December 12, 2012. Nuclear Vault, National Security Archive, George Washington University.
Burr, William, and Hector L. Montford, eds. "The Making of the Limited Test Ban Treaty, 1958–1963." National Security Archive, George Washington University, August 8, 2003. http://www.gwu.edu/~nsarchiv/NSAEBB/NSAEBB94/.
Bush, Vannevar, and James B. Conant. Vannevar Bush–James B. Conant Files. Office of Scientific Research and Development, S-1, National Archives.
Buzhievskaya, Tamara I., Tatiana L. Tchaikovskaya, Galina G. Demidova, and Galina N. Koblyanskaya. "Selective Monitoring for a Chernobyl Effect on Pregnancy Outcome in Kiev, 1969–1989." *Human Biology* (August 1995).
Byers, Nina. "Fermi and Szilard." *arXiv*, June 3, 2003. http://arxiv.org/html/physics/0207094.
Cain, Susan. "Shyness: Evolutionary Tactic?" *New York Times*, June 26, 2011.
Caldicott, Helen. "After Fukushima: Enough Is Enough." *New York Times*, December 2, 2011.
———. "Unsafe at Any Dose." *New York Times*, April 30, 2011.
Calloway, Larry. "The Trinity Test: Eyewitnesses." *Albuquerque Journal*, July 16, 1995. http://larrycalloway.com/category/new-mexico-southwest-history.
Cantwell, Alan R. Jr., M.D. "The Human Radiation Experiments—How scientists secretly used US citizens as guinea pigs during the Cold War." *New Dawn* 68 (September–October, 2001).
Cardwell, Diane. "Renewable Sources of Power Survive, but in a Patchwork." *New York Times*, April 10, 2012.
Carroll, Dennis J. "Trinity Eyewitness Recalls World's First Atomic Blast." Reuters, August 10, 2010.
Carson, Cathryn, and David Hillinger, eds. "Reappraising Oppenheimer: Centennial Studies and Reflections." *Berkeley Papers in History of Science.* Berkeley: University of California Press, 2005.
Chandler, David L. "Explained: Rad, Rem, Sieverts, Becquerels: A Guide to Terminology About Radiation Exposure." *MIT News*, March 28, 2011.
"Chernobyl: Assessment of Radiological and Health Impacts." The Organisation for Economic Co-operation and Development's Nuclear Energy Agency, 2002. http://www.oecd-nea.org/rp/chernobyl/.
Cho, Adrian. "At DOE, Body Blows to Fusion, Nuclear Physics, and Particle Physics." *Science Insider*, February 13, 2012.
Clynes, Tom. "The Boy Who Played with Fusion. Taylor Wilson always dreamed of creating a star. Now he's become one." *Popular Science*, February 14, 2012.
Cockburn, Alexander. *Whiteout: The CIA, Drugs, and the Press.* New York: Verso, 1998.
"Cold War Strategy." NuclearFiles.org. http://www.nuclearfiles.org/menu/key-issues/nuclear-weapons/history/cold-war/strategy/index.htm.
Collins, Martin, ed. *After Sputnik: 50 Years of the Space Age.* New York: Smithsonian/Collins, 2007.

Columbia Center for Oral History, Columbia University. http://library.columbia.edu/indiv/ccoh.html.
Conant, Jennet. *109 East Palace.* New York: Simon & Schuster, 2005.
———. *Tuxedo Park.* New York: Simon & Schuster, 2002.
Cook, Gareth. "Why Did Japan Surrender?" *Boston*, August 2011.
Cowan, George A. *Manhattan Project to the Santa Fe Institute.* Albuquerque: University of New Mexico Press, 2010.
Cox, Patrick. "Hirohito's Nukes." *TCS Daily*, March 31, 2003.
Crick, Francis. *Life Itself, Its Origin and Nature*. New York: Simon & Schuster, 1981.
Cronin, J. W. *Fermi Remembered*. Chicago: University of Chicago Press, 2004.
Crowell, William P. "Remembrances of Venona." CIA Headquarters, July 11, 1995. http://www.nsa.gov/public_info/declass/venona/remembrances.shtml.
"Cuban Missile Crisis." AtomicArchive.com. http://www.atomicarchive.com/Docs/Cuba/index.shtml.
"Cuban Missile Crisis." Wilson Center Digital Archive. http://digitalarchive.wilsoncenter.org/collection/31/Cuban-Missile-Crisis.
Cumings, Bruce. "Korea: Forgotten Nuclear Threats." *Le Monde Diplomatique*, December 2004.
Curie, Ève. *Madame Curie.* New York: Doubleday, 1937.
Curie, Marie. *Cher Pierre que je ne reverrai plus (Journal 1906–1907).* Paris: Editions Odile Jacob, 1996.
———. *Pierre Curie*. New York: Macmillan Company, 1923.
Curie, Marie, and Irène Curie. "Radium." *Encyclopædia Britannica*, 1926.
Curie, Mme. Skłodowska. "Radium and Radioactivity." *Century*, January 1904.
Curie, P., Mme. P. Curie, and G. Bémont, "Sur une nouvelle substance fortement radioactive, contenue dans la pechblende." *Comptes rendus de l'Académie des Sciences* (Paris) 127 (December 26, 1898).
Curry, Andrew. "Why Is This Cargo Container Emitting So Much Radiation?" *Wired*, October 21, 2011.
Dam, H. J. W. "The New Marvel in Photography." *McClure's*, April 1896.
"Dan Hornig Interview." *The Story with Dick Gordon.* WUNC North Carolina Public Radio, October 30, 2006.
Dannen, Gene. *Leo Szilard Online.* http://www.dannen.com/szilard.html.
David, Leonard. "Fifty Years of Nuclear-Powered Spacecraft: It All Started with Satellite Transit 4A." SPACE.com, June 29, 2011. http://www.space.com/12118-space-nuclear-power-50-years-transit-4a.html.
Dawidowicz, Lucy S. *The War Against the Jews: 1933–1945.* New York: Bantam, 1986.
Dean, Stephen O. "Fifty Years of U.S. Fusion Research—An Overview of Programs." *Nuclear News*, July 2002.
Demetriou, Danielle. "Wild Monkeys to Measure Radiation Levels in Fukushima." *Telegraph*, December 12, 2011.
Dempsey, Judy. "How Merkel Decided to End Nuclear Power." *New York Times*, August 13, 2011.
———. "Merkel Pays a Price for Her Energy Policy Shift." *New York Times*, May 28, 2012.
Diaz, Jesus. "How Japan's Tsunami Massive Debris Plume Will Hit California and Hawaii." *Gizmodo*, July 13, 2011.
"Discovery of Fission." *Moments of Discovery*. Center for History of Physics, Niels Bohr Library and Archives, American Institute of Physics.
Dobbs, Michael. *One Minute to Midnight: Kennedy, Khrushchev, and Castro on the Brink of Nuclear War*. New York: Alfred A. Knopf, 2008.
Donadio, Rachel. "Looking for Leonardo, with Camera in Hand." *New York Times*, August 26, 2011.
Donn, Jeff. "NRC and Industry Rewrite Nuke History." Associated Press, June 28, 2011.

Dor-Ner, Zvi, executive producer. "Fallout, 1945–1995." *The People's Century*. BBC and WGBH, undated.
Douple, Evan B., et al. "Long-Term Radiation-Related Health Effects in a Unique Human Population: Lessons Learned from the Atomic Bomb Survivors of Hiroshima and Nagasaki." *Disaster Medicine and Public Health Preparedness* 5, suppl. 1 (2011).
Dower, John W. *Ways of Forgetting, Ways of Remembering: Japan in the Modern World*. New York: New Press, 2012.
Drain, Margaret, director. "Meltdown at Three Mile Island." *The American Experience*, February 22, 1999. http://www.pbs.org/wgbh/amex/three/filmmore/transcript/index.html.
Dvorak, Phred, Juro Osawa, and Yuka Hayashi. "Japanese Declare Crisis at Level of Chernobyl." *Wall Street Journal*, April 12, 2011.
Dyson, Freeman. "Memoirs." *Radiations* (Spring 2004).
———. "Oppenheimer: The Shape of Genius." *New York Review of Books*, August 15, 2013.
Dyson, George. *Turing's Cathedral: The Origins of the Digital Universe*. New York: Pantheon, 2012.
Earns, Lane R. "Reflections from Above: An American Pilot's Perspective on the Mission Which Dropped the Atomic Bomb on Nagasaki." *Crossroads: A Journal of Nagasaki History and Culture* (Summer 1995). http://www.uwosh.edu/faculty_staff/earns/olivi.html.
Edge, Dan, writer, producer, and director. "Inside Japan's Nuclear Meltdown." *Frontline*, February 28, 2012.
Edwards, Paul N. "Entangled Histories: Climate Science and Nuclear Weapons Research." *Bulletin of the Atomic Scientists*, July 13, 2012.
Eilperin, Juliet. "The Clean Tech Meltdown." *Wired*, February 2012.
Einstein, Albert. "Space-Time." *Encyclopædia Britannica*, 1926.
Eisenhower, Dwight D. "Atoms for Peace." Presented at the 470th Plenary Meeting of the United Nations General Assembly, Tuesday, December 8, 1953, at 2:45 p.m.
Elliott, Justin. "Japan's Nuclear Danger Explained." *Salon*, March 18, 2011.
Engel, Richard. "One Year After Disaster at Fukushima Nuclear Plant, Town Remains Frozen in Time." *Rockcenter*, MSNBC, March 7, 2012.
Eyewitness Accounts. Record Group 227, OSRD-S1 Committee, box 82, folder 6, "Trinity." National Archives.
Fackler, Martin. "Atomic Bomb Survivors Join Nuclear Opposition." *New York Times*, August 6, 2011.
———. "Devastation at Japan Site, Seen Up Close." *New York Times*, November 12, 2011.
———. "Fukushima's Long Link to a Dark Nuclear Past." *New York Times*, September 5, 2011.
———. "Japan's Former Leader Condemns Nuclear Power." *New York Times*, May 28, 2012.
———. "Japan Split on Hope for Vast Radiation Cleanup." *New York Times*, December 6, 2011.
———. "Nuclear Disaster in Japan Was Avoidable, Critics Contend." *New York Times*, March 9, 2012.
Falk, Jim. *Global Fission: The Battle over Nuclear Power*. New York: Oxford University Press, 1983.
Faulconbridge, Guy. "Russia Faced Major Nuclear Disaster in 2011: Report." Reuters, February 14, 2012.
Federation of American Scientists' Nuclear Information Project. http://www.fas.org/programs/ssp/nukes/index.html.
Feichtenberger, Klaus, writer and director. "Radioactive Wolves." *Nature*, October 19, 2011.
Feld, Bernard T., and Gertrud Weiss Szilard, eds. *Collected Works of Leo Szilard: Scientific Papers*. Cambridge, MA: MIT Press, 1972.
Felt, Jonathan S., producer and director. "The Men Who Brought the Dawn." Greenwich Workshop/Smithsonian Networks, 1997.
Fermi, Enrico. "The Development of the First Chain-Reacting Pile." *Proceedings of the American Philosophical Society* 90 (1946).

———. *Enrico Fermi Papers. Bulletin of the Atomic Scientists* files. Joseph Regenstein Library, University of Chicago.

———. "Physics at Columbia University: The Genesis of the Nuclear Energy Project." *Physics Today*, November 8, 1955.

Fermi, Laura. *Atoms in the Family.* Chicago: University of Chicago Press, 1954.

"Fermi Facts, Fables: Colleagues and Friends Share Memories." Argonne National Laboratory. http://www.anl.gov/Media_Center/logos20-1/fermi01.htm.

Ferro, Shaunacy. "Physicists Create Crystals That Are Nearly Alive." *Popular Science*, February 1, 2013.

Feynman, Richard P. *"Surely You're Joking, Mr. Feynman!"* New York: W. W. Norton, 1985.

Firestein, Stuart. *Ignorance: How It Drives Science.* New York: Oxford University Press, 2012.

Fowler, Tom. "Toll on Shellfish Takes a While to Become Clear." *Wall Street Journal*, April 12, 2012.

Fox, John F. Jr., FBI Historian. *In the Enemy's House: Venona and the Maturation of American Counterintelligence.* Presented at the Symposium on Cryptologic History, October 27, 2005.

Frank, Barney, guest. *Real Time with Bill Maher.* HBO, October 26, 2012.

Frankel, Max. "150th Anniversary: 1851–2001; Turning Away from the Holocaust." *New York Times*, November 14, 2001.

Franklin, Jane. *Cuba and the United States: A Chronological History.* Sydney: Ocean Press, 1966.

Freedman, Michael. "Can We Unlearn the Bomb?" *Atlantic*, May 6, 2011.

Friedrich, Thomas. *Hitler's Berlin: Abused City.* New Haven, CT: Yale University Press, 2012.

Frisch, Otto. *What Little I Remember.* Cambridge: Cambridge University Press, 1980.

Frisch, Otto Robert, and Rudolf Peierls. "Memorandum on the Properties of a Radioactive Super-Bomb"; "On the Construction of a 'Super-Bomb' Based on a Nuclear Chain Reaction in Uranium." March 1940. Atomic Archive. http://www.atomicarchive.com/Docs/Begin/FrischPeierls.shtml.

Fuhrmann, Matthew. "Nuclear Inertia—How do nuclear accidents affect nuclear power-plant construction? I built a giant database to find out." *Slate*, April 26, 2011.

Fursenko, Aleksandr, and Timothy Naftali. *Khrushchev's Cold War.* New York: W. W. Norton, 2007.

Gaddis, John Lewis. *The Cold War: A New History.* New York: Penguin Press, 2005.

———. *George F. Kennan: An American Life.* New York: Penguin Press, 2011.

Gagnon, Geoffrey. "Brave Thinkers." *Atlantic*, November 2012.

Galbraith, Kate. "A New Urgency to the Problem of Storing Nuclear Waste." *New York Times*, November 27, 2011.

Gale, Jason. "Fukushima Radiation May Cause 1,300 Cancer Deaths, Study Finds." *Businessweek*, July 17, 2012.

Gann, Carrie. "Tobacco Companies Knew of Radiation in Cigarettes, Covered It Up." ABC News, September 29, 2011.

Garber, Stephen J. "Multiple Means to an End: A Reexamination of President Kennedy's Decision to Go to the Moon." *Quest: The History of Spaceflight Quarterly* 7, no. 2 (Summer 1999).

Gavett, Gretchen. "Did This Man Predict the Tsunami at Fukushima?" pbs.org, January 17, 2012.

Gil, Brenda. "Bourgeois of the Air." *New Yorker*, May 9, 1953.

Gilinsky, Victor. "Behind the Scenes of Three Mile Island." *Bulletin of the Atomic Scientists*, March 23, 2009.

———. "Indian Point: The Next Fukushima?" *New York Times*, December 16, 2011.

Glanz, James. "Almost All in U.S. Have Been Exposed to Fallout, Study Finds." *New York Times*, March 1, 2002.

Gleick, James. "After the Bomb, a Mushroom Cloud of Metaphors." *New York Times*, May 21, 1989.

Goetzman, Keith. "Chernobyl Death Toll: 4,000 or 1 Million?" *Utne Reader*, May 4, 2010.
Goldman, Eric F. "The Wrong Man from the Wrong Place at the Wrong Time." *New York Times*, January 5, 1969.
Goldstein, Joshua S. *Winning the War on War.* New York: Dutton, 2011.
Gonyeau, Joseph. "The Virtual Nuclear Tourist: Nuclear Power Plants Around the World." http://www.nucleartourist.com/contents.htm.
Gonzalez, Robert, and Cyriaque Lamar. "How Will We Dispose of Spent Nuclear Fuel Rods for Centuries to Come?" *Gizmodo*, March 29, 2011.
Goodman, Amy, and David Goodman. "Hiroshima Cover-Up: How the War Department's Timesman Won a Pulitzer." *Common Dreams*, August 10, 2004.
Gopnik, Adam. "Decline, Fall, Rinse, Repeat." *New Yorker*, September 12, 2011.
Görtemaker, Heike B. *Eva Braun: Life with Hitler.* New York: Alfred A. Knopf, 2011.
Gottlieb, Andrew C. "Why I Made Myself Radioactive." *Bellevue Literary Review* (Fall 2011).
Grady, Denise. "Countering Radiation Fears with Just the Facts." *New York Times*, March 26, 2011.
———. "Radiation Is Everywhere, but How to Rate Harm?" *New York Times*, April 4, 2011.
Gray, Theodore. *The Elements*. New York: Black Dog & Leventhal, 2009.
Grimberg, Sharon, producer/writer. "Race for the Superbomb." *American Experience*, 1999. http://www.pbs.org/wgbh/amex/bomb/filmmore/index.html.
Griswold, Eliza. "The Fracturing of Pennsylvania." *New York Times*, November 17, 2011.
Grunberg, Slawomir. "Forty Years of Nuclear Contamination in Chelyabinsk, Russia." *Wentz*. http://www.wentz.net/radiate/cheyla/.
Gueret, Eric, director, and Sophie Parrault, producer. *Waste: The Nuclear Nightmare*. 2009.
Haffner, Sebastian. *Defying Hitler.* New York: Picador, 2003.
Hahn, Otto. *Otto Hahn: My Life*. New York: Herder and Herder, 1970.
Hall, Kenji, and Julie Makinen. "Workers Suffer Hardships in Effort to Stabilize Fukushima Plant." *Los Angeles Times*, March 29, 2011.
Hargittai, István. *Judging Edward Teller.* Amherst, MA: Prometheus, 2010.
———. *Martians of Science.* New York: Oxford University Press, 2006.
Harriman, W. Averell, and Elie Abel. "We Can't Do Business with Stalin." *American Heritage*, August 1977.
Harris, Gardiner. "Chernobyl Study Says Health Risks Linger." *New York Times*, March 17, 2011.
Harris, Michael. *The Atomic Times: My H-Bomb Year at the Pacific Proving Ground.* New York: Word One International, 2005.
Hartung, William D. "Don't Forget Nuclear Weapons." *Salon*, July 9, 2012.
Harvey Mudd College Oral History Project on the Atomic Age. Claremont Colleges.
Harvie, David I. *Deadly Sunshine: The History and Fatal Legacy of Radium*. Stroud, Gloucestershire: Tempus, 2005.
Hawkins, Helen, G. Allen Greb, and Gertrud Weiss Szilard, eds. *Toward a Livable World.* Cambridge, MA: MIT Press, 1987.
Heisenberg, Werner. Two Letters from Werner Heisenberg to Robert Jungk, the author of *Brighter Than a Thousand Suns.* http://werner-heisenberg.unh.edu/Jungk.htm.
Herbert, Bob. "Tears for Teddy." *New York Times*, May 24, 2008.
Herken, Gregg. *Brotherhood of the Bomb.* New York: Henry Holt, 2002.
———. *The Winning Weapon: The Atomic Bomb in the Cold War, 1945–1950.* New York: Alfred A. Knopf, 1980.
Hersey, John. *Hiroshima*. New York: Alfred A. Knopf, 1946.
Hessler, Peter. "The Uranium Widows." *New Yorker*, September 13, 2010.
Hibbs, Mark, and James M. Acton. "Fukushima Could Have Been Prevented." *New York Times*, March 9, 2012.
Hiss, Alger. *Recollections of a Life.* New York: Arcade, 1988.
Hiyama, Atsuki, et al. "The Biological Impacts of the Fukushima Nuclear Accident on the Pale Grass Blue Butterfly." *Nature*, August 2012.

Hoffman, David E. *The Dead Hand*. New York: Doubleday, 2009.

Holton, Gerald, F. James Rutherford, Fletcher G. Watson, and Harvard Project Physics, producers. "The World of Enrico Fermi." *People and Particles.* Holt, Rinehart and Winston, 1972.

Holton, W. C., C. A. Negin, and S. L. Owrutsky. *Report NP-6931: The Cleanup of Three Mile Island Unit 2. A Technical History: 1979 to 1990.* Electric Power Research Institute, 1990.

Hoshiko, Eugene, and Mari Yamaguchi. "Radioactive Water Leaks from Crippled Japan Plant." Associated Press, April 2, 2011.

Inajima, Tsuyoshi. "Fukushima Disaster Was Man-Made, Investigation Finds." *Bloomberg*, July 5, 2012.

Inglis-Arkell, Esther. "How Does Radiation Travel, and What Kinds of Damage Can It Do?" *io9*, March 16, 2011. http://io9.com/5782367/how-does-radiation-travel-and-what-kinds-of-damage-can-it-do.

"Intelligence Operations in the Cold War." Wilson Center Digital Archive. http://digitalarchive.wilsoncenter.org/collection/45/Intelligence-Operations-in-the-Cold-War.

International Thermonuclear Experimental Reactor—ITER. http://www.iter.org/.

Isaacs, Eric D. "Forget About the Mythical Lone Inventor in the Garage." *Slate*, May 18, 2012.

Isaacson, Walter. *Einstein: His Life and Universe.* New York: Simon & Schuster, 2007.

Iversen, Kristen. *Full Body Burden: Growing Up in the Nuclear Shadow of Rocky Flats.* New York: Crown, 2012.

———. "Nuclear Fallout." *New York Times*, March 10, 2012.

Iyer, Pico. "Heroes of the Hot Zone." *Vanity Fair*, January 2012.

Jackson, Shirley Ann. "Greatest Engineering Achievements of the 20th Century." http://www.greatachievements.org.

Jamail, Dahr. "Gulf Seafood Deformities Alarm Scientists—Eyeless shrimp and fish with lesions are becoming common, with BP oil pollution believed to be the likely cause." *Al Jazeera*, April 18, 2012.

Johnson, Ben. "Radioactive Bluefin Tuna Have Reached U.S. Waters in Wake of Fukushima Disaster." *Slate*, May 29, 2012.

Johnson, George. "Radiation's Enduring Afterglow." *New York Times*, March 26, 2011.

Johnson, Kirk. "A Battle over Uranium Bodes Ill for U.S. Debate." *New York Times*, December 26, 2010.

Joint European Torus—JET—and the European Fusion Development Agreement. http://www.efda.org/jet/.

Joliot-Curie, Irène. "Artificial Production of Radioactive Elements." Nobel Lecture, December 12, 1935. http://nobelprize.org/nobel_prizes/chemistry/laureates/1935/joliot-curie-lecture.html.

Joliot-Curie, Irène, and George Boussières. "Polonium." *Encyclopaedia Britannica*, 1949.

Jones, Maggie. "Joan Hinton: True Believer: From the Manhattan Project to Maoism in One Lifetime." *New York Times*, January 2, 2011.

Jungk, Robert. *Brighter Than a Thousand Suns: A Personal History of the Atomic Scientists.* New York: Harcourt Brace, 1958.

Kagan, Robert. "Against the Myth of American Decline." *New Republic*, January 17, 2012.

Kaplan, Fred. "What the Cuban Missile Crisis Should Teach Us." *Slate*, October 10, 2012.

———. "What Robert Caro Got Wrong." *Slate*, May 31, 2012.

———. *Wizards of Armageddon.* Palo Alto, CA: Stanford University Press, 1983.

Kaufman, Brian, writer and director. "A Is for Atom, B Is for Bomb." *NOVA*, January 22, 1980.

Kaufman, Leslie. "Mutated Trout Raise New Concerns Near Mine Sites." *New York Times*, February 22, 2012.

Kean, Sam. "Cobalt: It Makes the Dirtiest of Dirty Bombs." *Slate*, July 27, 2010.

———. "Hafnium: Building the Doomsday Device of Tomorrow." *Slate*, July 28, 2010.
———. "Radium: Cures Gout! (Warning: Also Causes Cancer)." *Slate*, July 29, 2010.
———. "Thorium: The Nuclear Fuel of the Future?" *Slate*, July 23, 2010.
Kelly, Cynthia C., ed. *The Manhattan Project: The Birth of the Atomic Bomb in the Words of Its Creators, Eyewitnesses, and Historians.* New York: Black Dog & Leventhal, 2007.
Kennedy, John F. "Special Message to the Congress on Urgent National Needs." Delivered to a joint session of Congress, May 25, 1961. http://www.jfklibrary.org/Historical+Resources/Archives/Reference+Desk/Speeches/JFK/003POF03NationalNeeds05251961.htm.
Kent, Arthur, writer. *America's Lost Bombs: The True Story of Broken Arrows.* 2001.
Khrushchev, Nikita. *Khrushchev Remembers: The Last Testament.* New York: Little, Brown, 1974.
Khrushchev, Sergei. "The Cold War Through the Looking Glass." *American Heritage*, October 1999.
Kim, Margaret G., executive producer. "USS *Scorpion* Lost at Sea." *Vanishings!*, October 4, 2003.
Kinkead, Eugene. "Notes and Comment: The Talk of the Town." *New Yorker*, August 18, 1945.
Kivi, Rose. "A History of Nuclear Power Plant Disasters." *Bright Hub*, September 24, 2010. http://www.brighthub.com/environment/science-environmental/articles/13602.aspx.
Klein, Gunther, Stefan Brauburger, and Guido Knopp, writers and producers. *The Cuban Missile Crisis Declassified.* Itaga Filmproduktion GMBH/ZDF, 2002.
Kline, Martin, et al., eds. *Collected Papers of Albert Einstein*. Princeton, NJ: Princeton University Press, 1987.
Kloor, Keith. "The Pro-Nukes Environmental Movement—After Fukushima, is nuclear energy still the best way to fight climate change?" *Slate*, January 14, 2013.
Kolbert, Elizabeth. "Burning Love." *New Yorker*, December 5, 2011.
———. "Indian Point Blank—How worried should we be about the nuclear plant up the river?" *New Yorker*, March 3, 2003.
———. "The Nuclear Risk." *New Yorker*, March 28, 2011.
Krafft, Fritz. "Lise Meitner: Her Life and Times." *Angewandte Chemie International Edition: A Journal of the Gesellschaft Deutscher Chemiker* 17 (1978).
Kramer, Andrew W. "Nuclear Industry in Russia Sells Safety, Taught by Chernobyl." *New York Times*, March 22, 2011.
Krauss, Lawrence. "How Close Are We to Doomsday? The *Bulletin of the Atomic Scientists* says we're five minutes to midnight." *Slate*, January 15, 2013.
Krulwich, Robert. "Don't Come to Stockholm! Madame Curie's Nobel Scandal." National Public Radio, December 14, 2010.
Kubrick, Stanley, producer and director. *Dr. Strangelove or: How I Learned to Stop Worrying and Love the Bomb.* Columbia Pictures, 1964.
Kumar, Manjit. *Quantum*. New York: W. W. Norton, 2010.
Kunkin, Art. "Healing at a Radioactive Mine in Montana." *Desert Star Weekly*, June 14–20, 2012.
Lamphere, Robert J., and Tom Shachtman. *The FBI-KGB War.* New York: Random House, 1986.
Lang, Daniel. "A Reporter at Large: The Top Top Secret." *New Yorker*, October 27, 1945.
Langewiesche, William. *The Atomic Bazaar.* New York: Farrar, Straus and Giroux, 2007.
Lanouette, William, with Bela Silard. *Genius in the Shadows: A Biography of Leo Szilard, the Man Behind the Atomic Bomb.* New York: Charles Scribner's Sons, 1992.
Lapidos, Juliet. "Atomic Priesthoods, Thorn Landscapes, and Munchian Pictograms—How to communicate the dangers of nuclear waste to future civilizations." *Slate*, November 16, 2009.
Larrabee, Eric. *Commander in Chief: Franklin Delano Roosevelt, His Lieutenants, and Their War.* Washington: Naval Institute Press, 2004.

Laurence, William L. Archives. *New York Times*, 1939–1945. http://www.nytimes.com/2007/10/29/science/3manharchive.html?_r=0.

———. "Drama of the Atomic Bomb Found Climax in July 16 Test." *New York Times*, September 26, 1945.

———. "Eyewitness Account of Atomic Bomb over Nagasaki." Bureau of Public Relations, War Department, September 9, 1945.

———. "Vast Power Source in Atomic Energy Opened by Science." *New York Times*, May 5, 1940.

Lehren, Andrew W. "Walking the Streets of a Nuclear Ghost Town." *New York Times*, May 25, 2012.

LeMay, Curtis E. *Daily Diary*. Curtis LeMay Papers, Library of Congress.

———. *Mission with LeMay: My Story.* New York: Doubleday, 1965.

Lepore, Jill. "The Force." *New Yorker*, January 28, 2013.

LeRoy, Mervyn, director. *Madame Curie.* 1943.

Levi, Michael. "The Devil We Know—What backpedaling on nuclear power would mean for the rest of U.S. energy policy." *Slate*, March 21, 2011.

Liebert, Rachel, and Tony Hardmon, producers and directors. *Semper Fi: Always Faithful.* MSNBC Films, February 24, 2012.

Lifton, Robert Jay. "Fukushima and Hiroshima." *New York Times*, April 15, 2011.

Light, Michael. *100 Suns.* New York: Alfred A. Knopf, 2003.

Lindley, David. *Uncertainty*. New York: Doubleday, 2007.

Liszewski, Andrew. "Radioactive Used Cars Are Being Illegally Sold in Japan." *Gizmodo*, October 27, 2011.

Loader, Jayne, Kevin Rafferty, and Pierce Rafferty, directors. *Atomic Café.* Archives Project, 1982.

Lomborg, Bjørn. "Yes, Nukes: The tragedy in Japan shouldn't cause us to abandon nuclear power." *Slate*, April 13, 2011.

Los Alamos Oral History. Los Alamos Historical Archives. http://www.losalamoshistory.org/pods.htm.

Lovett, Ian. "Troubles at a 1960s-Era Nuclear Plant in California May Hint at the Future." *New York Times*, July 4, 2012.

Lucas, Paul. "Shadows of the Soviet Space Race." *Strange Horizons,* May 3, 2004.

Lumet, Sidney, director. *Fail-Safe.* Columbia Pictures, 1964.

Macmillan, Leslie. "Uranium Mines Dot Navajo Land, Neglected and Still Perilous." *New York Times*, March 31, 2012.

Maday, Charlie, and Gerald W. Abrams, producers. "Keep Out." *Modern Marvels*, 2010.

Madrigal, Alexis. *Powering the Dream: The History and Promise of Green Technology.* New York: Da Capo, 2011.

Mahaffey, James. *Atomic Awakening*. New York: Pegasus, 2009.

Malley, Marjorie C. *Radioactivity*. New York: Oxford University Press, 2011.

Manhattan Engineer District Records. National Archives.

Manhattan Project and Metallurgical Laboratory. Special Collections Research Center, University of Chicago.

Maremont, Mark. "Nuclear Waste Piles Up—in Budget Deficit." *Wall Street Journal*, August 9, 2011.

Mason, Bobbie Ann. "Fallout: Paducah's Secret Nuclear Disaster." *New Yorker*, January 10, 2000.

Massachusetts Institute of Technology. *MIT150 Infinite History.* http://mit150.mit.edu/infinite-history.

Matsumoto, Soji. *The Unforgettable Fire*. New York: Pantheon, 1981.

McDougall, Walter A. *. . . the Heavens and the Earth: A Political History of the Space Age.* New York: Basic Books, 1985.

McInerney, Michael, producer/director. "Shot Down: U2 Spyplane." *Man Moment Machine.* Edelman Productions, 2005.

McNamara, Robert, Secretary of Defense. "Mutual Deterrence Speech." San Francisco, September 18, 1967.

McNeill, David, and Lucy Birmingham. *Strong in the Rain: Surviving Japan's Earthquake, Tsunami, and Fukushima Nuclear Disaster.* New York: Palgrave Macmillan, 2012.

McPhee, John. "The Curve of Binding Energy." *New Yorker*, December 3, 1973.

Medvedev, Zhores A. *The Legacy of Chernobyl.* New York: W. W. Norton, 1990.

Mee, Charles L. Jr. "A Good Way to Pick a Fight." *American Heritage*, August 1977.

Mehra, Jagdish. *The Solvay Conference on Physics.* Dordrecht, Netherlands: D. Reidel, 1975.

Meitner, Lise. Meitner Collection. Churchill College Archives Centre.

———. "Status of Women in the Professions." *Physics Today*, August 1960.

Meitner, Lise, and Otto Frisch. "Disintegration of Uranium by Neutrons: A New Type of Nuclear Reaction, January 16, 1939." *Nature* 143 (1939).

"Meitner Interview." 1963. American Institute of Physics Archives. http://www.aip.org/history.

Menand, Louis. "Fat Man." *New Yorker*, June 27, 2005.

———. "Getting Real." *New Yorker*, November 14, 2011.

Mettler, Fred. "Chernobyl's Living Legacy." *Bulletin of the International Atomic Energy Agency*, Autumn 1996. http://www.iaea.org/Publications/Magazines/Bulletin/Bull472/htmls/chernobyls_legacy.html.

Miklós, Vincze. "The Tragic Story of the Semipalatinsk Nuclear Test Site." *io9*, March 5, 2013. http://io9.com/5988266/the-tragic-story-of-the-semipalatinsk-nuclear-test-site.

Miller, Judith. "New Look at Stopping Nuclear War." *New York Times*, April 17, 1982.

Mingle, Jonathan. "A Dangerous Fixation." *Slate*, March 12, 2013.

Mirsky, Steve. "Nuclear Experts Explain Worst-Case Scenario at Fukushima Power Plant." *Scientific American*, March 12, 2011.

Mizin, Victor. "Russia's 'Nuclear Renaissance.'" *Journal of International Security Affairs* 14 (Spring 2008).

Moffett, Cleveland. "The Röntgen Rays in America. *McClure's*, April 1896.

Monbiot, George. "Evidence Meltdown." *Guardian,* April 5, 2011.

Moran, Michael. "The Reckoning: The Future of American Power." *Slate*, November 7–23, 2011.

Morris, Errol, director. *The Fog of War.* Sony Pictures Classics, 2003.

Mosher, Dave. "Signs of Destroyed Dark Matter Found in Milky Way's Core." *Wired*, October 26, 2010.

Mueller, John. *Atomic Obsession: Nuclear Alarmism from Hiroshima to Al-Qaeda.* New York: Oxford University Press, 2010.

Myhrvold, Nathan. "After Fukushima: Now, More Than Ever." *New York Times*, December 2, 2011.

Nash, Philip. *The Other Missiles of October: Eisenhower, Kennedy, and the Jupiters, 1957–1963.* Chapel Hill: University of North Carolina Press, 1997.

National Security Archive. George Washington University. http://www.gwu.edu/~nsarchiv/.

Neuman, William, and Florence Fabricant. "Screening the Day's Catch for Radiation." *New York Times*, April 5, 2011.

Nevada Test Site Oral History Project. University of Nevada at Las Vegas. http://digital.library.unlv.edu/ntsohp/.

Nichols, Kenneth D. *The Road to Trinity.* New York: William Morrow, 1987.

Niels Bohr Library & Archives. American Center for Physics.

Niemeyer, Kyle. "Chain Reaction: The (slow) revival of US nuclear power." *arstechnica*, March 19, 2012. http://arstechnica.com/science/news/2012/03/chain-reaction-the-slow-revival-of-us-nuclear-power.ars.

Nitske, W. Robert. *The Life of Wilhelm Conrad Röntgen, Discoverer of the X Ray.* Tucson: University of Arizona Press, 1971.

Nocera, Joe. "Chernobyl's Lingering Scars." *New York Times*, July 11, 2011.
"Nuclear Accidents." http://www.luvnpeas.org/understand/nukes.html.
"Nuclear Proliferation." Wilson Center Digital Archive. http://digitalarchive.wilsoncenter.org/collection/63/Nuclear-proliferation.
Nuclear Technology in the American West Oral History Project. American West Center, University of Utah. http://www.amwest.utah.edu/?pageId=2542.
O'Brien, Miles, and Jon Palfreman, writers. "Nuclear Aftershocks." *Frontline*, January 17, 2012.
Onishi, Norimitsu. "Culture of Complicity Tied to Stricken Nuclear Plant." *New York Times*, April 26, 2011.
———. "Japan Held Nuclear Data, Leaving Evacuees in Peril." *New York Times*, August 8, 2011.
———. "'Safety Myth' Left Japan Ripe for Nuclear Crisis." *New York Times*, June 24, 2011.
Operation Crossroads, Nuclear Weapons Test at Bikini Atoll, 1946, Oral History. Navy Historical Center. http://www.history.navy.mil/faqs/faq87-6b.htm.
Oppenheimer, J. Robert. *J. Robert Oppenheimer Papers*. Manuscript Division, Library of Congress.
Orear, Jay. *Enrico Fermi: The Master Scientist.* Ithaca, NY: Laboratory of Elementary-Particle Physics, Cornell/First University Press, 2004.
Osborn, Andrew. "Chernobyl: The Toxic Tourist Attraction." *Telegraph*, May 28, 2012.
Osgood, Kenneth. *Total Cold War: Eisenhower's Secret Propaganda Battle at Home and Abroad.* Lawrence: University Press of Kansas, 2006.
Osnos, Evan. "Aftershocks: A nation bears the unbearable." *New Yorker*, March 28, 2011.
———. "The Fallout. Seven Months Later: Japan's Nuclear Predicament." *New Yorker*, October 17, 2011.
Overbye, Dennis. "Recalling a Fallen Star's Legacy in High-Energy Particle Physics." *New York Times*, January 17, 2011.
———. "Trillions of Reasons to Be Excited." *New York Times*, November 2, 2010.
Pais, Abraham. *A Tale of Two Continents*, Princeton, NJ: Princeton University Press, 1997.
Palfreman, Jon, writer, producer, and director. "Nuclear Reaction." *Frontline* #1511, April 22, 1997.
Palmer, Brian. "Radiation Exposure—How to know if you're at risk." *Slate*, March 14, 2011.
———. "Sievert, Gray, Rem, and Rad: Why are there so many different ways to measure radiation exposure?" *Slate*, March 28, 2011.
Paret, Peter, ed. *Makers of Modern Strategy*. Princeton, NJ: Princeton University Press, 1986.
Pasternack, Alex. "The Thorium Dream." *Vice*, November 23, 2011. http://motherboard.vice.com/read/motherboard-tv-the-thorium-dream.
"Patent on World's First Reactor Was a Long Time Coming." *Media Center News: Argonne at 50.* Argonne National Laboratory. http://www.anl.gov/Media_Center/News/History/news960518.html.
Payne, Stanley G. "Italian Fascism: Selections from the Fry Collection." University of Wisconsin–Madison, July–September 1998.
Perkins, Sid. "Earth Still Retains Much of Its Original Heat." *Science*, July 17, 2011.
Perrin, Jean, Paul Langevin, and Emma Jeanne Desfosses. *Testimonials.* Archives, Ecole Supérieure de Physique et de Chimie Industrielles de la Ville de Paris.
Peterson, Cass. "A Decade Later, TMI's Legacy Is Mistrust." *Washington Post*, March 28, 1989.
Petit, Charles. "It's Scary, It's Expensive, It Could Save the Earth—Nuclear power risking a comeback." *National Geographic*, April 2006.
Pickover, Clifford. *The Physics Book: From the Big Bang to Quantum Resurrection, 250 Milestones in the History of Physics.* New York: Sterling, 2011.
Pillitteri, Carl. "None of You Are Getting Out of Here." *Salon*, March 9, 2012.
Pinker, Steven. *The Better Angels of Our Nature: Why Violence Has Declined.* New York: Viking, 2012.

Pollack, Andrew. "After Japan Crisis, New Urgency for Radiation Drugs." *New York Times*, March 31, 2011.

———. "Japanese Revisit Nuclear Zone While They Can." *New York Times*, April 21, 2011.

Pool, Robert. "Searching for Safety." *Beyond Engineering*. New York: Oxford University Press, 1997.

Popham, Peter. "Mussolini's Former Home to Be Turned into a Roman Holocaust Museum." *Independent*, April 27, 2004.

Powell, Bill, and Hideko Takayama. "Fukushima Daiichi: Inside the Debacle." *Fortune*, April 20, 2012.

Prager, Stewart C. "How Seawater Can Power the World." *New York Times*, July 10, 2011.

Preuss, Paul. "What Keeps the Earth Cooking?" News Center, Lawrence Berkeley National Laboratory (Berkeley Lab), July 17, 2011.

"Primary Resources: The First Nuclear Test in New Mexico." *American Experience*, http://www.pbs.org/wgbh/americanexperience/features/primary-resources/truman-bombtest/.

Quinn, Susan. *Marie Curie*. Reading, PA: Perseus Books, 1995.

"Radiation Exposure and Cancer." American Cancer Society. http://www.cancer.org/cancer/cancercauses/othercarcinogens/medicaltreatments/radiation-exposure-and-cancer.

Ravilious, Kate. "Chernobyl Birds' Defects Link Radiation, Not Stress, to Human Ailments." *National Geographic News*, April 18, 2007.

Redniss, Lauren. *Radioactive: Marie & Pierre Curie: A Tale of Love and Fallout*. New York: It Books, 2010.

Reed, Stanley. "Report Casts Doubt on Britain's Nuclear Electricity Strategy." *New York Times*, March 4, 2013.

Rhodes, Richard. *Arsenals of Folly: The Making of the Nuclear Arms Race*. New York: Alfred A. Knopf, 2007.

———. "The Complementarity of the Bomb." *Journal of Chemical Education* (May 1989).

———. *Dark Sun: The Making of the Hydrogen Bomb*. New York: Simon & Schuster, 1995.

———. " 'I Am Become Death . . .': The Agony of J. Robert Oppenheimer." *American Heritage*, October 1977.

———. *The Making of the Atomic Bomb*. New York: Simon & Schuster, 1986.

———. "A Matter of Risk." *Nuclear Renewal*. New York: Penguin, 1993.

Richelson, Jeffrey T. *Spying on the Bomb*. New York: W. W. Norton, 2006.

Rife, Patricia. *Lise Meitner and the Dawn of the Nuclear Age*. Boston: Birkhauser, 1999.

Rittenmeyer, Nicole, and Seth Skundrick, executive producers. *Third Reich: Rise and Fall*. New Animal Productions, December 14, 2010.

"Rocky Flats Historical Public Exposures Studies." Colorado Department of Public Health and Environment. http://www.cdphe.state.co.us/hm/rf/rfhealth/index.htm.

Rose, Kenneth D. *One Nation Underground: The Fallout Shelter in American Culture*. New York: New York University Press, 2001.

Rosenbaum, Ron. *How the End Begins*. New York: Simon & Schuster, 2011.

———. "Radioactivity—Why is this blight different from all other blights?" *Slate*, March 18, 2011.

———. "Six Questions About the Nuclear Crisis in the Middle East." *Slate*, March 14, 2012.

———. "The Subterranean World of the Bomb." In *The Secret Parts of Fortune*. New York: Random House, 2000.

Rosenberg, David Alan. "A Smoking Radiating Ruin at the End of Two Hours: Documents on American Plans for Nuclear War with the Soviet Union, 1954–1955." *International Security* (Winter 1981/1982).

Rosenthal, Elisabeth. "Experts Find Reduced Effects of Chernobyl." *International Herald Tribune*, September 6, 2005.

Roy, Susan. *Bomboozled: How the U.S. Government Misled Itself and Its People into Believing They Could Survive a Nuclear Attack*. New York: Pointed Leaf Press, 2011.

Rozenthal, Stefan, ed. *Niels Bohr: His Life and Work as Seen by His Friends and Colleagues.* New York: Wiley & Sons, 1967.

Rutherford, Ernest. "Radioactivity." *Encyclopædia Britannica*, 1910.

Sacks, Ethan. "Chernobyl Study: Birds in the Radiated Region Have on Average 5% Smaller Brains." *New York Daily News*, February 10, 2011.

Sahota, Maninderpal, director. "Meltdown in Chernobyl." *Seconds from Disaster.* Darlow Smithson Productions, August 17, 2004.

Sakharov, Andrei. *Memoirs*. New York: Vintage, 1992.

Saletan, William. "Nuclear Overreactors—Let's cool the political meltdown over Japan's damaged nuclear power plants." *Salon*, March 14, 2011.

———. "Shaken to the Core—I'll feel safer about nuclear power when the industry looks more shaken by what happened in Japan." *Slate*, April 1, 2011.

Samuels, Lennox. "A Japanese Firefighter Talks Surviving the Nuclear Reactors." *Newsweek*, April 3, 2011.

Sanger, David E., Matthew L. Wald, and Hiroko Tabuchi. "U.S. Calls Radiation 'Extremely High,' Sees Japan Nuclear Crisis Worsening." *New York Times*, March 16, 2011.

Savranskaya, Svetlana. "Cuba Almost Became a Nuclear Power in 1962: The scariest moment in history was even scarier than we thought." *Foreign Policy*, October 10, 2012.

Sayare, Scott. "Wishing Upon an Atom in a Tiny French Village." *New York Times*, February 2, 2012.

Sayen, Jamie. *Einstein in America*. New York: Crown, 1985.

Schell, Jonathan. "The Fate of the Earth." *New Yorker*, February 1, 8, 15, 1982.

Schmidt, Chris, writer, producer, and director. "Hunting the Elements." *NOVA,* April 3, 2012.

Schwartz, Ariel. "Was a Nuclear Renaissance Possible Before the Japan Disaster?" *Fast Company*, March 25, 2011.

Schwartz, Benjamin. "The Real Cuban Missile Crisis." *Atlantic*, January/February 2013.

Schwartz, Stephen I., ed. *Atomic Audit: The Cost and Consequences of U.S. Nuclear Weapons Since 1940.* Washington: Brookings Institution Press, 1998.

Seaborg, Glenn T. *Seaborg by Seaborg.* https://isswprod.lbl.gov/Seaborg/bio.htm.

Segaller, Stephen, writer and director. *Rain of Ruin.* 1995.

Segrè, Emilio. *Enrico Fermi, Physicist*. Chicago: University of Chicago Press, 1970.

———. *From X-Rays to Quarks: Modern Physicists and Their Discoveries.* Milan: Mondadori, 1976.

Seife, Charles. "Fusion Energy's Dreamers, Hucksters, and Loons—Bottling up the power of the sun will always be 20 years away." *Slate*, January 3, 2013.

———. *Sun in a Bottle: The Strange History of Fusion and the Science of Wishful Thinking.* New York: Viking, 2009.

Shapiro, Fred C. "Nuclear Waste." *New Yorker*, October 19, 1981.

Sheehan, Jason. "The Birth of the Atomic Cheeseburger." *GiltTaste*, October 10, 2011.

Sherrod, Robert. Untitled manuscript on the history of NASA. Sherrod Archives, NASA Headquarters.

Shukman, Henry. "Chernobyl, My Primeval, Teeming, Irradiated Eden." *Outside*, March 2011.

Shultz, George P., William J. Perry, Henry A. Kissinger, and Sam Nunn. "A World Free of Nuclear Weapons." *Wall Street Journal*, January 4, 2007.

Silverstein, Ken. "The Radioactive Boy Scout." *Harper's*, November 1998.

Sime, Ruth Lewin. *Lise Meitner: A Life in Physics.* Berkeley: University of California Press, 1996.

Simpson, Robert. "The Infinitesimal and the Infinite." *New Yorker*, August 18, 1945.

Singer, Lynette, writer. *Japan's Atomic Bomb.* August 16, 2005.

Slabbert, N. J. "The Lost Prestige of Nuclear Physics." *New Atlantis*, Summer 2009.

Sledge, Matt. "Ted Turner Says Coal, Oil Industries Need 'A Good A** Kicking.'" *Huffington Post*, May 25, 2011.

Slowiczek, Fran, and Pamela M. Peters. "The Discovery of Radioactivity: The Dawn of the Nuclear Age." *Access Excellence*. National Health Museum. http://www.accessexcellence.org/AE/AEC/CC/radioactivity.php.
Smith, Alice Kimble and Charles Weiner, eds. *Robert Oppenheimer: Letters and Recollections*. Cambridge: Harvard University Press, 1980.
Smith, Elliot Blair and Gopal Ratnam. "$35B Missile Defense Misses Bullet with Bullet." *Bloomberg*, August 3, 2011.
Smith, Jean Edward. *Eisenhower in War and Peace*. New York: Random House, 2012.
Smith, P. D. *Doomsday Men*. New York: St. Martin's, 2007.
Smyth, Henry DeWolf. *Atomic Energy for Military Purposes: The Official Report on the Development of the Atomic Bomb Under the Auspices of the United States Government*. Aka the Smyth Report. Princeton, NJ: Princeton University Press, 1945. http://www.atomicarchive.com/Docs/SmythReport/index.shtml.
Society of Nuclear Medicine. "History of Nuclear Medicine." http://interactive.snm.org/index.cfm?PageID=1107&RPID=924.
Socolow, Robert. "Reflections on Fukushima: A time to mourn, to learn, and to teach." *Bulletin of the Atomic Scientists*, March 21, 2011.
Sorensen, Kirk. "Can Thorium End Our Energy Crisis?" TED Talks, April 22, 2011.
Sorensen, Ted. *Counselor*. New York: Harper, 2008.
Southern, Terry. "Check-Up with Doctor Strangelove." *Filmmaker*, Fall 2004.
Soviet Archives, Library of Congress.
Sparberg, Esther. "Study of the Discovery of Fission." *American Journal of Physics* 32 (1964).
Specter, Michael. "A Wasted Land: A Special Report—10 Years Later, Through Fear, Chernobyl Still Kills in Belarus." *New York Times*, March 31, 1996.
Staff. "Absorbent Yet to Soak Up Radioactive Water at Fukushima Plant. *Kyodo News*, April 3, 2011.
Staff. "Atom: Einstein, the Man Who Started It All." *Newsweek*, March 10, 1947.
Staff. "A Big Laser Runs into Trouble." *New York Times*, October 6, 2012.
Staff. "The Bloated Nuclear Weapons Budget." *New York Times*, October 29, 2011.
Staff. "The Complete Japan Crisis Timeline." *Gizmodo*, May 12, 2011. http://gizmodo.com/5780998/the-definitive-japan-crisis-timeline.
Staff. "Did You Know? Kodak Park Had a Nuclear Reactor—Uranium cache was spirited away more than four years ago." *Rochester Democrat and Chronicle*, May 11, 2012.
Staff. "Enrico Fermi Dead at 53; Architect of Atomic Bomb." *New York Times*, November 28, 1954.
Staff. "Greenpeace Penetrates French Nuclear Plant." *Al Jazeera*, December 5, 2011.
Staff. "In the Wake of Fukushima." *New York Times*, July 23, 2011.
Staff. "Iranians 'Confess' to Nuclear Scientist Murders on State Television." *Guardian*, August 6, 2012.
Staff. "Iran's First Nuclear Power Plant Goes into Operation." Reuters, September 4, 2011.
Staff. "Japan: Butterfly Mutations Found Near Damaged Nuclear Plant." Associated Press, August 16, 2012.
Staff. "The 'Jewish Spirit' in Science." *Nature*, April 30, 1938.
Staff. "The Oppenheimer Years, 1943–1945." *Los Alamos Science* (Winter/Spring 1983).
Staff. "Peace, Love and Understanding: Barack Obama Proposes a World Free of Nuclear Weapons." *Economist*, April 6, 2009.
Staff. "Prince Hints Saudi Arabia May Join Nuclear Arms Race." Associated Press, December 6, 2011.
Staff. "Prof Takes Girl for a Ride, Walks Home." *Berkeley Gazette*, February 14, 1934.
Staff. "Radiation in Japan Seas: Risk of Animal Death, Mutation?" *National Geographic News*, March 11, 2011.
Staff. "Radium Greatest Find of History." *Chicago Daily Tribune*, June 21, 1903.

Staff. "Swedish Man Arrested After Trying to Split Atoms in His Kitchen; Says It Was Only a Hobby." Associated Press, August 3, 2011.
Staff. "Utility Says It Underestimated Radiation Released in Japan." Reuters, May 24, 2012.
Staff. "Word for Word: Chernobyl Witnesses; After Meltdown, Unsung Heroes Talk of Rads, Duty and Vodka." *New York Times*, April 21, 1996.
Staff. "Workers Bid Ill-Fated Chernobyl a Bitter Farewell." *New York Times*, December 15, 2000.
Staff. "Would a New Nuclear Plant Fare Better than Fukushima?" *National Geographic Daily News*, March 23, 2011.
Staff, with Robert S. McNamara, Defense Secretary. "Is Russia Slowing Down in Arms Race?" *U.S. News & World Report*, April 12, 1965.
Stern, Mark Joseph. "Did Chernobyl Cause the Soviet Union to Explode? The nuclear theory of the fall of the USSR." *Slate*, January 25, 2013.
Stern, Philip M., and Harold P. Green. *The Oppenheimer Case—Security on Trial*. New York: Harper & Row, 1969.
Stern, Sheldon N. *The Cuban Missile Crisis in American Memory: Myths versus Reality*. Stanford, CA: Stanford University Press, 2012.
Stimson, Henry L., Chairman. *Notes on the Interim Committee*. May–November 1945. http://www.nuclearfiles.org/menu/key-issues/nuclear-weapons/history/pre-cold-war/interim-committee/index.htm.
Stone, Richard. "The Long Shadow of Chernobyl—Twenty years after a nuclear reactor exploded, blanketing thousands of square miles with radiation, the catastrophe isn't over." *National Geographic*, April 2006.
Stranahan, Susan Q. "A More Likely Nuclear Nightmare." *Huffington Post*, May 11, 2011.
Strauss, Lewis L. *Men and Decisions*. New York: Doubleday, 1962.
Sublette, Carey. *The Nuclear Weapons Archive*. http://nuclearweaponarchive.org/index.html.
Sullivan, Robert. "A Slight Chance of Meltdown." *New York*, November 14, 2011.
Sykes, Tom. "Queen's Speech in Event of Nuclear War Revealed." *Daily Beast*, August 1, 2013.
Szilard, Leo. *Leo Szilard Papers*. Mandeville Department of Special Collections, Central University Library, University of California at San Diego.
Tabuchi, Hiroko. "Citizens' Testing Finds 20 Hot Spots Around Tokyo." *New York Times*, October 14, 2011.
———. "A Confused Nuclear Cleanup." *New York Times*, February 10, 2012.
———. "In Japan, a Painfully Slow Sweep." *New York Times*, January 7, 2013.
———. "Japan Courts the Money in Reactors." *New York Times*, October 10, 2011.
———. "Japanese Tests Find Radiation in Infant Food." *New York Times*, December 6, 2011.
———. "Japan's Nuclear Disaster Severs Town's Economic Lifeline, Setting Evacuees Adrift." *New York Times*, April 2, 2011.
———. "Japan to Nationalize Fukushima Utility." *New York Times*, May 9, 2012.
———. "Radiation-Tainted Beef Spreads Through Japan's Markets." *New York Times*, July 18, 2011.
———. "Radioactivity in Japan Rice Raises Worries." *New York Times*, September 24, 2011.
Tabuchi, Hiroko, and Ken Belson. "Japan Releases Low-Level Radioactive Water into Ocean." *New York Times*, April 4, 2011.
Tabuchi, Hiroko, Keith Bradsher, and Matthew L. Wald. "Japan Faces Potential Nuclear Disaster as Radiation Levels Rise." *New York Times*, March 14, 2011.
Tabuchi, Hiroko, and Matthew L. Wald. "Japan Scrambles to Avert Meltdowns at Two Crippled Nuclear Reactors." *New York Times*, March 13, 2011.
Tanikawa, Miki. "Japan Gets Electricity Wake-Up Call." *New York Times*, October 26, 2011.
Tannenwald, Nina. "Stigmatizing the Bomb: Origins of the Nuclear Taboo." *International Security* 29, no. 4 (Spring 2005).
Taubman, Philip. "No Need for All These Nukes." *New York Times*, January 7, 2012.

Teller, Edward. *Memoirs*. New York: Perseus, 2001.

Teller, Edward, and Allan Brown. *The Legacy of Hiroshima*. New York: Doubleday, 1962.

Thomson, Sir William (Lord Kelvin). "The Wave Theory of Light." Lecture delivered at the Academy of Music, Philadelphia, under the auspices of the Franklin Institute, September 29, 1884.

Thrall, Nathan, and Jesse James Wilkins. "Kennedy Talked, Khrushchev Triumphed." *New York Times*, May 22, 2008.

"Three Mile Island Emergency." Community Studies Center, Dickinson College. http://www.threemileisland.org.

Timmer, John. "Dept. of Energy Signs Agreements to Develop Small Nuclear Generators." *arstechnica*, March 3, 2012.

"To Fermi—with Love." Argonne National Laboratories. http://osulibrary.oregonstate.edu/specialcollections/coll/pauling/peace/audio/energy1167a-stagg1.html.

"Trinity." Record Group 227, OSRD-S1 Committee, National Archives.

Truman, Harry S. "The Decision to Drop the Atomic Bomb." Harry S. Truman Library & Museum, 1945–64. http://www.trumanlibrary.org/whistlestop/study_collections/bomb/large/index.php.

———. *Memoirs*. New York: Doubleday, 1955.

Tuchman, Barbara W. *The Proud Tower: A Portrait of the World Before the War, 1890–1914*. New York: Macmillan, 1966.

United States Atomic Energy Commission. *In the Matter of J. Robert Oppenheimer*. Washington: Government Printing Office, 1954.

United States Department of Energy. "Enrico Fermi: Audio/Video Clips." *DOE R&D Accomplishments*. http://www.osti.gov/accomplishments/fermiAV.html.

United States Nuclear Regulatory Commission. "Backgrounder on the Three Mile Island Accident." http://www.nrc.gov/reading-rm/doc-collections/fact-sheets/3mile-isle.html.

United States Strategic Bombing Survey. "The Effects of Atomic Bombs on Hiroshima and Nagaski." Washington: Government Printing Office, 1946. http://www.ibiblio.org/hyperwar/AAF/USSBS/AtomicEffects/index.html.

Upton, John. "Fusion Experiment Faces New Hurdles." *New York Times*, June 24, 2011.

Vick, Charles P. "CIA/CIO Declassifies N1-L3 Details." Global Security.org. http://www.globalsecurity.org/intell/library/imint/4_n1_1.htm.

"Voice of Hibakusha." *Hiroshima Witness*. Hiroshima Peace Cultural Center/NHK, August 1990. http://www.inicom.com/hibakusha/.

Volkman, Ernest. *Science Goes to War: The Search for the Ultimate Weapon, from Greek Fire to Star Wars*. New York: John Wiley, 2002.

Wade, Mark. "What Did the CIA Know and When Did They Know It?" *Encyclopedia Astronautica*. http://www.astronautix.com/articles/whanowit.htm.

Wald, Matthew L. "If Indian Point Closes, Plenty of Challenges." *New York Times*, July 13, 2011.

———. "Japan Nuclear Crisis Revives Long U.S. Fight on Spent Fuel." *New York Times*, March 23, 2011.

———. "Nuclear Power's Death Somewhat Exaggerated." *New York Times*, April 10, 2012.

———. "Studies Clash on the Impact of Closing Indian Point." *New York Times*, October 17, 2011.

———. "U.S. Engineers Cite Lengthy Cleanup in Japan." *New York Times*, April 19, 2011.

Waldman, Katy. "Me, Myself, and Iodine—What's in the radioactive vapors leaking from the damaged Japanese nuclear power plant, and how dangerous is it?" *Slate*, March 12, 2011.

Walker, Gregory. Trinity Atomic website. http://www.abomb1.org.

Walsh, Bryan. "The Worst Kind of Poverty: Energy Poverty." *Time*, October 11, 2011.

Waltz, Kenneth N. "Why Iran Should Get the Bomb—Nuclear Balancing Would Mean Stability." *Foreign Affairs* (July/August 2012).

Weart, Spencer, and Gertrud Weiss Szilard, eds. *Leo Szilard: His Version of the Facts*. Cambridge, MA: MIT Press, 1978.

Weaver, J. Ellsworth III. "A Brief Chronology of Radiation and Protection." Idaho State University Department of Physics, 1995. http://www.physics.isu.edu/radinf/chrono.htm#top.

Weinberg, Steven. "The Crisis of Big Science." *New York Review of Books*, May 10, 2012.

Weiner, Tim. *Enemies: A History of the FBI.* New York: Random House, 2012.

Weinstein, Adam. "Nuclear Weapons on a Highway Near You." *Mother Jones*, February 15, 2012.

———. "We're Spending More on Nukes Than We Did During the Cold War?!" *Mother Jones*, November 9, 2011.

———. "World May Face New Nuclear Arms Race." *Mother Jones*, November 7, 2011.

Weller, George. *A Nagasaki Report* (censored by the War Department). September 1945. http://www.nuclearfiles.org/menu/key-issues/nuclear-weapons/history/pre-cold-war/hiroshima-nagasaki/weller_nagasaki-report.htm.

Welsome, Eileen. *The Plutonium Files: America's Secret Experiments in the Cold War.* New York: Dial Press, 1999.

Whisker, James B. "Italian Fascism: An Interpretation." *Journal for Historical Review* 4, no. 1 (Spring 1983).

Whyte, Chelsea. "Nuclear Weapons Labs Repurposed by Climate Scientists." *International Science Times*, July 15, 2012.

Wilford, John Noble. "With Fear and Wonder in Its Wake, *Sputnik* Lifted Us into the Future." *New York Times*, September 25, 2007.

Wilford, John Noble, and Matthew L. Wald. "A Flash, and an Uncontrolled Chain Reaction." *New York Times*, October 1, 1999.

Wilkins, Alasdair. "Why a Nuclear Reactor Will Never Become a Bomb." *i09*, March 16, 2011. http://io9.com/5782349/why-a-nuclear-reactor-will-never-become-a-bomb.

Wilson, Chris. "How to Fix the Nuclear Crisis in Japan: A Physicist Breaks Down the Best Reader Ideas for How to Fix Fukushima." *Slate*, March 28, 2011.

Wilson, Jane S., and Charlotte Serber, eds. *Standing By and Making Do: Women of Wartime Los Alamos.* Los Alamos, NM: Los Alamos Historical Society, 2008.

Wilson, Ward. *Five Myths About Nuclear Weapons.* New York: Houghton Mifflin Harcourt, 2012.

———. "The Myth of Nuclear Necessity." *New York Times*, January 13, 2013.

Wood, Graeme. "My Atomic Holiday." *Atlantic*, September 2012.

Yamaguchi, Mari. "Records Show Japan Government Knew Meltdown Risk Early." Associated Press, March 9, 2012.

Yokota, Takashi, and Toshihiro Yamada. "Heroes of Japan's Nuclear Disaster All but Forgotten." *Newsweek*, March 4, 2012.

York, Herbert F. *The Advisors: Oppenheimer, Teller, and the Superbomb*. Palo Alto, CA: Stanford University Press, 1989.

Yourgrau, Tug, producer. "Secrets, Lies, and Atomic Spies." *NOVA*, PBS, February 5, 2002. http://www.pbs.org/wgbh/nova/transcripts/2904_venona.html.

Zaitchik, Alexander. "Inescapable, Apocalyptic Dread: The Terrifying Nuclear Autumn of 1983." *Salon,* September 29, 2013.

Zakaria, Fareed. "Fareed Zakaria Answers Your Questions on Nuclear Weapons." CNN, March 27, 2012.

Zoellner, Tom. *Uranium: War, Energy, and the Rock That Shaped the World.* New York: Penguin, 2010.

Zubok, Vladislav, and Constantine Pleshakov. *Inside the Kremlin's Cold War: From Stalin to Khrushchev*. Cambridge, MA: Harvard University Press, 1996.

PHOTO CREDITS

If not listed below, the photographer is unknown and the rights are public domain.

1: AIP Emilio Segrè Visual Archives, Lande Collection
3: AIP Emilio Segrè Visual Archives, Physics Today Collection
5: Samuel Goudsmit, AIP Emilio Segrè Visual Archives, Goudsmit Collection
6: AIP Emilio Segrè Visual Archives
7: US Department of Energy, ID 2017562
8: US National Archives & Records Administration, ID 326-PV-4(4)
9: US Department of Energy, ID 2017709
10: US Department of Energy, ID C939c10
11: US National Archives & Records Administration, ID unknown
12: US Department of Energy, ID 2020381
13: US National Archives & Records Administration, ID 77-BT-115
14: US National Archives & Records Administration, ID unknown
15: US National Archives & Records Administration, ID 1242
16: US National Archives & Records Administration, ARC ID 6234452
17: US National Archives & Records Administration, ID 434-RF-10(2)
18: US National Archives & Records Administration, ARC ID 6234466
19: Courtesy of the author
20: Ralph Morse, Time & Life Pictures
21: US National Archives & Records Administration, ID 311-D-15(7)
22: Courtesy of the author
23: US National Archives & Records Administration, ID 412-DA-8666
24: AP Photo/Volodymir Repik, April 25, 2007
25: Air Photo Service Co. Ltd., Japan

INDEX

Abelson Philip, 144–45, 162
A-bombs. *See* atomic bombs
Academic Assistance Council, 80, 81, 92
Academy of Sciences (Académie des Sciences), France, 14, 27, 34, 37, 188, 325
Academy of Sciences, Sweden, 34, 96
Acheson, Dean, 220, 296
Aczel, Amir, 61
Adams, Henry, 33
Adamsky, Victor, 268
Adamson, Keith R., 119
Aeby, Jack, 200, 203
AEC. *See* Atomic Energy Commission
Agnew, Harold, 132, 225–26
Allen, Michael, 346
Allison, Sam, 165, 199–200
Alvarez, Luis, 376
  atomic bomb research and, 215, 261, 262, 221
  Lawrence and, 141–42
  nuclear fission research of Meitner and Frisch and, 144–45
  thermonuclear fusion research of, 212, 228, 229
Amaldi, Edoardo, 59, 62, 64, 185
americium-241, 3, 90, 363, 377
ammonia, 372–73
Anderson, Herbert
  injuries suffered by, 167, 226
  Los Alamos and, 158
  nuclear fission research at Columbia and, 103, 104–05, 133
  nuclear reactor (CP-1) installation and testing by, 126–27, 131, 132–33, 136
  uranium fission research at Chicago and, 110–11, 113, 119, 124, 225, 233
Anderson, Philip, 221
Andropov, Yuri, 327, 328
"Andy's Atomic Adventures" (comic book), 375
Appell, Paul, 41, 43, 47, 48
Argonne reactor complex, Chicago, 129–30, 138, 139, 160, 165, 166, 188, 227, 229, 243, 303. *See also* Chicago Pile-1 (CP-1) nuclear reactor
Arnold, Hap, 221, 281
Arrhenius, Svante, 48
Arzamas-16, Russia (nuclear research site), 233, 237, 255, 275
Association of Los Alamos Scientists, 220
atomic bombs
  Bradley's "psychological" use of, 235, 236–37, 278
  Fermi as father of, 266–67
  first use of term, 79
  understanding more to fear less about, 378–79
  Wells's novel on, 79–80
Atomic Energy Commission (AEC)
  arms control and, 236, 245–46
  creation of, 229
  image of atomic energy and, 230–31, 341
  labs and production facilities of, 282
  miners working for, 336
  nuclear detection system and, 234–35
  Oppenheimer and, 229, 257, 258, 259–61
  radiation experiments by, 270, 271
  reactor designs and, 304
  thermonuclear fusion (Super bomb) and, 235, 247, 251, 252, 254, 258, 259
  uranium sources and, 270
  weapons testing by, 332
Atoms for Peace, 304, 305, 331, 340, 341, 363
Ayers, Eben, 237

Baca, Rowena, 204
Baggott, Jim, 192
Bainbridge, Kenneth, 199
Barnes, Sidney, 82
Barnet, Richard, 239–40

Baruch, Bernard, 230
Baudino, John, 155
Bay of Pigs invasion, Cuba, 286, 293–94
Becquerel, Alexandre-Edmond, 14
Becquerel, Antoine César, 14
Becquerel, Antoine Henri, 14–15, 22, 25, 26, 27, 34, 35, 37
Belarus, 314, 334
  Chernobyl disaster's effect on, 312–13, 321, 322, 323, 324, 364, 369
  nuclear arsenal in, 292, 337
Bell, Daniel W., 162
Berg, Moe, 185–86, 190
Bergeron, Ken, 345–46, 347
Beria, Lavrenti, 193, 194, 232, 233–34, 237, 254, 269
beryllium, 53, 62, 81, 82, 104, 126–27, 167, 197, 253
Bethe, Hans, 228
  atomic bomb testing and, 168, 169, 196–97, 199, 203, 204
  concern about German progress by, 184
  energy production in stars and, 363
  on Fermi, 60–61, 62–63, 64
  Heisenberg's involvement in German bomb project and, 182–83
  Los Alamos and, 150, 151, 152, 157, 227
  nuclear explosives research of, 147
  Oppenheimer and, 142, 262
  Szilard and, 73
  Teller and, 157, 158, 169, 236, 264
  thermonuclear weapons and, 236, 249, 250, 252, 255
Beveridge, Sir William, 80
Bhabha, Homi Jehangir, 339
Bikini, tests on, 272–73, 274, 275, 340, 356
Bitter, Gustav, 209–10
black holes, 265, 266
Blair, Bruce G., 280, 292
Blandy, W. H. P., 272
Blayais nuclear power plant, France, 351
Bloch, Felix, 65
Bloomfield, Louis, 344
Bohr, Aage, 174, 175
Bohr, Harald, 173
Bohr, Margrethe, 94, 95, 120, 173, 175, 180, 181, 190
Bohr, Niels, 59, 66, 210
  Cold War fears of, 194
  death of, 269
  Einstein on, 94
  emigration to United States by, 83, 174, 191–92
  evacuation from Denmark of, 120–21, 173–74
  Fermi's Nobel Prize and, 69, 71
  German surveillance of, 179
  Hahn's research and, 96, 97, 103
  Heisenberg's involvement in German bomb project and, 178, 179–82, 188
  Los Alamos research of, 175–76, 191–92, 225
  Manhattan Project and, 148
  Meitner's support from, 93, 94–96
  nuclear arms race concerns of, 192, 194
  Oppenheimer and, 191–92, 257
  personality of, 174–75
  Planck's research and, 88
  research on nucleus by, 99, 100–101, 102
  research secrecy and, 113, 254
  Szilard's research and, 82
  Teller and, 94, 175
  uranium fission research and, 103, 105, 112, 113, 120, 144, 146
Boltzmann, Ludwig, 85
Borden, William, 260
Borel, Marguerite, 43, 47, 48
Borisevich, Valentin, 312–13
Born, Max, 58, 257
Bosch, Carl, 93, 94, 95
Bothe, Walther, 113, 183
Bourgeois, Henri, 45, 47
Bradbury, Norris, 227, 231, 251
Bradley, Omar, 235, 236–37, 278
Bravo thermonuclear tests, 272–73, 275
breeder reactors, 362–63
Brenner, David, 377–78
Bretscher, Egon, 146
Briggs, Lyman, 118, 119, 121, 147
Briggs Advisory Committee on Uranium, 118–19, 121–22
Brighter than a Thousand Suns (Jungk), 178, 180, 181
Brixner, Berlyn, 200, 203
Brodie, Bernard, 278, 282–83, 372
Brooks, Linton, 371
Brun, Jomar, 177
Brzezinski, Zbigniew, 293
Bundy, McGeorge, 294, 297
Bunn, Matthew, 335
Burchett, Wilfred Graham, 214
Bush, George H. W., 333, 334

Bush, Vannevar
arms race and, 188, 230, 252
atomic bomb research and, 147, 149, 206
German atomic bomb research and, 176
Office of Scientific Research and Development and, 122
Oppenheimer's support by, 262
Roosevelt and, 121, 146
test-ban agreement of, 255
thermonuclear research and, 125–26, 128
Byers, Eben, 38
Byrnes, James, 206, 207, 222, 230, 232, 334

Caldicott, Helen, 369
Campbell, Charles, 155
Canada, 34, 74, 117, 172, 235, 303, 306
cancer
iodine causing, 308, 336, 350, 365
nuclear medicine for treatment of, 5, 6, 13, 82, 363, 377, 379
radiation exposure and risks for, 3, 4, 5, 6, 215, 266, 268, 271, 272, 291, 320, 337, 358, 365, 366, 369, 378
radium treatments for, 35, 50
radon causing, 188
X-rays and, 13
carbon-14, 377, 378
Carter, Jimmy, 292, 293, 309, 327
Castle series of thermonuclear tests, 273, 274
Castro, Fidel, 298, 373
Centers for Disease Control (CDC), 270, 272, 336, 337
cesium, 271, 274, 322, 346, 350, 356, 357–58
Chadwick, James, 36, 52, 61, 152, 159, 173, 174, 225
Chalmers, Thomas, 82
Chernobyl, Ukraine, 314, 325
Chernobyl Forum, 323, 364, 365
Chernobyl nuclear reactor disaster, Ukraine, 5, 314–26
containment attempts in, 321–22
deaths after, 317–18, 320, 323, 361, 364, 365, 369
emergency state after, 318–20
evacuation after, 321, 322, 324
fear of, 324
financial cost of, 323
fire with toxic fallout after, 317–18
health effects of, 323–24, 349, 350, 364–65
as laboratory for scientists, 215, 325–26
local belief in Biblical prophecy about, 314
location of, 314
political effects of, 324, 359, 361–62, 369
radiation released in, 336, 356, 357, 365
ranking of, among other accidents, 322, 365
residents who stayed after, 322
safety test precipitating, 315–17
tourism to site of, 325
wildlife sanctuary in years after, 7, 325
workers (liquidators) at, 320, 347
Chesser, Ron, 326
Chevalier, Haakon, 142, 261, 262
Chicago Daily Tribune, 34–35
Chicago Pile-1 (CP-1) nuclear reactor, 129–38, 160, 173, 233, 266–67
Chicago Pile-2 (CP-2) nuclear reactor, 160, 163
Chicago reactor, Argonne, 129–30, 138, 139, 160, 165, 166, 188, 227, 229, 243, 303
China, 221, 231, 236, 243, 244, 245, 287, 288, 332, 334, 337, 361, 362, 365, 373
China Syndrome, The (movie), 309–10
Churchill, Winston, 71, 211, 232, 276, 279, 372
CIA, 148, 270, 282, 285, 287, 293, 294, 327, 340
Civil Defense Administration, 287
Coatesworth, Janet, 116
cobalt, 82, 153, 257, 267, 271, 326, 377
Cohen, Bernard, 365
Cohen, Lona (Helen), 171
comics, radiation theme in, 275, 367, 375
Comprehensive Test Ban Treaty, 336
Compton, Arthur
atomic bomb development and, 206, 207, 225, 228
Chicago Met Lab under, 126, 128, 147, 148, 149
CP-1 nuclear reactor and, 133, 137, 138
Fermi's work with, 129, 165
Oppenheimer's work with, 147, 148
postwar research and, 225, 228
Compton, Betty, 129
Comte, Auguste, 16
Conant, James Bryant, 138
arms race and, 230
atomic bomb research and, 147, 199, 202, 206, 207

Conant, James Bryant (*cont.*)
Briggs Committee and, 121
Chicago nuclear research and, 126, 128, 160
Oppenheimer's support by, 262
thermonuclear fusion and, 228
Copenhagen (Frayn), 181
Corbino, Orso Mario, 58–59, 64, 193
corium, 319–20, 321–22
Corona surveillance-satellite program, 287
Coster, Dirk, 94, 95
Counter-Intelligence Corps (Creeps), Manhattan Project, 154–55
Cowan, George, 126–27, 273–74
Crocker, William, 141
Cronkite, Walter, 308–09
Crookes, Sir William, 9, 32
Cuba, Bay of Pigs invasion of, 286, 293–94
Cuban Missile Crisis, 267, 268, 294–300
Curie, Dr. Eugène, 25, 31, 41, 43, 53
Curie, Eve, 18, 19, 39, 40, 50, 53
Curie, Irène. *See* Joliot-Curie, Irène
Curie, Jacques, 22
Curie, Marie (born Marja (Manya) Skłodowska), 5, 7, 341, 378, 379
courtship and marriage to Marie, 23–25
daughter Irène's marriage and work and, 51–52, 53
death of, 53
effect of death of Pierre on, 41–43
Einstein on, 43
family and childhood of, 16, 18–20
first meeting between Pierre and, 22–23
funding for, 141
health of, 52, 53
as hero in popular culture, 369
Langevin's affair with, 43–45, 46, 47–48, 49
Meitner's admiration for, 85
move to Paris by, 20
Nobel Prize to Pierre and, 34, 47–48
relationship with government of, 251
research collaboration with Pierre, 26–37
success and fame of, 35–39, 43, 98, 369
US tour of, 189
Universal Exposition visit of, 33–34
uranium source of, 36, 114, 194
work in Poland of, 16–18
Curie, Pierre, 353, 369
background of, 21–22
courtship and marriage to Marie, 23–25
death of, 40–41
early research of, 22–23, 25
first meeting between Marie and, 22–23
health of, 38, 39
Nobel Prize to Marie and, 34, 47–48
research collaboration with Marie, 26–37
success and fame of, 35–39
Universal Exposition visit of, 33–34
curietherapy, 35, 43

Daiichi power plant, Japan. *See* Fukushima Daiichi power plant disaster, Japan
Dallet, Joe, 172
Dally, Clarence Madison, 13
David's Sling, 334
Day After, The (TV movie), 299, 329
Dead Hand defense strategy, 292–93, 372
Debye, Peter, 119, 176
Decapitation defense strategy, 292, 328, 372
de Gaulle, Charles, 337
de Hevesy, George, 54, 376
Department of Energy (DOE), 335, 336, 341, 353, 356, 363, 374–75
Desfosses, Emma Jeanne, 44–46, 47, 48–49
Detinov, Nikolai, 299–300
deuterium, 113, 120, 147, 228, 237, 250, 253, 255
Dirac, Paul, 94
dirty bombs, 335
Dluski, Bronisława (Bronya) Skłodowska, 16, 18, 19, 20, 29, 38, 39, 42, 48, 51, 53
Dluski, Casimir, 18, 20, 29
Dobrynin, Anatoly, 297–98
Dr. Strangelove (movie), 6, 267, 291–92, 293
Donne, John, 189
Doolittle, James, 227, 255
Doomsday Machine, 291–92
Döpel, Robert, 183–84
Douple, Evan, 366
downwinders, 271–72, 336
Dubovsky, B. G., 233
DuBridge, Lee, 247
Dulles, John Foster, 283
Dunning, John R., 104–05, 133
DuPont, Robert, 369–70
DuPont and Company, 131, 149, 164, 282
Dyson, Freeman, 225, 231, 262, 265, 267, 305

earthquakes, in Japan, 5, 7, 342–45, 352–53
Ebermayer, Erich, 93

Edison, Thomas, 13
Edmundson, James, 245
Eifler, Carl, 185
Einstein, Albert, 5, 34, 46, 244, 341
  on atomic bomb use, 220–21
  on Bohr, 94
  Brownian motion explanation of, 34
  childhood interest in science of, 84–85
  Emergency Committee of Concerned Scientists of, 190
  emigration to the United States by, 68, 80, 115
  Fermi on, 66
  German denunciations of, 67–68, 92, 176
  on Haber, 373
  as hero in popular culture, 369
  on Langevin, 44
  letters between Roosevelt and, 116–18, 119, 153, 176, 220, 369
  Los Alamos consultations by, 153
  Manhattan Project and, 83, 122
  on Marie Curie, 43, 48
  on Meitner, 75, 89
  need for atomic bomb research advocated by, 114–15
  nuclear energy and, 112, 369
  nuclear warfare and 299
  Oppenheimer and, 261
  Planck's research and, 87–88
  on scientists and war efforts, 89
  Szilard and, 77, 78, 127, 163
  theory of general relativity of, 67
  unified field theory of, 83
  US military research and, 122
Eisenhower, Dwight D., 340, 370
  arms control (Open Skies) plan of, 284, 287
  arms race and, 277, 282, 286, 304
  Atomic Energy Commission and, 260, 267
  Atoms for Peace and, 304, 305, 331, 363
  ballistic missile tracking and, 330
  concern about nuclear arms use and, 255, 370
  defense policy of, 283
  first US nuclear power plant and, 304
  military-industrial complex of, 252
  missile defense shield concept and, 330
  nuclear merchant ship commissioned by, 304–05
  Oppenheimer and, 259–60
  presidency of, 246, 252, 283
  Rosenbergs' execution and, 242
  Soviet threat and, 245, 255, 256, 259
  test-ban agreement and, 267
  use of atomic bomb in Japan and, 208, 222
Eltenton, George, 261
Elugelab atoll, Mike thermonuclear tests on, 250, 251, 252–53
Emergency Committee of Concerned Scientists, 190
ENIAC computer, 248, 249
Eniwetok Island, Mike thermonuclear tests on, 250, 251, 252–53
Environmental Protection Agency (EPA), 128, 308, 310, 363
Esau, Abraham, 105
Ester, Julius, 26

Falk, Jim, 369
fallout shelters, 286, 376
Fate of the Earth, The (Schell), 299, 328
Fat Man bomb, 154, 197–98, 211, 216, 217, 230, 237, 371
FBI
  Bohr and, 174–75
  Einstein and, 122
  Fermi and, 123
  Manhattan Project and, 149, 155, 171
  Oppenheimer and, 155, 170, 257, 258, 260, 261, 262
  spying detected by, 238–39, 240, 241, 242, 363
  Szilard and, 149
Feather, Norman, 146
Fermi, Enrico, 54, 55–72, 157, 369, 379
  atomic bomb design and, 168, 170, 266–67
  atomic bomb testing and, 201–02, 206, 207, 229, 266–67
  childhood of, 56–57
  death of, 265–66, 269
  decision to emigrate to the United States by, 64–65, 67, 68–72
  education of, 57–58
  element discovered by, 84
  family life of, 123
  German nuclear program and, 185
  Hahn's research on nucleus and, 103–04
  Hanford reactor and, 164–65
  irradiation research of, 90
  legacy of, 55–56, 266, 369

Fermi, Enrico (*cont.*)
Los Alamos and, 158–59, 160, 225, 228, 247
Manhattan Project and, 148, 154, 155
marriage of, 66–67, 123, 159
need for atomic bomb research advocated by, 114, 119
Nobel Prize to, 69–70, 71, 83, 97, 110, 128
nuclear fission experiment at Columbia and, 104, 105, 133
nuclear reactor design by, 129–39, 149, 160, 163, 172, 185, 233
Oak Ridge reactor and, 163
on Oppenheimer, 261
personality of, 165, 193, 225–26
political campaigns and power of, 251
postwar research at Chicago by, 225
research approach of, 123–25
research secrecy and, 112–13, 254
on scientific advances, 379
subatomic particle research of, 226
Szilard and, 229
Teller and, 147, 157, 158, 227, 228, 229, 247, 253, 265, 330
thermonuclear fusion research and, 228, 235, 247, 254
US military research and, 122–23
University of Rome research on uranium of, 58–64, 91, 266
uranium fission research of, 109–13, 124–25, 129, 266
wartime status as enemy alien, 128–29
Wigner on, 109
Fermi, Giulio, 55, 69, 71–72, 128, 226
Fermi, Laura, 55, 63–64, 65, 66–67, 69, 70, 71–72, 83, 115–16, 123–24, 126, 128, 129, 159, 170, 176, 225–26, 265, 266
Fermi, Nella, 55, 65, 70, 71–72, 123, 128, 226–27, 266, 269
Fermilab, 266
Fermi paradox, 76, 337
fermium, 253, 266
Forsmark nuclear power plant, Sweden, 312
Ferraby, Tom, 212
Feynman, Richard (Dick), 151, 152, 164, 166, 202, 205, 247
films, radiation theme in, 274–75, 310–11, 367
First Lightning weapons test, 233–34, 235, 238, 239, 373
Fischer, Emil, 86, 88
fishermen
thermonuclear testing affecting, 273, 340
tsunami affecting, 344–45
Fitch, Val, 196, 203
Flerov, Georgi, 171–72, 193–94
Flexible Response strategy, 283, 287, 372
Fokker, Adriaan, 94, 95
Fonda, Jane, 310
France, 351, 364
Franck, James, 89, 91, 93, 207, 225
Frank, Barney, 334–35
Frayn, Michael, 181
French Academy of Sciences (Académie des Sciences), 14, 27, 34, 37, 188, 325
Frisch, David, 150
Frisch, Otto Robert, 91, 121
atomic trigger device designed by, 166–67
background of, 91
Meitner's relationship with, 97, 189, 190
move to England by, 120
move to Los Alamos by, 152–53
nuclear fission research of Meitner and, 99–102, 103, 105, 144
thermal diffusion and, 162–63
Frisén, Jonas, 378
Fromm, Friedrich, 183
Fuchs, Elizabeth, 170–71
Fuchs, Klaus
background of, 170–71
Los Alamos research by, 152, 169–70, 227, 240
spying by, 170, 171, 172, 173, 194, 237, 238, 239, 240, 251, 259
Fukushima Daiichi power plant disaster, Japan, 5, 340–60
deaths from, 360
earthquake causing, 342–45
evacuations after, 346, 349–50, 356
evidence of ancient tsunami near, 342
government regulators and, 352–53, 354
health effects of, 350, 358
heat generation from fission fragments in, 343–44, 345–46, 356
information withheld in, 350–51, 354–55
location of, 341–42
plant damage after, 347–49, 351–52, 356, 359–60
political effects of, 357, 359
private industry's relationship with government regulators in, 353

radiation released in, 336, 355, 357, 358
rehabilitation of area and resettlement after, 358–59
steam venting in, 346–47, 349
TEPCO's responsibilities after, 351–52
tsunami damage in, 344–45, 352–53
US consultants on, 349, 351, 355, 356
workers at, 346–48, 349, 352, 353–55, 356–57
Fuller, Loie, 33–34

Gabor, Dennis, 75, 79
Gale, Robert, 319, 349
Gamow, George, 94, 99, 248, 282
Gard, Robert, 334
Gardner, Meredith, 238, 242
Garson, Greer, 51
Garwin, Richard, 225, 247–48, 251
Geiger, Hans, 74, 87, 89, 113
Geitel, Hans Friedrich, 26
genetic abnormalities, and radiation, 326
Giannini, Gabriel Maria, 59–60
Gibertson Company, 375
Giesel, Friedrich, 31, 35, 188
Girshfield, Viktor, 290
Glicksman, Maurice, 265
Gold, Harry, 170, 172–73, 240–41, 242, 251
Goldstine, Herman, 248
Goncharov, German, 255, 275
Goodyear Tire & Rubber Company, 132, 282
Gopnik, Adam, 337
Gorbachev, Mikhail
Chernobyl accident and, 318, 359, 361
International Thermonuclear Experimental Reactor (ITER) and, 364
on nuclear weapons, 328
proposed nuclear arms reduction treaty and, 329–30, 332–33
Reagan's Strategic Defense Initiative and, 331, 332
START treaty and, 333
Göring, Hermann, 183
Goudsmit, Samuel, 186–87
Gore, Albert Sr., 293, 294
Great War (World War I), 32–33, 36, 74, 89, 91, 193, 372
Greenglass, David, 171, 172, 173, 194, 237, 240–41, 251
Greenglass, Ruth, 171, 173, 241
Greenglass, Samuel, 171
Greenewalt, Crawford, 134, 138, 164
Greisen, Kenneth, 204
Groves, Leslie
Allied bombing targets recommended by, 186
atomic bomb and, 197, 198, 199, 206, 207, 202, 229
Bohr surveillance ordered by, 174–75
bombing of Japan as signal to Russia and, 222
bombing of Vemork plant, Norway, and, 178
intelligence missions of, 185–86
Los Alamos and, 151, 157, 169, 225, 228
Manhattan Project management by, 147–50, 156, 165, 175, 197–98, 199
Oak Ridge National Laboratory and, 161, 162–63
Oppenheimer and, 262
postwar atomic arms research and, 229, 230
Russian scientists and, 194
silver supply and, 161
Soviet threat and, 227
Szilard and, 127, 150, 207, 208
uranium source and, 154
Grubbe, Emil, 13

Haber, Fritz, 75, 89, 92, 372–73
Hahn, David, 362–63
Hahn, Edith, 93
Hahn, Otto
Allied capture and internment of, 186–87, 188–89, 190
atomic bomb research and, 113, 121, 183, 187
awards and recognition of, 190
Hitler's Jewish laws and, 92, 93
on Meitner, 86, 96, 98, 191
Meitner's research with, 64, 84, 85–87, 88–90, 98, 99, 100, 102, 103, 121, 189, 190
Meitner's treatment by, 92, 93, 96–97, 102–03, 189, 190–91
Nobel Prize to, 188–89, 190
as scientist in Nazi Germany, 190–91
Hall, Joan, 242
Hall, Theodore, 171, 194, 237, 242
Hanford, Washington, reactor complex, 159, 160, 164–65, 166, 168, 175, 188, 225, 282
Harding, Warren G., 50
Harris, Michael, 253, 272

Harrison, Richard Stewart, 144
Haukelid, Knut, 178
Havenaar, Johan, 323, 324
Hawkins, David, 192
Hawks, H. D., 12–13
H-bombs. *See* hydrogen (Super) bombs
health
 downwinders and, 271–72, 336
 radiation experiments and, 270–71
 thermonuclear testing's long-term effects on, 340, 356
Heisenberg, Werner, 93
 Allied capture of, 186–87
 atomic bomb research and, 113, 183
 Bohr and, 94, 178, 179–82, 188
 early research of, 58, 59
 German bomb project and, 176, 177, 178–83, 185–88, 191
 German denunciations of, 176
 Groves's surveillance plans for, 185–86
 personality of, 66
 reactor design of, 175, 176–77, 183–84, 186
 rumors of death of, 184
 Teller and, 77
Hendrix, John, 161
Hersey, John, 215–16
Heslep, Charter, 258
Hess, Kurt, 92, 93, 95
Hilberry, Norman, 134–35, 135–36
Himmler, Heinrich, 94, 176
Hinckley, John Jr., 327
Hinton, Joan, 150, 203
Hirohito, Emperor, 222, 351
Hiroshima, Japan, bombing, 211–16
 American reactions to, 116, 219, 220
 attitudes toward nuclear weapons after, 6, 167, 187, 202, 230, 254, 271, 313, 330, 336, 363, 367, 369
 biological and health effects of, 7, 365–66
 bombing flight in, 216, 221, 222, 371
 damage caused by, 210, 211, 213, 214, 371
 effectiveness of bombing of, 222
 Japanese reactions to, 215–16, 219, 222, 350–51, 359
 number of deaths from, 214–15, 366, 371
 peace memorial for, 216
 recovery of city during decades after, 374
 Soviet reaction to, 232
 survivors (*hibakusha*) of, 7, 213, 214, 215, 219, 350–51, 359, 365–66, 371
 technology of, 54, 154, 166, 211, 212–15
 US planning for, 221, 237
 US support for, 228, 310
Hiss, Alger, 242–43
Hitler, Adolf, 69
 concern about nuclear arms use by, 153, 175, 176, 178, 180, 184, 208, 221, 227, 231
 German émigrés' hatred of, 112, 184, 187–88, 189, 192
 Jews and scientists exiled under, 91, 92–93, 98
 Planck's conversation with, 91
 Speer's support of nuclear research and, 183
 Szilard's concerns about, 80, 82, 114
 von Stauffenberg's assassination attempt on, 186
Hooper, Don, 266
Hooper, Stanford, 114
Hoover, J. Edgar, 122, 123, 241, 257, 260
Hornig, Dan, 198–99
Horthy, Miklós, 74, 77
House Un-American Activities Committee, 257
Howard, Charles Henry George "Mad Jack," 120
Humphrey, Hubert, 256
Hutchins, Robert, 129
hydrogen (Super) bombs
 accidental dropping of, 305
 long-term health effects of testing of, 340, 356
 Oppenheimer's opposition to, 228, 258, 259, 262, 264
 research secrecy on, 254
 Soviet testing of, 275
 Teller's research on, 125, 158, 227–28, 235, 236, 239, 247, 249–50, 252, 253, 258, 259, 262, 264, 275, 372
 testing of Bravo at Bikini, 272–73, 275, 340, 356
 testing of Castle series of, 273, 274
 testing of Mike at Elugelab, 250, 251, 252–54, 282–83, 372
 Ulam's design for, 248–50, 252, 253, 262, 264, 265, 275

Ignatenko, Lyudmilla, 318
India, 227, 337, 339, 361, 365, 373

Institute for Biological Studies, 268
Institute for Nuclear Power Engineering, Minsk, Belarus, 312–13
Interim Committee, 206, 208, 211, 228, 371
International Atomic Energy Agency (IAEA), 305, 338, 356, 364
International Thermonuclear Experimental Reactor (ITER), 364
"In the Matter of J. Robert Oppenheimer," 261–65
iodine, 188, 215, 271, 308, 336, 350, 356, 365, 377
Iran, 6, 338, 373
Israel, 334, 337–38, 360, 370

Jaczko, Gregory, 355
Japan, 364
  demand for surrender of, 211
  earthquakes in, 342–45, 352–53
  firebombing of, 209–11
  nuclear power plants in, 340–41, 342
  nuclear research in, 210
  regulation of nuclear technology in, 341
  thermonuclear testing affecting fishermen from, 273, 340
  See also Fukushima Daiichi power plant disaster, Japan; Hiroshima, Japan, bombing; Nagasaki, Japan, bombing
Joachimsthal, Czechoslovakia, mine, 36, 195, 231
Johnson, Lyndon, 211, 337
Joliot-Curie, Irène, 61, 82, 376
  death of, 188
  deuterium source for, 120
  education of, 49–50
  family background of, 25, 30, 31, 39, 41, 48
  fame of, 98, 369
  Fermi's research and, 90
  Hahn and Meitner on errors in research of, 90
  man-made radiation research of, 61, 105
  marriage of, 51–52
  Meitner's research and, 103
  nuclear fission research of, 112, 117, 120, 142, 193
  uranium sources and, 154
  work with husband Fred, 52–53, 54, 61, 62, 90, 103, 112
Joliot-Curie, Jean-Frédéric (Fred), 376
  deuterium source for, 120
  French resistance work of, 188
  health of, 188
  marriage of, 51–52
  nuclear fission research of, 142
  uranium sources and, 154
  work with wife Irène, 52–53, 54, 61, 62, 90, 103, 112
Jungk, Robert, 178, 180, 181

Kahn, Herman, 290–91, 292
Kan, Naoto, 346, 351, 352, 357
Karle, Isabella, 127
Kaufman, Irving, 241–42
Kazakhstan, atomic tests in, 233, 235, 336, 372
Kelvin, Lord William Thomson, 22–23, 39
Kennan, George, 256–57
Kennedy, John F., 265
  arms race and, 276–77, 282, 289, 376
  atomic satellite programs and, 305–06
  Bay of Pigs invasion and, 286, 293–94
  Cuban Missile Crisis and, 294–99
  fallout shelters and, 286
  joint moon mission proposal and, 284
  Khrushchev and, 285–86
  missile gap and, 282, 285–86, 293–94
  nuclear air power programs and, 305
  nuclear defense strategies and, 288–89
  Nuclear Test Ban Treaty and, 374
  presidential campaign against Nixon by, 282, 284–85
Kennedy, Robert, 297
KGB, 172, 238, 240, 292, 293, 327
Khan, Abdul Qadeer, 338
Khrushchev, Nikita, 370–71
  arms race and, 245, 254, 267, 276–77, 284–85, 373
  Bay of Pigs invasion and, 286, 293–94
  Cold War and, 300
  Cuban Missile Crisis and, 294–95, 296–97, 298, 299
  joint moon mission proposal and, 284
  rise to power by, 254
  Szilard's meeting with, 267
  US perception of threats from, 256
Khrushchev, Sergei, 246, 256, 284, 290, 294–95, 297, 298
Killing a Nation defense strategy, 277, 292, 293, 372
Kim Il Sung, 243, 373
King, Ernest, 221–22
Kissinger, Henry A., 282, 335
Kistiakowsky, George (Kisty), 169, 171, 173, 196, 197–98, 199, 202, 228

Klaproth, Martin, 25
Knuth, August, 130
Kolbert, Elizabeth, 360–61
Korea. *See* North Korea; South Korea
Korean War, 236, 241, 245, 246, 279, 370, 376
Kowalski, Joseph, 40
Kubrick, Stanley, 267, 291–92
Kurchatov, Igor Vasilievich, 192, 193, 194, 232–34, 237, 268–69

Lamb, Willis, 102, 103
Lamphere, Robert, 238, 239, 240
Langevin, Emma Jeanne Desfosses, 44–46, 47, 48–49
Langevin, Michel, 49
Langevin, Paul, 51, 53
  affair with Marie Curie, 43–45, 46, 47–48, 49
  background and early life of, 32, 38, 40
  Irène Curie's career and, 49, 53
  marriage with Jeanne, 44–46, 48–49
Langevin-Joliot, Hélène, 49
Langewiesche, William, 213
Lanouette, William, 122
Larionov, Nikolay, 255
Laurence, William, 189, 204, 217–18
Lawrence, Ernest Orlando, 174, 369
  atomic bomb and, 199, 206, 207, 261
  big science of, 141, 252
  cyclotron at Berkeley built by, 140–41, 145, 193
  Livermore lab and, 252, 262, 282
  Manhattan Project and, 147–48, 154, 220
  Nobel awarded to, 145
  Oppenheimer and, 141–42, 262
  plutonium discovery by, 146
  postwar research and, 225, 228
  thermonuclear fusion research and, 228, 229, 251–52
Lawrence, John, 141, 145
Lawrence, Molly, 142, 220, 282
Lawrence Livermore National Laboratory, California. *See* Livermore Laboratory, California
Leahy, William, 222
Lehman, Joe, 151
LeMay, Curtis Emerson
  aerial reconnaissance and, 279–80
  civilian oversight of armed forces and, 245–46, 279
  Cuban Missile Crisis and, 295, 296, 298, 299
  firebombing of Japan by, 208, 209–12, 221, 371
  nuclear weapons and, 278, 279
  personality of, 208–09
  satellite development and, 281, 284
  Soviet Union and, 227, 244, 285
  Strategic Air Command and, 279, 287–88
Lenard, Philipp, 67
Lenin, V. I., 193, 314
Lenin Atomic Power Station, Ukraine
  location of, 314
  tourism to, 325
  use of unaffected reactors at, 324
  See also Chernobyl nuclear reactor disaster, Ukraine
Lewis, Robert, 213
Lilienthal, David, 229, 230, 235–36, 257
Limerick reactor, Pennsylvania, 310–11
Lindbergh, Charles, 116, 145
Lippmann, Walter, 38
Little Boy bomb, 154, 166, 168, 208, 211–12, 230, 371
Litvinenko, Alexander, 54
Livermore Laboratory (later Lawrence Livermore National Laboratory), California, 252, 262, 282
  Bravo thermonuclear testing by, 273–74
  nuclear weapons designed at, 282
  power generation through fusion at, 364
  Reagan's visit to, 330
  test-ban agreement and, 267
Lochbaum, David A., 348, 353
Loomis, Alfred Lee, 141
Los Alamos National Laboratory, New Mexico, 150–53, 282
  atomic trigger device designed at, 166
  Bohr's move to, 175–76
  Bravo thermonuclear testing by, 273–74
  Fermi's move to, 158–59
  move by scientists to, 150–51, 152–53, 165–66
  Oak Ridge reactor and, 164
  Oppenheimer's discovery of site for, 144, 150
  Oppenheimer's work at, 151, 153, 164, 261–62
  plutonium supply for, 160
  quality of life at, 151–52, 153, 155–56, 157, 159–60

refusal by Meitner and Rabi to move to, 152, 153
research at, 159–60
spying at, 239, 240, 242
Teller's postwar return to, 231, 252
thermonuclear fusion research at, 247–51, 252
uranium production for, 165
Loubet, Émile, 37
Loubet, Marie Louise, 37
Lyman, Edwin, 355

MacArthur, Douglas, 243, 244–45
Maclean, Donald, 172, 238
Madame Curie (Eve Curie), 49
Madrid, Jim, 204–05
Mahaffey, James, 137, 161, 310, 367, 370
Manhattan Project
anti-Russian feelings in, 194
atomic bomb advice to Truman from, 207
concern about German progress by, 184–85
Einstein and, 83, 122
espionage and, 238
funding for, 126
headquarters building of, 149
naming of, 149
staffing of, 147–49
surveillance of, 154–55
uranium source for, 153–54
MANIAC computer, 248
Manin, Louis, 40
Manzotti, Luigi, 33
Mao Zedong, 6, 244, 373
Marshak, Ruth, 153
Marshall, George, 186, 206–07, 221, 222
Marshall, Leona, 164
Marshall Islands, tests on, 250, 251, 252–54, 270, 273–74, 340, 356
Massive Retaliation defense strategy, 283, 288, 372
Mattson, Roger, 308
MAUD Committee, 121, 147, 166
Max Planck Institute, 190
May, Alan Nunn, 172
Mayak Plutonium Facility, Russia, 323
McCarthy, Joseph, 240, 261, 264
McCone, John, 294
McEnaney, Laura, 287
McKellar, Kenneth, 161
McKibben, Joseph, 150, 199, 203
McMillan, Edwin M., 201
McNamara, Robert, 211, 280, 289, 294, 296, 297, 298–99
medicine, 375, 376
isotopes in, 106, 141, 363–64
radiation for diagnostics in, 5, 13, 30, 54, 82, 89, 137, 266, 377
tracers in, 5, 54, 188, 377
Meitner, Auguste, 97
Meitner, Lise, 54, 59, 75, 379
Bohr's support for, 94–96, 120–21
childhood interest in science of, 84–85
death of, 190
departure from Germany by, 94–96, 97, 189
education of, 85
Einstein on, 75, 89
Eleanor Roosevelt's interview of, 189
Frisch on, 99, 100
Hahn on, 86, 96, 98
Hahn's research with, 64, 84, 85–87, 88–90, 98, 99, 100, 102, 103, 121, 189, 190
Hahn's treatment of, 92, 93, 96–97, 102–03, 189, 190–91
Hitler's Jewish laws and, 91–92, 93–95
move to Los Alamos rejected by, 152, 153
Nobel Prize and, 189
nuclear fission research of Frisch and, 99–102, 103, 105, 144
Planck's support for, 87, 88, 92, 94
posthumous recognition of, 191
story about "fleeing Jewess" with atomic secrets about, 189
Szilard and, 79
uranium fission research of, 121, 189
US tour of, 189
meitnerium, 191
Meloney, Marie (Missy), 50
Menand, Louis, 300
Met Lab (Metallurgical Laboratory), University of Chicago, 127–29, 147, 148, 149, 207, 305
Metropolitan Edison Company (Met Ed), 306, 307, 308
Mettler, Fred, 323–24, 367
Mike (thermonuclear device) testing, at Elugelab, 250, 251, 252–54, 372
Minoura, Koji, 342
Minutemen ICBMs, 280–81, 288, 292
Missile Defense Agency (MDA), 333–34

Moffett, Cleveland, 11
Molander, Roger C., 291
Molotov, Vyacheslav, 232
Morrison, Emily, 155
Morrison, Phil, 61, 155, 185, 202
Moulder, John, 366
Mourning Journal (Marie Curie), 41–42, 43
movies, radiation theme in, 274–75, 310–11, 367
Murakami, Haruki, 359
Mussolini, 67, 69, 128, 226
Mutual Assured Destruction (MAD) strategy, 189, 289, 290, 331, 332, 372, 376
"My Trial as a War Criminal" (Szilard), 268

Nader, Ralph, 341
Nagasaki, Japan, bombing, 216–19
  American reactions to, 116, 221, 278
  attitudes toward nuclear weapons after, 6, 167, 187, 190, 208, 230, 336, 364, 367, 369
  biological and health effects of, 7, 213–15, 364, 365–66
  bombing flight in, 216–18, 221, 371
  damage caused by, 210, 211, 213–14, 218–19, 371
  effectiveness of bombing of, 222
  Japanese reactions to, 219, 222, 350–51, 359
  number of deaths from, 366
  survivors (*hibakusha*) of, 7, 213–14, 215, 219, 350–51, 359, 365–66
  technology of, 154, 168, 196, 211
  US planning for, 237
  US support for, 310
Nakagawa, Koichi, 354–55
NASA, 122, 186, 266, 270, 306, 377
National Defense Research Committee (NDRC), 121–22
National Ignition Facility (NIF), 364
NATO, 283, 293, 296, 297, 299, 327, 328
Nature journal, 82, 99, 101, 102, 103, 112
Nautilus submarine, 303–04, 305
Neddermeyer, Seth, 168, 169
Nelson, Steve, 172
Nevada, possible nuclear waste repository in, 374–75
Nevada Test Site (NTS), 254, 270, 271–72, 287, 332, 336, 337, 373, 375
Newton, Isaac, 5, 67, 381
New York Times, 11, 50, 105, 111–12, 121, 204, 217, 221, 241, 266, 369
nickel-63, 377
Nixon, Richard, 327
  arms race and, 277
  détente policies of, 327
  Kennedy's campaign against, 282, 284–85
  nuclear defense strategies of, 289–90
Nobel, Alfred, 215, 376
Nobel Prize laureates, 265
  antimissile strategy protest of, 334
  in Chemistry, 47–48, 52, 54, 93, 146, 188–89, 190, 191, 372
  exile from Germany of, 91
  in Physics, 12, 34, 39, 47, 52, 60, 61, 63, 69, 70, 71, 75, 79, 83, 88, 91, 97, 110, 126, 128, 140, 145, 363
  for Peace, 7, 376
Nobel Prizes
  Curies and, 34, 47–48
  Einstein and, 88
  Fermi and, 69–70, 71, 83, 97, 110, 128
  Haber and, 372–73
  Hahn and Meitner and, 188–89, 190, 191
Noda, Yoshihiko, 359
Noddack, Ida, 90
Non-Proliferation Treaty (NPT), 337, 338
NORAD, 330–31
Nordau, Max, 33
North Korea, 6, 243, 244–45, 337, 338, 339, 373
NS *Savannah* submarine, 305
nuclear accidents
  Blayais nuclear power plant, France, 351
  China Syndrome concept of, 309–10, 364
  Forsmark nuclear power plant, Sweden, 312
  general public's memories of, 367
  Mayak Plutonium Facility, Russia, 323
  Rocky Flats, Colorado, plant fire, 323
  See also Chernobyl nuclear reactor disaster, Ukraine; Fukushima Daiichi power plant disaster, Japan
nuclear power
  benevolent qualities of, 376–77
  myths and popular images of, 274–75, 310–11, 367
  problems in public perception of, 369–70, 374–75
nuclear power plants
  in Blayais, France, 351

design improvements in, 362–63
first Soviet, Atom Mirny-1 reactor, 313–14
first US, in Colorado, 304
in Forsmark, Sweden, 312
government support for, 361–62
in Minsk, Belarus, 312–13
number of countries using, 360
popular support for, 360–61
problems in public perception of, 370
Rocky Flats, Colorado, 323
storage facilities for waste from, 374–75
See also Fukushima Daiichi power plant disaster, Japan

Nuclear Regulatory Commission (NRC), 341
Davis-Besse plant, Michigan, and, 353
Emergency Response Plan and, 363
Fukushima disaster and, 351, 355, 356
private industry's relationship with, 353
safety procedures and, 309, 363
Three Mile Island reactors and, 307–08, 309, 311

Nuclear Test Ban Treaty, 374

Nuclear Utilization Target Selection (NUTS) strategy, 289, 290

nuclear weapons
American attitudes toward, 299–300, 370–72
arsenals left behind after dissolution of USSR, 335, 338–39
Bush's elimination of, 333
civilian oversight of, 279–80
current US spending on, 334–35
false alarms about, 293
Gorbachev's proposed nuclear arms reduction treaty and, 332–33
great myth of, 371
Killing a Nation strategy for, 277, 292, 293, 372
Massive Retaliation strategy for, 283, 288, 372
potential terrorist use of, 336
public perception of, 375–76
Reagan's Strategic Defense Initiative against, 331–32
secret launch codes for, 280–81
Single Integrated Operational Plan (SIOP-62) for, 288–89, 292
Sunday Punch strategy for, 279–79, 283, 288, 292, 372
understanding more to fear less about, 378–79
US Triad Doctrine on, 288

Nunn, Sam, 335

Oak Ridge National Laboratory, Tennessee, 160, 161–64, 165, 166, 175, 188, 230, 282, 303, 376

Obama, Barack, 334

Office of Civil Defense (OCD), 286

Office of Scientific Research and Development (OSRD), 122, 252

Ogasawara, Haruko, 213–14

Oliphant, Mark, 121

Onda, Katsunobu, 344

On Thermonuclear War (Kahn), 290–91, 292

Open Skies arms control plan, 284, 287

Oppenheimer, Frank, 141, 142, 144, 149, 196, 203, 257, 261–62

Oppenheimer, J. (Julius) Robert (Oppie), 36
arms race concerns of, 230, 235, 243–44, 278
atomic bomb design and, 168
atomic bomb testing and, 196, 198, 199, 200, 205, 229
atomic bomb recommendations from, 206, 207, 208, 371
Atomic Energy Commission and, 229, 257, 258, 259–61
awards and recognition of, 265
background of, 141, 144
Bohr and, 175, 176, 191–92, 257
Chicago Met Lab work of, as Coordinator of Rapid Rupture, 147, 148, 150
Communist past of, 149
concern about German progress and, 184
death of, 265, 269
eccentricity of, 143
firebombing of Japan and, 210
hearing on security clearance for, 261–65
hydrogen bomb opposition of, 228, 258, 259, 262, 264
influence of, 142
Livermore lab opposition of, 262
Los Alamos research and, 144, 150, 155, 158, 166, 175, 225, 227, 261–62
Manhattan Project and, 147–49, 155, 156, 157, 158, 184, 220
marriage of, 144
nuclear detection system and, 234
nuclear explosives research by, 147, 163

Oppenheimer, J. (Julius) Robert (Oppie) (*cont.*)
  nuclear power plant dangers and, 341
  Oak Ridge reactor and, 164
  personality of, 141–43, 257–48
  research secrecy and, 259
  research with Lawrence at Berkeley by, 141–42, 145
  Roosevelt's death and, 194–95
  Strauss and, 257–48
  suspicions about and surveillance of, 155, 170, 185, 256–57, 258, 259, 260
  Teller and, 170, 220, 227, 231, 258, 262–64
  thermal diffusion and, 163
  thermonuclear fusion research and, 228, 229, 250, 258, 259
Oppenheimer, Kitty, 144, 149, 155, 156, 172, 198
Orear, Jay, 104, 125
Overbeck, Wilcox, 135

Paderewski, Ignaz, 20
Page, Katherine, 143–44
Page, Winthrop, 143–44
Pais, Abraham, 261
Pakistan, 6, 337, 339, 373
Palladino, Eusapia, 32
Panasyuk, I. S., 233
Parsons, Deke, 156, 212
Partial Test Ban Treaty, 299
Pash, Boris, 185, 186
Pasteur, Louis, 29, 33, 141
Pasteur Institute, 43
Pauli, Wolfgang, 58, 83, 94, 95
Pearson, Mike, 376
Pegram, George, 110, 112, 114, 125, 146
Peierls, Gennia, 159, 170
Peierls, Rudolf, 120, 121, 152, 159, 169, 170
Perle, Richard, 333
Perrin, Henriette, 34, 44
Perrin, Jean, 9, 34, 41, 44, 46, 47, 49, 52
Perry, William J., 335
Persico, Enrico, 56
Petrzhak, Konstantin, 171
Philby, Kim, 238
Phillips, Melba, 143
Photography magazine, 11–12
Picasso, Pablo, 33–34
Picture Parade (comic book), 375
Pillitteri, Carl, 343, 344
Planck, Erwin, 186
Planck, Max, 46, 59, 74, 75
  Einstein's research and, 87–88
  Hitler's laws against Jews and, 91, 92
  Meitner's support from, 87, 88, 92, 93, 94
  wartime bombing of house of, 186
plutonium
  Mayak Plutonium Facility, Russia, accident and release of, 323
  Rocky Flats, Colorado, plant fire and release of, 323
Poincaré, Henri, 14
polonium, 4, 26, 28, 30, 51, 52, 53, 90, 188, 197, 230, 253
popular culture, radiation theme in, 274–75, 310–11, 367
Pousson, Jens, 177
Power, Thomas, 288
pregnancy, effects of radiation on, 215, 271, 308, 324
Preuss, Paul, 4
Pripyat, Ukraine, 314, 316, 319–20, 321, 324, 325
Project Alsos, 185, 186, 187
Public Awareness of Nuclear Science, 369
Pulitzer, Elinor, 155
Putin, Vladimir, 54

Rabi, Isidor Isaac, 102, 156, 188, 331
  ballistic missile tracking and, 330
  concern about German progress by, 184
  Hungarian background of, 77
  move to Los Alamos rejected by, 153
  Oppenheimer's support by, 142, 260–61, 262, 264–65
  Szilard and, 77, 110, 112
  on Teller, 147, 264
  thermonuclear weapons development and, 235, 236
  uranium fission research of, 110, 112
Racah, Giulio, 69
radiation
  cancer from exposure to, 3, 4, 5, 6, 215, 266, 268, 271, 272, 291, 320, 337, 358, 365, 366, 369, 378
  effects of, after Chernobyl disaster, 323–24
  fear of, 324, 370
  medical use of, 5, 13, 30, 54, 82, 89, 137, 266, 377
  as popular culture theme, 274–75, 310–11, 367
  pregnancy and, 215, 271, 308, 324

research on health effects of, 365–66
Teller's concepts of effects of, 331
as two-faced miracle, 368, 379
understanding more to fear less about, 378–79
Radiation Effects Research Foundation, 365–66
Radiation Exposure Compensation Act, 336
radiation sickness, 5, 215, 320, 354
radioisotopes, 141, 257, 370, 376–77
radioisotope thermoelectric generators (RTGs), 306
radiophobia, 324, 370
Radithor (radium water), 36, 38
radium, 3, 26, 27, 28, 29
burns from, 35, 38, 39, 53, 127
Curies' research on, 30–34, 39, 47, 49, 54, 63
medical use of, 35–36, 43, 54, 82
public enthusiasm for, 34–35, 36–37, 38
Radium Chemical Company, 127
Radium Institute, Paris, 43, 82, 193
radium isotopes, 97, 98
radon, 3, 4, 30, 54, 60, 62, 82, 104, 127, 188, 311, 336
RAND, 281–83, 288, 290, 372
Rasetti, Franco, 56, 57–58, 59, 62, 64
Ray, Maud, 121
Reagan, Ronald, 7, 299, 376
fear of atomic bomb threats by, 327–28, 329
Flexible Response strategy of, 283
problems in defense strategies and, 330–31, 334
proposed nuclear arms reduction treaty and, 329–30, 332–33
Strategic Defense Initiative (SDI; Star Wars) and, 5, 330, 331–32, 333, 335, 364
Reese, Willy Peter, 133
Regaud, Claude 43
Reich Research Council, 105, 183
Rhodes, Richard, 237, 316
Richardson, Owen, 120–21
Rickover, Hyman, 288, 303–04
Ridgway, Matthew, 245
Riken Institute, Saitama, Japan, 210
Rittenmeyer, Nicole, 91
Robb, Roger, 261, 262
Rockwell, Theodore, 162
Rocky Flats, Colorado, plant, 323
Rodin, Auguste, 34
Ronnenberg, Joachim, 178
Röntgen, Anna Bertha, 10
Röntgen, Wilhelm, 9, 85, 97
public enthusiasm for work of, 11–12, 34
X-ray discovery by, 10–13, 14
Röntgen rays, 11–12, 13, 15, 16, 35, 126
Röntgen Society, 13
Roosevelt, Eleanor, 68, 115, 189
Roosevelt, Franklin D., 68, 369
Bohr's letter to, about nuclear arms race, 192
letters between Einstein and, 115, 116–18, 119, 153, 176, 220, 369
Manhattan Project and, 147, 161, 303
Nazi development of nuclear weapons and, 176
Oak Ridge National Laboratory and, 161
Oppenheimer and, 192, 194–95
uranium fission research and, 121, 125, 146
Rosbaud, Paul, 95
Rosenberg, Ethel, 171, 238, 240, 241–43, 251
Rosenberg, Julius, 171, 238, 240–43, 251
Rosenbluth, Marshall, 273
Rosenfeld, Léon, 101–02
Rossi, Bruno, 69
Rostow, Elspeth, 376
Rotblat, Joseph¸ 222
Roy, Susan, 286
Royal Swedish Academy of Sciences, 34, 69, 96
Rubens, Heinrich, 85
Rupp, Arthur, 163
Rusk, Dean, 293, 297
Rutherford, Ernest, 29–30, 38, 59, 81, 82, 86–87, 112

SAC. *See* Strategic Air Command
Sachs, Alexander, 115, 116, 118, 176
Saffer, Tom, 271
Sagan, Carl, 55–56, 266
St. Joachimsthal, Czechoslovakia, mine, 36, 195, 231
Sakharov, Andrei, 211, 237, 250, 254–55, 268, 317
Saletan, William, 361
Savannah submarine, 305
Sawachika, Hiroshi, 214
Schell, Jonathan, 299, 328
Scherrer, Paul, 93, 361

Schlesinger, James, 289–90
Schrödinger, Erwin, 88, 91, 168
Scranton, William III, 307
Seaborg, Glenn, 145, 146, 161, 207, 228, 377
Segawa, Kikuno, 214
Segrè, Emilio, 90, 140, 205
  atomic bomb testing and, 200, 205
  on Bohr, 175
  early career of, 59, 60, 62, 64
  Fermi and, 55, 58, 68, 104, 265
  Hanford reactor output and, 168
  Manhattan Project and, 148, 154
  medical imaging research of, 377
  Los Alamos and, 152, 153, 159–60, 175
  nuclear reactor (CP-1) installation and, 132, 138
  Oak Ridge reactor and, 163–64
  on Oppenheimer, 142–43
  plutonium discovery by, 146
  postwar research of, 225, 228
  thermonuclear weapons development and, 236
Sengier, Edgar, 153–54
Serber, Robert (Bob), 144, 147, 155, 165–66, 200–201, 264
Shevardnadze, Eduard, 323
Shoup, David, 288
Shultz, George P., 330, 333, 335
Siegbahn, Manne, 95, 96, 98, 189
Silard, Bela, 74, 80, 82, 113
Sime, Ruth Lewin, 103
Simon, Walter, 164–65
Single Integrated Operational Plan (SIOP-62), 288–89, 292
Skardon, William, 239
Skłodowska, Bronisława (Bronya), 16, 18, 19, 20, 29, 38, 39, 42, 48, 51, 53
Skłodowska, Marja (Manya). *See* Curie, Marie
Smith, Francis, 176
Smyth, Henry D., 194
Snow, C. P., 299
Society for the Protection of Science and Learning, 80
Socolow, Robert, 6, 343
Soddy, Frederick, 29–30, 34, 80
Solvay, Ernest, 46
Somervell, Brehon, 147
Sorbonne, 15, 16, 20, 21, 34, 37, 43, 44, 47, 48, 50
Sorensen, Ted, 295
South Korea, 327, 338, 360
Speer, Albert, 183
Spinoza, Baruch, 78
Sputnik, 256, 261, 282, 285, 286
Stalin, Joseph, 172
  arms race concerns about, 208, 230, 267
  atomic weapons program and, 194, 221, 232–33, 235, 245
  death of, 245, 254
  mass murders of, 362, 373
  North Korea and, 243
  nuclear bomb threat under, 6, 373
  Truman and, 232
  US relations with, 193, 235
Stallings, Richard, 374
Stark, Johannes, 67, 92, 176
START and START II treaties, 333
Star Wars (Strategic Defense Initiative, SDI), 5, 330, 331–32, 333, 335, 364
Steen, Charles, 270
Stern, Mark Joseph, 361–62
Stevenson, Adlai, 252
Stevenson, Robert Louis, 36
Stimson, Henry, 148, 150, 206, 207–08, 210–11, 221, 228, 370
storage facilities, for nuclear waste, 374–75
Strategic Air Command (SAC)
  aerial reconnaissance by, 279–80
  civilian oversight of, 279, 280
  headquarters of, 281
  LeMay as chief of, 279, 287–88
  power of, 287–88
  secret missile launch codes and, 280–81
Strauss, Lewis, 267
  arms race and, 255, 258–59
  image of atomic energy promoted by, 230–31
  nuclear detection system and, 234
  Oppenheimer and, 257–61, 263, 264
  support for Szilard from, 82
  test-ban agreement and, 267
Strassmann, Fritz, 83–84, 90, 92, 96–97, 98, 100, 102–03, 186, 190
Strategic Defense Initiative (SDI; Star Wars), 5, 330, 331–32, 333, 335, 364
Sunday Punch defense strategy, 279–79, 283, 288, 292, 372
Super thermonuclear bombs. *See* hydrogen (Super) bombs
Suzuki, Tomohiko, 352
Swedish Academy of Sciences, 34, 69, 96
Sweeney, Chuck, 216, 217

System for Nuclear Auxiliary Power (SNAP), 306
Szilard, Trude (Gertrude), 82, 268
Szilard, Leo, 54, 222, 379
- arms control and, 267–68
- assistance to exiles in London by, 80–81, 92
- atomic bomb petition from, 207–08
- on atomic bombs, 127–28
- childhood of, 73–74
- cobalt experimentation by, 83, 153, 257, 267
- Cold War fears of, 194, 206
- concern about German use of nuclear arms and, 139, 153, 184
- Cuban Missile Crisis and, 268
- death of, 268, 269
- education of, 74–75
- Einstein and, 77, 78, 127, 220
- Fermi's work with, 83, 149
- Groves's suspicions about, 149–50, 207, 208
- Hungarian background of, 75, 76, 77
- Institute for Biological Studies work of, 268
- inventions of, 77–78
- Met Lab research in Chicago by, 149–50
- morality of atomic bomb use and, 211
- move to New York by, 82–83
- need for atomic bomb research advocated by, 114–16, 118, 119
- neutron-triggered chain reaction research of, 81–82, 105
- nuclear fission research and, 105
- nuclear reactor design and, 131, 133, 135, 137, 139, 149, 188
- nuclear research oversight and, 251
- Oak Ridge reactor and, 163
- political campaigns and power of, 251
- postwar research by, 229
- research secrecy and, 112–13, 208, 254
- Teller's radiation concepts and, 331
- US military research and, 122–23
- uranium fission research of, 106, 109–13, 126, 127–28, 129, 177
- Wells's concept of atomic bomb and, 79–80, 105
- Wigner on, 76

Szilard, Louis, 73
Szilard, Tekla, 73

Taniguchi, Sumiteru, 219
Tanimoto, Kiyoshi, 216
Tatlock, Jean, 133, 171, 198
technetium, 140, 377
Teflon, 161
Telegdi, Valentine, 66, 225, 226, 261
Teller, Edward, 225, 268, 291, 379
- arms race with Soviets and, 258–59
- atomic bomb design and, 168, 169, 199
- atomic bomb use in Japan and, 207
- Bohr and, 94, 175
- concern about German progress by, 184, 185
- concerns about Soviet power by, 230, 231
- exile visas for family of, 267
- FBI's suspicions about spying by, 240
- on Fermi's research, 90, 170
- fusion-bomb usage proposals of, 305
- Hungarian background of, 75, 77, 231
- hydrogen bomb research of, 125
- Livermore lab and, 252, 330, 331
- Los Alamos and, 151, 157, 158, 227, 228, 231, 252
- need for atomic bomb research advocated by, 114, 116, 118, 119
- nuclear defense strategy and, 234, 288
- nuclear explosives research of, 147
- nuclear power reactor safety and, 310
- on Oppenheimer, 220, 258, 259, 262–64
- personality of, 157–58, 248–49, 250, 251
- political campaigns and power of, 251
- professional relationships of, 262, 265
- Reagan's Strategic Defense Initiative and, 330, 331–32, 333, 364
- research at Chicago by, 229, 242
- test-ban agreement and, 267
- thermonuclear fusion (Super bomb) design of, 125, 158, 227–29, 235, 236, 237, 239, 247, 249–50, 255, 258, 259, 262, 264, 275, 372
- thermonuclear testing (Mike) at Elugelab and, 250, 251, 252–54, 282–83
- Ulam's relationship with, 231, 250, 265
- US military research and, 122
- uranium fission research of, 111

Teller, Mici, 157, 159
test-ban agreements, 255, 267, 299, 336, 374
Thatcher, Margaret, 328
thermonuclear (Super) bomb. *See* hydrogen (Super) bomb
Thompson, Silvanus, 27
Thomson, William, 1st Baron Kelvin, 22–23, 39

thorium breeder reactors, 362
Thornburgh, Richard, 308, 309
Three Mile Island Nuclear Generating Station (TMI), Pennsylvania, 5, 306–10, 312, 346, 353, 369
federal officials on, 307–09, 311
mechanism of accident at, 306–07
public reaction to, 309–10
radiation released in, 336
Tibbets, Enola Gay, 212
Tibbets, Paul Warfield, 211–12, 213
Todreas, Neil, 345
Tokyo, 359
firebombing of, 209–10, 221, 222, 371
radiation fallout affecting, 346, 349, 350, 351, 355, 356, 358, 359
Tokyo Electric Power Company (TEPCO), 341–42, 344, 345, 346, 347, 349, 350, 351, 352–53, 354–55, 357, 359, 366
Trabacchi, G. C., 60, 62
Tracerlab, 234–35
tracers, 5, 54, 188, 377
Treaty on the Non-Proliferation of Nuclear Weapons (NPT), 337, 338
Triad Doctrine, 288
Truman, Harry, 211
advice from scientists on atomic weapon use to, 206–07
aerial reconnaissance and, 279
arms race with Soviet Union and, 235–36, 238, 246
Atomic Energy Commission and, 229
Cold War and, 244
future of nuclear weapons and, 206, 235
MacArthur replaced by, 245
Meitner's visit to, 189
Stalin and, 232, 235
thermonuclear weapons development and, 236, 237, 247, 252
use of atomic bombs in Japan and, 208, 215, 216, 221, 232, 311
tsunami, in Japan, 342, 344–45, 352–53
Tuchman, Barbara, 33, 46
Turing, Alan, 248

Udall, Stewart, 336
Ukraine, 314, 334
Chernobyl disaster's effect on, 322, 323, 324–25, 364, 369
nuclear arsenal in, 291, 337
nuclear reactors in, 313, 360
Ulam, Françoise, 188, 249–50
Ulam, Stanislaw, 76, 150, 282, 379
move to Los Alamos by, 150
personality of, 249
Teller's relationship with, 231, 250, 265
thermonuclear fusion research by, 247, 248–50, 252, 253, 262, 264, 265, 275
Union of Concerned Scientists, 353
United Nations (UN), 79, 189, 207, 220, 230, 267, 297, 304, 364
Atomic Energy Commission, 230
International Atomic Energy Agency (IAEA), 305, 338, 356, 364
Scientific Committee on the Effects of Atomic Radiation (UNSCEAR), 364–65
U.S. Radium Corporation, 38
Universal Exposition (Paris, 1900), 33
University of Chicago
Met Lab at, 127–29, 147, 148, 149, 207, 305
nuclear reactor (CP-1) installation in Stagg Field at, 129–34
University of Rome, 57, 58–64, 266
uranium
common uses of, 25–26
Curies' joint research on, 29–31
emission of rays from, 15, 26, 27
half-life of, 5, 30
Marie Curie's initial research on, 27
naming of, 25
Pierre Curie's research on, 26–27
processing of ore for, 28–29
sources of, 4, 6
Uranium Club (Uranverein), 105, 176, 177, 178, 179, 183, 186–87
uranium-235, 4, 120, 122, 123, 161, 166
Urey, Harold, 94, 122, 123, 147, 225
nuclear reactor design by, 185
uranium production and, 160–61, 162, 170, 238
USS *Nautilus* submarine, 303–04, 305
U-2 aerial surveillance program, 279, 282, 285, 288, 294, 297–98

Velikhov, Yevgeny, 331, 332
Vernadski, Vladimir I., 193
Verne, Jules, 79, 303
Viner, Jacob, 278
viruses, computer, 338
von Bahr-Bergius, Eva, 96, 98
von Braun, Wernher, 79, 186, 291
von Halban, Hans, 120

von Kármán, Theodore, 75, 76, 77
von Laue, Max, 75, 79, 88, 92, 93, 186, 188
Von Laue, Theo, 188
von Neumann, Janos (John), 77, 80, 290, 291
  atomic bomb design and, 168, 168, 220
  background of, 168–69
  on computers in the future, 248
  Hungarian background of, 75, 76, 231
  Manhattan Project and, 148, 155
  nuclear defense strategy and, 288
  Oppenheimer's support from, 262
  political campaigns and power of, 251
  preemptory nuclear strike and, 244
  thermonuclear research by, 239, 248–49, 250
von Neumann, Klari, 168–69, 248
von Stauffenberg, Claus, 186
von Weizsäcker, Carl Friedrich, 118, 178, 180–81, 183, 187

Walker, Andrew, 155
Wall Street Journal, 37
Watras, Stanley, 310–11, 312
Wattenberg, Albert, 130–31
Wefelmeier, Wilfrid, 99
Weil, George, 134, 135, 136, 137
Weimar Republic, 74, 77, 79, 91, 225, 251
Weinberger, Caspar, 283
Weisband, Bill, 238
Weiss, Trude, 82, 268
Weisskopf, Victor, 236
Welsome, Eileen, 271
Wells, H. G., 7, 79–80, 81, 105
Wheeler, John, 102, 164, 244, 265
Wiesner, Jerome, 277–78
Wigner, Eugene, 228
  early career of, 76, 77, 80, 82
  on Fermi, 109
  Hanford reactor and, 164
  Hungarian background of, 75
  need for atomic bomb research advocated by, 114–16, 118, 119
  nuclear reactor (CP-1) testing and, 137–38
  Oak Ridge reactor and, 163
  on Szilard, 76
  on Teller, 229
  US military research and, 122
  uranium fission research of, 109
  von Neumann and, 76, 77, 168
Wilson, Anne, 205
Wilson, Jane, 159
Wilson, Richard, 324
Wilson, Robert (Bob), 159, 167, 205, 219, 220
Wilson, Taylor, 363–64
Wilson, Volney, 134–35
Wohlstetter, Albert, 283–84, 288, 290, 376
Woods, Leona, 124, 131, 137, 164
World Set Free, The (Wells), 79–80
World War I (the Great War), 32–33, 36, 74, 89, 91, 193, 372
World War II, 128, 194, 195, 206, 230, 232–33, 239
Wrye, William, 205

X-rays, 3, 5, 11–12, 13–14, 34, 35, 36, 49, 50, 57, 58, 79, 85, 126, 215, 230, 250, 309, 368, 377, 378
X-ray lasers, 332
X-ray satellites, 331

Yates, Sidney, 340
York, Herbert, 236, 250, 252, 281, 290
Yoshida, Masao, 346, 351–52
Yucca Mountain nuclear waste storage facility proposal, 373, 374

Zaitchik, Alexander, 329
Zakaria, Fareed, 339
Zinn, Walter, 233
  injuries suffered by, 167
  nuclear reactor (CP-1) installation and testing by, 131, 132–33, 135, 137
  uranium fission research of, 106, 124, 129
Zorawski, Bronka, 17, 18
Zorawski, Kazimierz (Casimir), 17, 18, 51
Zorawski, Stas, 17–18
Zorawski family, 17–18, 21

ABOUT THE AUTHOR

Craig Nelson is the author of the *New York Times* bestseller *Rocket Men*, as well as several previous books, including *The First Heroes*, *Thomas Paine* (winner of the Henry Adams Prize), and *Let's Get Lost* (short-listed for W. H. Smith's Book of the Year). His writing has appeared in *Vanity Fair*, the *Wall Street Journal*, *Salon*, *National Geographic*, *New England Review*, *Popular Science*, *Reader's Digest*, and a host of other publications; he has been profiled in *Variety*, *Interview*, *Publishers Weekly*, and *Time Out*. Besides working at a zoo and in Hollywood, and being an Eagle Scout and a Fuller Brush man, he was a vice president and executive editor of Harper & Row, Hyperion, and Random House, where he oversaw the publishing of twenty national bestsellers. He lives in Greenwich Village.